WORLD DEFORESTATION IN THE TWENTIETH CENTURY

Edited by

John F. Richards

and

Richard P. Tucker

Duke Press Policy Studies
Duke University Press
Durham and London 1988

Printed in the United States of America
on acid-free paper ∞
Second printing, 1989
Library of Congress Cataloging-in-Publication Data
World deforestation in the twentieth century / edited by John F. Richards and Richard P. Tucker.
p. cm.—(Duke Press policy studies)
Bibliography: p.
Includes index.
ISBN 0-8223-0784-7 cloth
ISBN 0-8223-1013-9 paper
1. Deforestation—History—20th century—Congresses.
SD418.W67 1988
333.75'11'0904—dc 19 87-31953

CONTENTS

Tables and Figures v
Preface ix
Introduction John F. Richards and Richard P. Tucker 1

PART 1 Forests in the Developing World

1 Deforestation in the Araucaria Zone of Southern Brazil, 1900–1983
John R. McNeill 15

2 Southern Mount Kenya and Colonial Forest Conflicts
Alfonso Peter Castro 33

3 The Impact of German Colonial Rule on the Forests of Togo
Candice L. Goucher 56

4 Deforestation and Desertification in Twentieth-Century Arid Sahelien Africa James T. Thomson 70

5 The British Empire and India's Forest Resources: The Timberlands of Assam and Kumaon, 1914–1950 Richard P. Tucker 91

6 Agricultural Expansion and Forest Depletion in Thailand, 1900–1975 David Feeny 112

PART 2 Linkages: The Global Timber Trade

7 Export of Tropical Hardwoods in the Twentieth Century
Jan G. Laarman 147

8 The North American–Japanese Timber Trade: A Survey of Its Social, Economic, and Environmental Impact Thomas R. Cox 164

PART 3 Forests of the Developed World

9 Changing Capital Structure, the State, and Tasmanian Forestry John Dargavel 189

10 The Death and Rebirth of the American Forest: Clearing and Reversion in the United States, 1900–1980 Michael Williams 211

11 Perspectives on Deforestation in the U.S.S.R. Brenton M. Barr 230

Notes 263
Index 309
Contributors 323

TABLES AND FIGURES

Tables

1.1 Changes in Forest Area 18
1.2 Coniferous Sawnwood Exports from Brazil, 1911–1981 21
1.3 Official Coniferous Sawnwood Production, 1905–1978 22
1.4 Argentinian Coniferous Sawnwood Imports from Brazil, 1935–1977 23
1.5 Cultivated Area in Paraná, Santa Catarina, and Rio Grande do Sul, 1920–1980 24
1.6 Area Approved for Reforestation in Brazil by Species and by State, 1967–1985 30
2.1 Compounded Forest Offenses, Southern Mount Kenya, August 1937 to March 1940 43
4.1 Woodstock Perspectives, Niger 85
4.2 Woodstock Perspectives, Upper Volta 87
4.3 Deforestation around Ouahigouya, Upper Volta 88
4.4 Woodstock Perspectives, Mali 89
6.1 Average Annual Rates of Change of Cropped Area, Paddy Area, Paddy Production, Paddy Yield, and Population in Thailand, 1905–1955 114
6.2 Paddy Area Harvested and Total Cropped Area Harvested in Thailand, 1950–1975 115
6.3 Average Annual Rates of Change in Cropped Area in Thailand, 1911–1976 116

6.4 Estimates of the Area under Forests and Rates of Change in the Area under Forests in Thailand, 1913–1980 118
6.5 Comparison of Changes in the Area under Forests and in the Cropped Area in Thailand, 1913–1975 121
6.6 Average Annual Production of Teak, Other Timber, Total Timber, Firewood, and Charcoal in Thailand, 1890–1978 122
6A-1 Whole Kingdom Paddy Area, Paddy Output, Paddy Yield, and Major Crop Area for Thailand, 1905/6–1955 131
6A-2 Area Planted in Rubber Trees and Total Area under Major Crops in Thailand, 1913–1955 134
6A-3 Various Estimates of the Thai Population, 1906–1956 137
6A-4 Average Annual Rates of Growth of Population for Thailand, 1906–1955 138
6A-5 Comparison of Rice Balance Sheet Estimate to Official Paddy Production Figures for Thailand, 1906–1955 139
6A-6 Regressions of the Ratio of the Rice Balance Sheet Estimates to the Official Production Figures on Time, Thailand, 1906–1955 140
7.1 Postwar Trends in Tropical Hardwood Production and Export 152
7.2 Direction of Trade of Major Forest Products Exported from the Tropics, 1983 154
8.1 Lumber Production in Japan, 1880–1960 166
8.2 Percentage of Forest-owning Japanese Households by Class Size, 1960–1980 169
8.3 Number of Sawmills in Japan According to Power Use, 1960–1977 173
8.4 Log Exports from Oregon and Washington to Japan, 1961–1974 182
8.5 Number of Sawmills in Oregon by Log Consumption Capacity, 1968–1976 183
10.1 Estimated Area of Commercial and Noncommercial U.S. Forest and Total Standing Saw-Timber Volume 214
10.2 Estimated Volume and Annual Growth and Drain in the United States, 1909–1977 217
10.3 Cleared Farmland in the United States, 1910–1979 223
10.4 Major Land-use Changes in Carroll County, Georgia, 1937–1974 225
11.1 Land-use Composition of State Industrial Forests, U.S.S.R. 236
11.2 Distribution of Forested Area and Characteristics of Industrial Stands, U.S.S.R. 240

11.3 Timber Harvesting in the U.S.S.R., 1913–1983 242
11.4 U.S.S.R. Forest Resource 249
11.5 U.S.S.R. Forest Management by Central State Forestry Agencies 250

Figures

2.1 Southern Mount Kenya Forest, Kirinyaga District, Kenya 35
7.1 World Exports of Tropical Hardwoods, 1900–1980 149
8.1 Supply of Wood and Wood Products in Japan, 1954–1978 167
8.2 Price of Domestic and Imported Timber in Japan, 1960–1978 168
8.3 Distribution of the National Forest in Japan 172
8.4 Housing Starts and Wholesale Price Index of Timber for Japan, 1973–1982 180
9.1 Land Transferred to Private Ownership in Tasmania 192
9.2 Pulpwood and Wood Chip Concession Areas in Tasmania by Date Mills Started 203
10.1 Annual Net Growth and Drain of Timber in the United States, 1800–2040 219
10.2 Annual Extent of Forest Fires in the United States and Average Size of Area Burned per Fire, 1910–1970 221
10.3 Abandonment of Agricultural Land to Forest and Conversion of Forest to Agricultural Land, Carroll County, Georgia, 1937–1974 224
10.4 Replanting and Reseeding of Land to Forest in 13 Southern States, 1925–1980 226
10.5 Per Capita Consumption of Major Timber Products in the United States, 1900–1980 228

PREFACE

The essays published in this volume were first presented and discussed at a research symposium held at the National Humanities Center, Research Triangle Park, North Carolina. The conveners (the editors of this volume) planned this scholarly meeting as a sequel to our original conference and subsequent volume on the world forests in the nineteenth century. As on the former occasion, we hoped to bring together scholars presenting case studies from different countries and world regions. We also hoped to detect larger generalizations and themes emerging from these individual contributions. We were especially concerned to notice divergences from the trends and issues posed in our review of the nineteenth-century world economy and world forests. Certainly, the meeting itself was successful in this regard. The revised papers included in this volume also fulfill this goal.

With only few exceptions, the contributions assembled here portray a continuing process of global deforestation. This process is pushed by newly urgent population loads in every part of the world and their accompanying food and resource needs. The global market and the nation-states have escalated their demands on the world's forests as the decades of the twentieth century have slipped past. Forests are a valued capital resource for the newly independent nations of the former colonial world. At the same time the land they cover is often seen as necessary for an expanded agricultural base. Thus in the twentieth century the linkage between local and regional resource and land needs is powerful and direct—perhaps more so than in the newly consolidated world economy of the previous century. Nevertheless the far-reaching grasp of the wealthy, developed countries has not relaxed, but intensified. The net result has been to convert and degrade the world's wooded lands at a rate and scale never seen before in human history.

It is our belief that the debate over current forest and environmental issues—such as the fate of those large tropical rain forests in Amazonia and Southeast Asia still uncut and undegraded—should be conducted with a longer-term historical perspective. It is our hope that in some measure this volume will help contribute to this perspective and understanding.

In closing, we wish to acknowledge the sponsorship of the Duke Center for International Studies for the original conference and for many of the expenses of preparing the work for publication. We are especially grateful to A. Kenneth Pye, who as Director of the Center made a special effort to obtain funds to meet our needs. We also wish to thank the National Humanities Center for the use of its facilities during our conference.

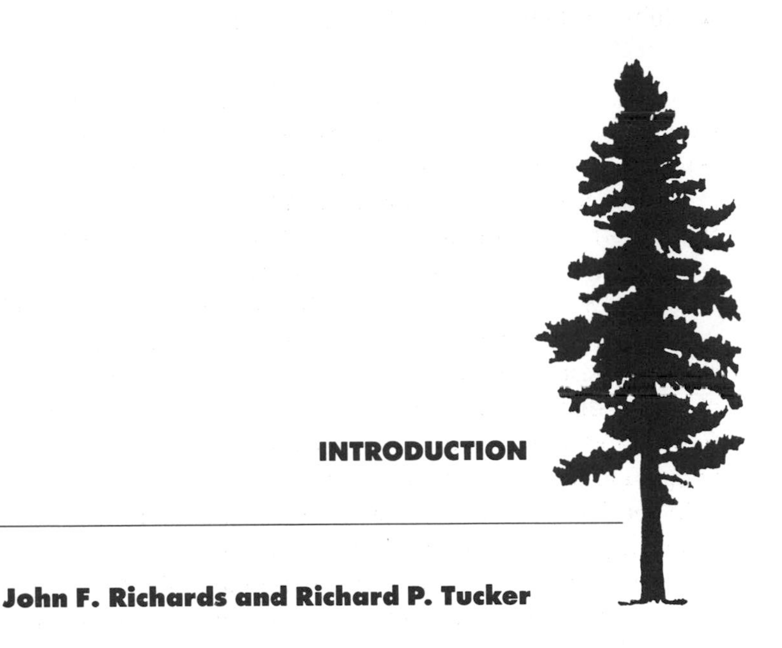

INTRODUCTION

John F. Richards and Richard P. Tucker

Since 1900 the forests of the world have come under increasing human pressure. In Asia, Africa, and Latin America forests have been cut at a faster rate than ever before. Those areas not converted to other uses have been increasingly degraded and impoverished. Net forest regrowth has not kept up with the extraction of wood and other products. Economic planners and resource managers, as well as botanists and other natural scientists, have become more and more disturbed at the trend. In Canada, Australia, and other parts of the developed world, forest depletion is a direct result of intensifying demands upon wood for pulp, plywood, construction wood, and a host of other uses. Even the vast forests of the U.S.S.R. have diminished in extent and have suffered from excessive use in the tumultuous events of the twentieth century. Whether the expansion of forested area reported since World War II is accurate is uncertain. It is certain that the Soviet-managed economy has depended on utilization of its forests for domestic and export applications. In short, for most of the land surface of the globe, conversion of forest lands to other uses and impoverishment of standing forests has been a consistent theme in the twentieth century.

Only in certain areas of the developed world has this trend not been evident. In the industrial countries of Europe (save the U.S.S.R.) forest cover has largely stabilized. By 1914 land law had been refined and the forestry profession had been largely integrated into socioeconomic life. Foresters had long experience in administering timberlands, and the economics of strongly organized and industrialized societies had made extensive reforestation of formerly arable land a reality. In a closely related trend, the transition to intensive agriculture and nonwood energy sources was considerably more advanced in industrialized Europe than in other countries by 1914. Only the effects of acid rain in both western and eastern Europe now threaten to disrupt this longstanding balance between forests and human economic demands. Japan, with its equally longstanding tradition of forest management, has had a similar history of forest area stabilization and careful management of its wooded resources. Like Europe, it has been able to import the timber and wood products necessary for its growing economy and population in the twentieth century.

In the United States forests have steadily increased in extent and in density for the past half-century. In large measure this fortunate circumstance results from the commanding ability of the American economy to import much of its specialized wood product needs at a favorable price. Much can be credited to the efficiency and long-term investment policies of the American forest products industry. Forest regrowth, however, has primarily resulted from the large-scale abandonment of farm land in twentieth-century America. Mechanization and intensification of U.S. agriculture has permitted the steady shrinkage of cropland with no loss in overall production. Finally, some regrowth can be attributed to successful conservation and preservation efforts for wilderness and forest areas.

Unfortunately, this national success story is deceptive. For the remainder of the world, deforestation in its twin aspects—outright land conversion and biotic impoverishment of standing stock—is the prevailing condition. Despite the best efforts of environmentalist groups, economists calculating the true costs of forest depletion, and concerned citizens of all countries, the demand for wood and land prevails. Since World War II (or in some countries earlier) steeply climbing human numbers have intensified demand for fuelwood, pulp, and land for grazing and crops. Prior to 1930 commodity demands for food, fiber, and timber expressed through the emergent world economy of the colonized world order were relatively more significant. Since 1945 pressures upon the world's forests have become considerably more severe. But both before and after this divide, various commodity demands, including that for timber, have played an important role in the exploitation of the world's wooded resources. Those countries and regions with the means to do so have paid for and extracted wood or agricultural products from every region of the world. The global timber trade is merely one aspect of a complex net of economic relationships that consume forests and other natural resources on a worldwide scale.

In short, the title of this volume accurately reflects a global reality. Our world continues to be subjected to a cycle of deforestation. We are now involved in the self-conscious management of those dwindling forests and wooded areas that have survived the nearly five-century-long process of forest harvest and agricultural expansion. Obviously, consumption of timber and land has been an important, if not essential, aspect of economic growth and material betterment in every society—from the United States to India. Whether the choices made and the overall result could have been more responsible in both economic and ecological terms is by this time virtually irrelevant. We are now dealing with an accomplished fact. The essays in this volume are aimed at documenting, describing, and analyzing aspects of this world trend in the twentieth century.

A fundamental transformation of the planet's vegetation had begun around 1500 when Europeans first extended their quest for wealth and resources to other continents. But Europe's power to remodel the arable, grazing, and timberlands of the globe had been severely restricted and selectively local in its impact until the Napoleonic Wars. Thereafter its trade networks rapidly expanded into many economic and ecological hinterlands, leading to widespread deforestation. Historians of European colonial systems have often analyzed the expansion of agriculture but have rarely paid explicit attention to the other two dimensions of the land: forests and grasslands. The essays in our previous volume traced colonialism's impact on the changing frontier between arable and forest lands from 1815 until the world's next great turning point in 1914; the present collection extends the coverage toward the 1980s.

Even in regions that were still largely autonomous economically, we note, similar processes of deforestation were afoot before 1900. Brenton Barr and Thomas Cox remind us that Russian territories—including those east of the Urals—and Japan were both deforesting rapidly for their own internal purposes during the nineteenth century. Candice Goucher discusses the extensive deforestation that has been going on for centuries in the Togo Gap of West Africa, where indigenous iron smelting used a technology and a marketing system long predating European incursions. And Peter Castro shows that forests in parts of Kenya were also cut in the nineteenth century to provide new food-producing lands for the expanding caravan routes of East Africa.

But the dominant theme of the nineteenth century concerned the increasingly unified global economy, controlled in many ways by British capital and technology and British imperial institutions. The impact of the global economy has continued to spread in the present century; one of its most pronounced consequences has been accelerating deforestation. This is often attributed primarily to rural population growth in the developing countries. But in this collection we see more clearly the impact of outside capital: industrial economies tapping the developing economies' timber resources to meet their consumption demand and private investors (in some cases in alliance with local commercial interests) cashing in on the high short-term profitability of timber exports from capital-starved countries. The consequences of this imbalance of power between industrialized and developing nations are the main concern of this volume.

Under conditions of rapid industrialization and urbanization, it has sometimes been possible for a society to reverse the deforestation trend. In the United States, as Michael Williams describes, forest resources underwent a great resurgence in the modern period, after the devastating deforestation of the early industrial era in the nineteenth century. After 1900 the United States was no longer a frontier of the global market economy or the unstable hinterland of modern land-use bureaucracies, as it had been before. Timber production intensified, and forests reclaimed tilled land where old farms were deserted by migrants to the cities. Thomas Cox's paper on the timber economies of the northwestern United States and Japan since 1945 shows two economically powerful countries, each able to trade timber internationally to its marketing advantage. His analysis of market factors suggests that the explanation of why Japan has remained one of the most heavily forested countries in the world (despite its dense population) lies in the relative costs of different transport systems. It has been cheaper to import forest products by sea from North America than to move them across Japan's mountainous terrain, and Japan's international trading firms have operated on a scale that has overwhelmed her many small and fragmented internal timber producers. This pattern underlies Jan Laarman's discussion of Japan as the dominant extractor of Southeast Asia's forest wealth now. As he writes, the future of that region's tropical forest cover is likely to depend in considerable degree on future Japanese markets.

For many parts of the developing world the trade in timber as a major commodity has been a far less important factor in forest depletion than has export agriculture. European colonial systems became adept at turning their colonies' lands to the production of crops for metropolitan markets. The expansion of these croplands at the expense of grasslands and forests has continued steadily from the nineteenth into the mid-twentieth century. John Dargavel, writing on Tasmania, indicates the vast changes that resulted from the introduction of hundreds of thousands of sheep for Britain's woolen industry. Richard Tucker describes the transformation of forest lands in northeastern India into the plantation economy of Assam tea. Peter Castro explains how British settlers claimed lands from the Kikuyu for their estates in Kenya. In a broad sense the French colonial system was similar as it took hold in West Africa after 1920. James Thomson describes the changes of savanna land from its former agropastoralism to the production of cotton and peanuts for European markets. The brief German period in Togo centered on parallel processes of experimenting with various export crops, including both food and rubber. In Brazil, even though the country had been formally independent of Portuguese political control since 1822, similar processes were afoot: large forested areas centering in São Paulo state were transformed into coffee plantations before 1900. Farther south, in the pine belt of Paraná state, as John McNeill describes, coffee was never grown extensively but the pine stands themselves were the major source of employment once railroads were built in the interior shortly before 1900. Indigenous lumber companies responded to the sudden global demand for sawnwood in World War I; they sustained high levels of export until the early 1970s, by which time the Paraná pine belt was reduced to a tiny fragment of its former extent.

International timber trade must be seen within this context of interactions between the globally dominant power of Western capital and regional markets controlled in part by local entrepreneurs responding to regional opportunities for profit. After the Boer War Tasmanian timber was used in British enterprise in South Africa. David Feeny shows that even independent Thailand's teak must be added to the list of timbers controlled by international markets: the height of its export to non-Asian markets came shortly after 1900, when the natural stands of the north were still vast and easily penetrated, and King Chulalongkorn's government had not yet gained effective control over the export economy. In the Soviet Union, Siberian timber exports were indirectly related to this global expansion; they began moving to Japan and elsewhere around the Northwest Pacific in the years before World War I. In this case, Western capital did not dictate the pattern of trade, though it was deeply implicated in the general expansion of the Pacific Rim's economy by that time.

The capital structures of colonial trade and its non-Western partners appear in these papers at several important junctures. Dargavel explores the economics of Australian timber firms in detail and shows that their expansion was intimately related to the development politics of government in both Tasmania state and the national Australian regime. Government support came through new systems of property law and tax concessions. Without that colonial substructure, in Tasmania, Kenya, and eastern India, the firms exporting forest products and plantation crops could not have functioned. In ex-colonial Brazil as well, the government of Paraná was heavily influenced by large-scale land development companies including Europeans in overriding the land claims of squatters.

Behind this lay governments' power to provide the transport infrastructure that was an essential precondition of large-scale commodity export. Forested areas around the world were particularly vulnerable to the penetration of railways supported by governments and financed in part by European and American capital, as shown in places as distant from each other as Paraná, Togo, India, and northern Thailand. The Soviet government followed suit in the 1920s and 1930s, with severe consequences for the forests of European Russia. In the 1950s the Japanese government systematically improved harbor facilities for import of North American and Southeast Asian wood products, in close support of its international trading concerns.

The links between corporate interests and governmental development strategies take us into the realm of political lobbying; this too must be traced in a survey of the forces that have transformed the planet's natural systems. The land companies of Paraná, the colonial planters of Kenya and Assam, and the loggers' lobby of Tasmania all used the power of the state as well as they could. Our most detailed analysis comes in Cox's discussion of the multifaceted timber lobby of the American Northwest as it has attempted to shield itself from economic downturns and Japanese competition.

In the colonial cases the transformation of vegetation systems by these processes has been closely associated with the construction of rural labor forces that uprooted peasants and tribal peoples from the land to produce export crops for minimal wages. Thomson presents our most detailed treatment of this strategy: in French West Africa cash taxes at levels too high for local people to pay forced many men to work for money on the export plantations. He traces the resulting disruption of their traditional primary production strategies and the elaborate kinship system that had sustained them: ecological dislocation and social exploitation were parallel and intimately connected. Tucker alludes to similar processes at work in the mobilization of semi-bonded labor for India's tea estates. In contrast, Castro indicates that a strategy proposed for the Mount Kenya area never succeeded in producing a concentrated rural proletariat. But other restrictions on the Kikuyus' traditional access to forests and grazing lands similarly disrupted the environmental adaptations of precolonial times.

When, if at all, did any awareness of the environmental costs of these trends arise? Indications are that until very recently this has occurred primarily in the industrialized, high-income, high-literacy nations, which are not surveyed in detail in this collection. In other regions, which are under far greater fiscal and developmental pressure, awareness of the ecological costs of deforestation has dawned more slowly. Several of these essays indicate that even in recent years there has been little awareness of anything beyond rising commodity costs. McNeill argues that the grave threat to sustained economic growth in southern Brazil is still not generally recognized. Barr suggests that Soviet planning for both European and Asian lands is still locked into an obsolete variant of Marxist bureaucratic thought, which assumes that problems of resource management merely reflect unresolved inefficiencies of resource extraction. The fact that many of the costs of timber extraction are subsumed under administrative budgets there, rather than being directly reflected in the market, only disguises resource scarcities further.

Williams describes a very different ideological drift relating to American forests at the turn of this century, when President Roosevelt and the founders of the Forest Service began predicting a timber famine in the country whose timber resources had been used most profligately in the previous century. Responding to this alarm, the young forestry profession—in both governmental and corporate employ—began to reforest vast areas of land that had been denuded by 1900. In addition the rapid urbanization of America society had already begun allowing old farms, some of their soils severely depleted, to return to second-growth timber stands. In this, the eastern states' growth of new forests has resembled many mountain areas of Japan and Europe whose forests have been less intensively cut as the rural populace has drifted into urban and industrial settings.

It might seem that the most dramatic example of the new ideology of forest preservation could be seen in the British Empire's forestry service, which was born in India in the 1850s. (These essays, it should be noted, do not cover the earlier rise of continental European forestry, from which both North American and colonial forestry professions were outgrowths. Nor do they touch the autonomous growth of Japanese forestry since the seventeenth century.) The picture that emerges here, however, is ambiguous. From the beginning India's foresters spoke of watershed preservation and sustained growth of the remaining forests, yet they also concurred with their superiors in the colonial regime that lands that could be productively plowed should be turned into croplands. Whether this was an agreement on principle or a tacit recognition of their subordinate status in the colonial regime is a moot point. Forestry management was perennially understaffed and underfinanced, and in the government's debates over fiscal and developmental priorities the advocates of expanded agriculture dominated.

Nonetheless, laws for forest preservation and sustained-yield timber cutting were designed and incorporated into the colonial land law system, in the form of Reserved Forests. The establishment of governmental forest reserves, though, brought colonial regimes directly into confrontation with traditional, and far less monetized, subsistence systems; this was often a vital factor in undermining indigenous forest-oriented societies. By restricting peasant and tribal peoples' access to their former sources of subsistence in forest and savanna lands, the effect of these laws was often to redefine custom as theft of government property. Peasant resistance movements were likely to arise as an expression of the conflict between the old localized systems and the new global strategies.

The most dramatic case in this collection was in Kenya, where early foresters assessed Kikuyu farming methods as disastrous to soil and vegetation but seem to have been much more complacent about European settlers' methods. Further, in demarcating lines between Africans' lands and government forests, they were partners of the planters' desire to transform local labor into rural wage labor. They also saw the usefulness of cutting off public access to mountain lands that might be used as a base for resistance. One origin of the bloody Mau Mau rebellion can be traced to these maneuvers.

Whatever the impact of modern forestry management on land and indigenous people, it is clear from several of these essays that it was closely tied to production of commercial timber. In many countries serious enforcement of forest laws did not begin until years after the first laws were passed. In Sahelien West Africa, though forest laws were eventually passed for Niger and its neighbors, almost all foresters were ultimately sent to the Ivory Coast to assist in marketing its rain forest rather than being stationed in the open dry forests farther north. Extraction of profits remained the dominant purpose of empire.

Reforestation must be taken into account, though, as an issue in this century. Considerable acreage in Europe, North America, and Japan is more heavily forested now than in 1900. Some of this has resulted from the reversion of abandoned farmland to new timber; some has come from systematic replanting by forestry agencies and timber corporations. In the developing world the pattern has been very different. The subject is not systematically assessed in this volume, partly because most of the reforestation efforts are so new that they lie outside the chronological scope of the essays and partly because the total acreage of reforestation as yet is not a major factor in overall vegetation patterns. In many countries budgets have been pointed far more toward tax collection, agricultural expansion, and social control than toward meeting the forest product needs of the rural poor. Moreover, most reforestation has been pointed toward regular cutting of the new stands for commercial marketing. Consequently, as on all continents of the globe, the species composition of most planted forests in the developing world has been radically simplified, emphasizing eucalypti, loblolly pine, and other fast-growing species instead of the far more varied original forest cover. In general the unplanned reversion to second growth, as in Japan and the northeastern United States, has tended to preserve botanical diversity while tree plantations, like their food and fiber crop counterparts, have tended toward monocultures.

The ecological balance sheet is far more complex than this collection can attempt to summarize. But several broad patterns emerge. One of the most frequent and emphatic depicts the ecological disruption of river systems and their watersheds. In southern Brazil the Iguaçu River, which runs through what used to be Paraná pine country, has responded to the demise of the pine forests by producing increasingly severe cycles of flooding and then low volume. In Thailand the depletion of lowland forests when they were replaced by vast rice fields was less disruptive than the felling of teak in the hills to the north and northeast, which threatens to disrupt not only the subsistence of the hill people but agriculture downriver as well. In the Indian Himalayas the combination of rising human and livestock populations, commercial timber operations, and colonial tea plantations has brought increasing flooding to the Indus, Ganges, and Brahmaputra basins, which are home to hundreds of millions of people.

Savanna lands follow different dynamics but are just as fragile, as we see in the example of the open acacia forests of the Sahel. The French colonial system, by dismantling traditional kinship and farming systems, forced an ever-accelerating trend of extensive farming that now is associated with the advance of the desert southward.

South of the Sahel's grasslands, the rain forest is similar to the tropical belts of South America and Southeast Asia. In the true tropics, as Laarman points out, the great shift in extraction patterns came after World War II. Not the least of the challenges that the rain forest presents to further exploitation is its great variety of species in each forest, only a few of which had commercial potential until recently. But with the vast expansion of world markets for wood pulp and the development of new technologies for processing a mixture of wood fibers, the rain forest has become steadily more "valuable." The greatest expansion of tropical logging has come in Southeast Asia, where bureaucratic institutions, transport infrastructure, and investment capital all were more highly developed by the late 1940s than in the tropical regions of Latin America and Africa. In Dutch Indonesia, British India and Burma, and independent Thailand alike, teak was the greatest prize for a century and more. But during the past twenty years or so, a far broader range of tropical species has fallen to local, American, European, and Japanese saws and axes. In recent years the states that depend on timber exports for an important portion of their national revenues have been struggling primarily over control of the wood processing industry. By now very little timber is exported from Southeast Asia in the form of logs; nearly all is processed in one way or another. In this also Southeast Asia leads the two other tropical regions in the development of its exploitation strategies; the ecological implications are open to debate.

How much is known with confidence about the rate of timber harvesting globally and the relative role of Western capital in the harvest? And how closely does this correlate with the expansion of agriculture in all its forms? This underlying problem of collecting adequately detailed, accurate, and comparable data from one time and country to another is well understood in forestry circles, but perhaps less so by social scientists newly interested in the field. Several of the present essays address problems of quantification and the institutional settings of available data. Thomson shows that Francophone West Africa has only recently begun collecting systematic land-use data, for forests as well as arable land. Barr defines the severe difficulty of extracting quantitative patterns from Soviet data. Williams discusses the complexities of gathering and interpreting statistics even for the United States.

The papers on Southeast Asia combine disciplines and use somewhat contrasting methodological strategies to confront the changing variety of data-gathering systems in that region. Feeny has previously analyzed several systems of data for rice production in Thailand; in this essay he relates that to forestry data for the same country. Laarman centers on data on timber cutting and marketing for the years since the late 1940s when it became somewhat systematized. None of the essays would claim a high degree of certainty about its statistical base; our collective effort to assess both the trends in global biomass and the structures of human activity that determine those trends is still in an early stage of development. This collection may help to refine the discussion.

In sum, can we make any conclusions or predictions from these essays? All the contributors are historians in one way or another. As such, their purpose is not to predict or prescribe but to show, first, how broad global trends developed, and second, how broad a range of issues must be considered in order to gain a full grasp of the subject. Most of the contributors, in surveying the political economy of deforestation in the developing world, remind us that the transformation of natural systems has been intimately parallel to the transformation of human systems; the two must be studied in tandem.

In probing the relation of the decline of forest cover to the processes of international trade, it is tempting to speculate on whether industrialization and its attendant social complex are likely to lead to reforestation in the developing countries, as they have in Europe, North America, and Japan. From the nineteenth century well into the present century the industrial countries could reach out to exploitable frontiers on other continents, finding vast untapped resources and no large-scale commercial competitors except in some cases each other. Developing countries today face a profoundly different and more difficult prospect. They have few exploitable forested hinterlands except for portions of the tropical rain forest, and they face intense competition from wealthy industrial societies that are still hungry for their resources and still maintain a high level of access to them. This basic character of the world economy, along with the inexorable expansion of the world's population, means that the recovery of the developing world's forest resources will have to proceed from a very different financial and organizational base.

PART 1

FORESTS IN THE DEVELOPING WORLD

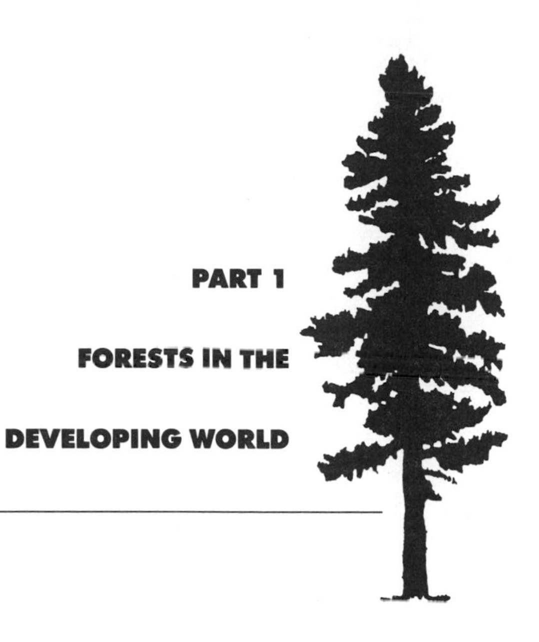

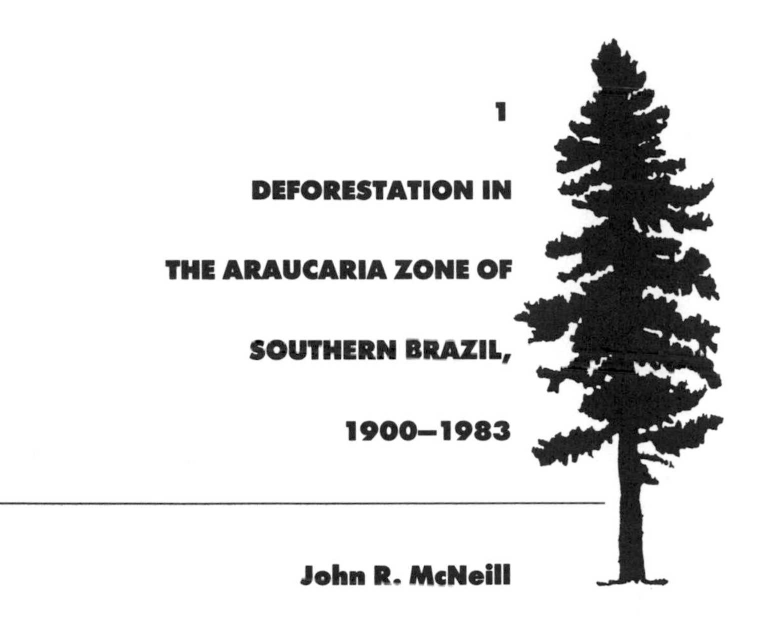

1 DEFORESTATION IN THE ARAUCARIA ZONE OF SOUTHERN BRAZIL, 1900–1983

John R. McNeill

Brazilians agree that God must have been in an especially generous mood when creating Brazil. Among its many bounties Brazil counts forest resources that rank second only to those of the Soviet Union. Slightly more than half of the nation's 851-million-hectare land area is covered with forest. The great majority of Brazilian forest lies in Amazonia, the earth's single largest forest zone. Although it is shrinking daily, and at an ever-climbing rate, the forest of Amazonia is by Brazilian standards untouched. The country's other two major forest zones, while much smaller to begin with, have undergone much more serious depletion. The tropical forest of the long Atlantic coast, from Pernambuco to São Paulo, has been under siege since the sixteenth century, and the largely coniferous forest of southern Brazil has all but disappeared in the span of two human generations. The incorporation of coastal Brazil into a growing world economy (1550 to 1850) encouraged the gradual depletion of the Atlantic coast forests. And the rapid inclusion of southern Brazil into a burgeoning world economy of 1920 to 1980 ensured the quick exhaustion of the coniferous forests of the South.

Only two species of conifer are native to Brazil, and only one, *Araucaria angustifolia,* or Paraná pine, ever existed in quantity. Until the twentieth century, araucaria forests dominated large parts of the three southernmost states of Brazil: Paraná, Santa Catarina, and Rio Grande do Sul. Together these three states cover about 580,000 square kilometers, of which perhaps 253,000 once contained araucaria formations.[1] Stricter definitions of forest land yield 197,000 square kilometers as the original domain of araucaria forest.[2] The state of Paraná contained the most extensive stands, about 40 to 45 percent of the total; the remaining area was divided evenly between Santa Catarina and Rio Grande do Sul. Santa Catarina, the smallest of the three states (95,000 square kilometers in area) was 70 percent covered by araucaria, while Rio Grande do Sul, the largest, featured araucaria on only 24 percent of its 285,300 square kilometers. Paraná state (199,900 square kilometers) originally supported araucaria on 58 percent of its land area. These three states lie between 22 and 33 degrees south latitude.

The natural habitat of the araucaria is roughly between 21 and 30 degrees south latitude—there were some stands in the state of São Paulo and in the Argentine province of Misiones. Generally the araucaria habitat lies in the gentle hills between 700 and 1,100 meters in altitude, although it is somewhat lower to the south in Rio Grande do Sul. It features rainfall of between 1,300 and 1,900 millimeters annually (about 30 percent higher in Rio Grande do Sul), well-distributed throughout the year. Median annual temperatures in the araucaria zone are 15 to 18 degrees celsius.[3]

Araucaria forest normally developed into three or four tiers, with the genus araucaria dominant. It generally grew in association with various subtropical hardwoods and amidst a host of undergrowth. The densest strands supported 65 araucaria per hectare, but the average was only about 26 per hectare. Much of the *campos* or grassland, of southern Brazil was dotted by the occasional araucaria, especially in the neutral strips between prairie and forest. The denser concentrations were mostly in Paraná, while araucaria parkland was most common in Rio Grande do Sul.[4]

The araucaria itself is a sturdy, heavy, and, according to Brazilians, an aesthetically delightful conifer. Normally it grows to a height of 25 to 30 meters, although occasionally it reaches 50 meters. Diameters vary between 50 centimeters and 90 centimeters in most cases, attaining 200 centimeters in exceptional ones. The trunk is normally knot-free and bare of branches for two-thirds of its length. Toward the top, slender branches curve upward from the trunk, presenting a shape reminiscent of a candelabrum. Its specific gravity ranges from 0.50 to 0.61; it grows slowly compared to most conifers; it is very sensitive to soils; it survives to an age of 200 to 300 years; it is fire resistant when young; it yields 2.5 to 3.0 cubic meters of sawnwood per tree;[5] and, in the words of Arturo Dias, "It is a fruit tree, it is an architectural column, it is a first class fuel, it produces a most useful rosin, it is the most beautiful shademaker in the vast plains it dominates, and it is above all a pleasure to the observing traveler, never mind how little of the poet and the artist there may be in his soul."[6]

With all these characteristics, the araucaria is suitable for countless purposes. Authorities have pronounced it superior (or at worst comparable) to Baltic pine. Brazilians have found it useful for the construction of floors, ceilings, furniture, packing cases, scaffolding, and more.[7]

The Disappearance of the Araucaria Forest

Of the 20 to 25 million hectares of araucaria forest once in southern Brazil, about 445,000 remained in 1980. Available data do not permit a confident reconstruction of the rates of deforestation, but the general outline of the forest's fate is clearly discernible.

In the early part of this century, from 1900 to 1930, deforestation proceeded gradually in Santa Catarina and Rio Grande do Sul, but rather more quickly in Paraná, where 9 percent of the state's total forest cover disappeared between 1920 and 1930 (see table 1.1). As early as 1907, a visitor noticed the scarcity of araucaria around Curitiba.[8] Between 1930 and 1945, the pace accelerated somewhat; after 1945 it quickened markedly, continuing to rise until about 1970. In 1950 the rate of deforestation in the state of Paraná amounted to 250,000 hectares per year, about 2.8 percent of the total forest cover annually. In the following decade, the deforestation rate mounted higher still, perhaps approaching 350,000 hectares per year. By 1953 there were very few araucaria stands left in the vicinity of Curitiba, and by 1960 the araucaria in eastern Paraná and all of Rio Grande do Sul had been nearly exhausted. By the 1970s the pace of deforestation had slowed in absolute terms, because little of the araucaria forest was left. Between 1973 and 1977, only about 80,000 hectares of araucaria disappeared, but that amounted to 8 to 10 percent of existing stands in 1973.[9]

Forests not dominated by araucaria have also disappeared in southern Brazil. Paraná state, originally about 89 percent forest, was only 45 percent forested in 1950, with most of the clearing (3.9 million hectares) taking place between 1930 and 1950. The processes that have brought about this deforestation have continued uninterruptedly

Table 1.1 Changes in Forest Area (1,000 hectares)

	Paraná		Brazil
Year	All forest	Araucaria	Araucaria
Origin	17,739	7,650	19,700
1920	16,035	—	—
1926	15,576	—	—
1930	14,359	3,958	—
1937	—	3,455	—
1950	8,940	2,750	—
1955	—	2,450	4,800
1964	[5,000]	1,800	3,500
1965	—	1,593	—
1973	—	434	1,985
1978	—	317	605
1980	[2,000]	[200]	445

Sources: Brazil, Ministério da Agricultura, Indústria e Comércio; Brepohl, "A Contribução económica," 349; Carlos Augusto Ribeiro Campos, *Atlas estatístico do Brasil* (Rio de Janeiro, 1941), 4; Diretoria Geral de Estatística, *Recenseamento do Brasil 1920. III. Agricultura*, 2 vols. (Rio de Janeiro, 1924), 2:vii; FAO, *Proyecto de evaluación*, 38–39, 48, 58; Reinhard Maack, "Plano de proteção des florestas do Paraná," *Anuário Brasileiro de Economia Florestal* 5 (1952): 64.

Note: Figures in brackets are author's estimates.

in southern Brazil and in the last five years have extended on a large scale across the Paraná River into Paraguay. The rapid deforestation of eastern Paraguay is a result of the expansion of Brazilian agriculture, Brazilian settlement (300,000 Brazilians have settled across the river), Brazilian capital, Brazilian markets, and Brazilian companies.[10] Neither national boundaries nor the Paraná River have slowed the frontier processes transforming the temperate forest zones of southeastern South America.

Sources of Deforestation in Southern Brazil

The frontier processes responsible for southern Brazilian deforestation are fairly typical of Latin American frontier zones in the twentieth century. They are, in order of importance, (1) land clearing for agriculture, (2) timber cutting, and (3) fires. Each of these, including fire, is to a significant degree linked to the rather sudden inclusion of southern Brazil into the national and world economy, beginning about the turn of the century. The "opening up" of Paraná, Santa Catarina, and Rio Grande do Sul has meant the disappearance of a large ecosystem peculiar to this corner of the world.[11]

Chronologically, fire made the first inroads on the stands of virgin araucaria. Well before the arrival of Europeans, indigenous cultivators burned forests in order to plant crops among the ashes. These traditional Amerindian methods of cultivation were generally adopted by the Europeans who first settled in the araucaria zone in the nineteenth century. Like the Indians, early white settlers did not hesitate to burn huge swatches of forest in order to cultivate a few hectares, whether for subsistence or eventually for marketing in the tiny towns growing up near the coast. In this way, a very few pioneers could destroy a considerable area of forest.[12]

Although few efforts could have been made to control fires, the impact of intentional burning on forest cover was limited by the preference of shifting cultivators to burn regrowth rather than virgin forest. Polish settlers in Paraná, for example, tried to burn vegetative growth before it reached three meters in height.[13] The labor involved accounts for this preference: much less work is required to cut and prepare for burning a second-growth forest. If population had remained static, nineteenth-century cultivators would simply have burned the same patches of forest over and over again—if fires had not burned out of control.

But intentional fires set by pioneer cultivators often raged far and wide, well beyond what was required for crops. In the early decades of the twentieth century, forest fires were considered a major problem in Paraná. Livestock raising, which became important in southern Brazil very early in the history of white settlement, also imperiled forest land through fire. Range burning, believed to improve forage, easily spread to adjacent forest land. In 1963, one such fire destroyed 2.02 million hectares in Paraná, more forest land than normally burns in a year in the whole United States.[14]

Fire has destroyed forests in southern Brazil for millennia, but in modern times the incidence of fire has increased through human efforts to create crop and pasture land. Because these efforts derive in part from market incentives, even destruction by forest fire is related, albeit indirectly, to the participation of southern Brazilians in the market economy.

Timber cutting in southern Brazil began on a small scale in the eighteenth century. Colonial administrators deplored the preference of the first white settlers for logging over farming. By the early nineteenth century, observers noticed that the leading item of export from southern Brazil was araucaria. Much of it floated down the Paraná River to Argentina, while the rest came down to the coast at Paranaguá. The scale of cutting, however, remained small, and it had no effect on forest land more than a kilometer or two from water transport. Furthermore, the market for araucaria timber was confined to the underpopulated regions of the River Plate and southern Brazil. At the turn of the century, Brazil still imported far more coniferous timber than it exported.[15]

In the 1880s the scope of timber cutting began to widen with the coming of the railway. The first railways came to the region in 1874; Paraná's first line opened in 1883. By 1914 the eastern quarter of Paraná was well served by rail. The locomotives used wood for fuel, so the forests adjacent to the tracks soon disappeared. More importantly, the railway encouraged commercial logging in areas that had previously been inaccessible. In 1908 the Paraná state railway carried 36,000 metric tons of araucaria sawnwood. More than 100 sawmills were in business, several of them steam powered. Sawmills became the single largest source of employment in Paraná, and the stands of araucaria in the eastern part of the state nearest to Curitiba disappeared. The pace of timber exploitation remained modest, however, because local demand did not warrant large-scale operations.[16]

World War I brought a sudden surge in foreign demand for araucaria timber, inaugurating a sixty-year epoch of intensive timber exploitation in the araucaria zone. Brazilian coniferous sawnwood exports increased fourteen-fold from the 1911 to 1914 period to the 1916 to 1918 period. Henceforth production and exports continued to rise, slowly at first, then rapidly in the late 1930s and again in the late 1940s. Exports peaked in the years 1956 to 1971, and have declined sharply thereafter (see table 1.2). In the early years of this epoch, exports accounted for a substantial share of total production: 41 percent in 1921, 32 percent in 1950, and 40 percent in 1965. But beginning in the late 1960s, the export market took increasingly small shares of coniferous sawnwood production. In 1970, 20 percent went for exports, in 1975 only 5 percent, and in 1978 less than 2 percent (see tables 1.2 and 1.3).

Table 1.2 Coniferous Sawnwood Exports from Brazil, 1911–1981

	1911–20	1921–30	1931–40	1941–50	1951–60	1961–70	1971–80	1981
Year 1	7,368	120,300	126,317	495,502	1,094,531	1,104,551	994,000	198,000
Year 2	6,239	168,293	131,867	550,861	645,204	838,290	836,000	
Year 3	19,926	239,216	136,990	478,832	941,606	821,725	688,000	
Year 4	9,701	188,555	178,645	498,807	809,750	1,052,000	382,000	
Year 5	51,301	160,059	218,353	431,575	1,123,459	1,153,000	307,000	
Year 6	118,780	133,498	240,810	793,160	648,075	1,188,708	177,000	
Year 7	76,341	148,281	342,788	836,628	1,364,343	1,032,000	192,000	
Year 8	253,875	133,299	359,957	955,291	1,121,702	1,312,000	178,000	
Year 9	119,607	151,833	514,016	647,346	805,102	1,022,000	239,000	
Year 10	141,758	141,990	412,562	751,278	933,652	927,000	187,000	
Totals	804,896	1,585,324	2,662,305	6,439,280	9,487,424	10,451,274	4,180,000	198,000
Decennial Averages	80,490	158,532	266,231	643,928	948,742	1,045,127	418,000	19,800

Sources: Associão Promotora de Estudos de Economia, *A Economia brasileira e suas perspectivas* 21 (1982), appendix I-11; FAO, Yearbook of Forest Products (Rome, 1981), table D-1.

Table 1.3 Official Coniferous Sawnwood Production, 1905–1978 (1,000 cubic meters)

Year	Paraná	Santa Catarina	Rio Grande do Sul	Brazil
1905	96	—	—	60
1910	—	—	—	—
1921	—	—	—	290
1945	537	486	369	1,415
1947	—	—	—	1,561
1950	1,243	—	—	2,319
1955	1,852	—	—	—
1960	1,145	—	—	—
1965	1,608	1,648	311	2,961
1970	2,946	1,944	372	4,535
1975	2,351	—	—	5,469
1978	2,707	—	—	6,952
1980	—	—	—	7,143

Sources: Aubreville, "A Floresta do pinho," 170; Brepohl, "A Contribução econômica," 348; Carvalho, *Brésil méridional*, 380; Centro Indústrial do Brasil, *O Brasil*, 751; *Anuário Brasileiro de Economia Florestal* 2 (1949), 540; Dammis Heinsdijk, *Forestry in Southern Brazil* (The Hague, 1970), 27; *Unasylva* (1948), 34; James W. Wilkie, ed., *Statistical Abstract of Latin America* 21 (1981), 216.

Note: Because sources are occasionally not in agreement, the Brazil total may be more or less than the sum of its parts.

The decline in araucaria exports since 1970 represents several different trends. As far as absolute quantities are concerned, araucaria exports have diminished because the cost of the product has risen. The necessity to penetrate into increasingly remote parts of the forest has raised costs, and the prospect of exhaustion of natural stands has raised the price further. The principal importers of araucaria have looked elsewhere. The Argentines, who relied heavily on araucaria after 1918, have turned to Chilean softwoods and their own araucaria stands in Misiones. When Buenos Aires was growing at its fastest rate, between 1900 and 1914 and again between 1947 and 1970, timber cutting in the Brazilian araucaria zone also grew rapidly (see table 1.4). In the 1970s Brazilian araucaria producers priced themselves out of their traditional markets and as a result the export trade has come almost to a close.

Table 1.4 Argentinian Coniferous Sawnwood Imports from Brazil, 1935–1977 (1,000 cubic meters)

1935	205	1943	361	1964	568	1972	343
1936	216	1944	374	1965	708	1973	259
1937	277	1945	326	1966	—	1974	161
1938	242	1946	530	1967	607	1975	142
1939	328	1947	599	1968	695	1976	34
1940	270	1948	789	1969	598	1977	75
1941	358	1949	510	1970	541		
1942	439	1950	401	1971	530		

Sources: Brazil, *Anuário Brasileira de Economia Florestal* (1951), 17; FAO, *Yearbook of Forest Products* (Rome, 1965–77).
Note: After 1977 the FAO figures are arranged in such a way that no data specific to the Brazil-Argentina trade are available.

The decline of araucaria exports as a proportion of total coniferous sawnwood production in Brazil represents both a decrease in araucaria production and the rapid rise of other softwoods, notably eucalyptus, loblolly pine, and other fast-growing plantation species. These plantation pines generate much higher yields than araucaria and can probably account for the rapid increase in total coniferous sawnwood production in the late 1970s (see table 1.3). The development of eucalyptus and plantation pines, it should be noted, took place mostly in southeastern Brazil, not in the araucaria zone at all.[17]

The rise of plantation pine at a time when coniferous sawnwood exports almost disappeared testifies to the rapid growth of the Brazilian domestic market. Araucaria, although comparatively scarce and expensive by the late 1970s, figured in this market. Population and housing construction have boomed in southern Brazil in recent decades (see table 1.5), and the large markets of Rio de Janeiro and São Paulo are nearby. As of 1979 the largest market for araucaria was within the state of Paraná; the second largest was the rest of Brazil; the export market, formerly so important, ran a very distant third.[18] Clearly it was the domestic market that in the late 1970s and early 1980s permitted the continued exploitation of increasingly scarce stands of araucaria.

All of these remarks are based on official statistics published by Paraná, Brazilian, and international agencies. They do not however, accurately cover the full extent of timber cutting in the araucaria zone. The Paraná River, the western border of the state of Paraná, is navi-

Table 1.5 Cultivated Area in Paraná, Santa Catarina, and Rio Grande do Sul, 1920–1980 (fallow included) (1,000 hectares)

Year	Paraná	Santa Catarina	Rio Grande do Sul
1920	223	117	756
1931	292	179	1,170
1940	619	343	1,323
1950	1,358	670	2,503
1960	3,471	1,013	3,796
1970	4,747	1,333	4,997
1975	5,544	1,425	5,897

Sources: Instituto Brasileiro de Geografia e Estatística, *Anuário estatístico do Brasil* (1946), 91–92; ibid. (1974), 170; ibid. (1978), 339; Instituto Brasileiro de Geografia e Estatística, Serviço Nacional de Recenseamento, *Sinopse preliminar do censo agrícola 1960* (Rio de Janeiro, 1963), 2–3; Ministério da Agricultura, Indústria e Comércio, Diretoria Geral de Estatística, *Recenseamento do Brasil 1920. III. Agricultura* (Rio de Janeiro, 1924), III-2, xiv.

gable for about 2,060 kilometers of its 4,700-kilometer length. Since the 1870s steamers have chugged upstream as far as Mato Grosso, the state immediately north of Paraná. Although most of the araucaria cut early in this century stood in the eastern part of Paraná and found its market via the railroads, since the 1940s a great deal of cutting has taken place in the southwestern part of the state. How much of this floated downstream to Argentina and Uruguay without making an appearance in official ledgers cannot be known. Today timber felled in eastern Paraguay crosses the Paraná to Brazil undetected by governments. The tradition of lumber smuggling is a venerable one in this part of the world.[19]

Official lumber exports account for the cutting of about 534,000 hectares between 1911 and 1981, assuming a yield of 65 cubic meters of sawnwood per hectare. Official araucaria production might account for three or at most four times as much—no more than 2 million hectares.[20] Thus the total area deforested by registered timber cutting between 1911 and 1981 can amount to no more than 2.5 million hectares, perhaps 10 to 15 percent of the original araucaria zone. Since more than 95 percent of the zone had been cleared as of 1981, this leaves a great deal of deforestation unaccounted for. Certainly illegal timber operations took a share, as did accidental fires from lightning and other sources. The rest of the forest, probably the majority of the araucaria zone, was either used for fuelwood or was cleared away for agriculture; either way, it went up in smoke.

The history of agriculture and settlement in the araucaria zone is also the history of its deforestation. This is of course true of many places around the globe, but in southern Brazil two matters stand out: human occupation has almost completely transformed a forest ecosystem unique in the world; and the transformation has been a very wasteful one, because the occupants derived much less economic benefit than they might have from the felling of a half billion valuable trees.

Settlement in the araucaria zone has proceeded along three main paths. The first of these emerged from the first white settlements along the coast and around Curitiba and was associated with food crop agriculture. The second path, also associated with food crops, carried a current of migration north from Rio Grande do Sul, first into Santa Catarina, and then by 1940 into Paraná. The third path brought a torrent of settlers into northern Paraná in the lee of the westward migration of the São Paulo coffee zone. This third path, which reached its apogee in the 1950s, affected the araucaria zone only marginally because most of the coniferous forest lay to the south of good coffee land.

Prior to 1860 the forest zones of southern Brazil supported very little agriculture. Most farming took place near the coast: Paraná was originally developed in order to help feed São Paulo. Only the grasslands were settled (mostly in Rio Grande do Sul), and the scattered inhabitants of the forest were loggers or gatherers of *erva mate* (*Ilex paraguariensis*), or Paraguayan tea, a favorite beverage throughout the region.[21] Settlement of the inland upland araucaria zone was hampered by the coastal escarpment especially prominent in Paraná. Rivers there that rise within sight of the Atlantic flow westward, eventually contributing to the Paraná or Uruguay rivers. The araucaria zone is for the most part the natural hinterland of Buenos Aires rather than any Brazilian port. Only the railways in the late nineteenth century reduced the formidable difficulty of communication between the coast and the interior.

As Curitiba grew, agriculture pierced the forest in order to feed the city. Settlers within perhaps 15 kilometers of the tracks could market their surplus. The governments of Paraná and Brazil took an active hand in promoting this development. Immigrants from Europe were recruited and settled in frontier districts throughout the region.

In eastern Paraná, Ukrainians, Poles, and Ruthenians figured prominently; in Santa Catarina, Germans formed the largest group; and in Rio Grande do Sul Italians led the way. In the two southernmost states, however, settlements normally were not made on forest land but in the grasslands. The process of state-sponsored colonization slowed after 1914, but these early immigrants, their descendants, and their ways of adapting to new environments made a strong and lasting imprint on the whole region.

When faced with abundant forest, these European immigrants practiced the same agricultural techniques known for millennia in the Americas. Swidden, or shifting, cultivation became their way of life. They cut and burned the forest, planted, and after a few years moved on, completely abandoning the traditions of settled agriculture they had known in Europe. One contingent of immigrants, the French from Algeria, were told in 1907 that there were no plows in the whole state of Paraná. Slash-and-burn cultivation agriculture seemed a "corruption" to many Brazilian officials, but they did not assist immigrants to move to plow agriculture once lands had been parceled out. By 1910 Poles and Ukrainians had pushed 50 kilometers west of Curitiba, slashing and burning as they went. As one contemporary observer put it, "colonization and deforestation are synonymous." Some of these cultivators eventually settled down, becoming permanent residents producing for market in Curitiba. But the majority became *caboclos*, as shifting cultivators are known in Brazil. Still, by 1908 colonists' lots occupied only about 1,000 square kilometers in Paraná, about 0.5 percent of the state's total land area.[22]

As these settlers pushed westward from Curitiba, another movement began, northward from the prairies of Rio Grande do Sul into the forests of Santa Catarina and eventually Paraná. This migration involved Italians and the descendants of German immigrants (German immigration had dwindled to a trickle after 1866 for reasons unconnected to Brazil). These migrants brought with them the custom of swine raising, but also engaged in food crop cultivation, generally of maize, beans, and manioc. They, like the Poles and Ukrainians, learned the traditional caboclo methods when they moved from grassland to forest. This movement began as early as the late nineteenth century but did not become important in Paraná until about 1940. In contrast to the first movement, this migration was composed entirely of unaided volunteers; neither state nor federal government played any role.[23]

The third major path of migration into the araucaria zone was a fringe current of the torrential flow of coffee and Paulistas (as residents of São Paulo are known) westward into northern Paraná. While this development looms very large in the history of the state of Paraná—several million hectares of forest have been replaced by coffee since 1930—it is of secondary importance for the araucaria zone. Coffee cannot flourish where frosts occur, and most of the araucaria zone lay too far to the south for coffee.[24]

Fire, timber cutting, and agriculture have reduced the araucaria forests of southern Brazil from about 20 million hectares to less than a half million hectares. The greatest part of this forest clearing, probably more than 90 percent of it, has taken place in the twentieth century. Inhabitants of southern Brazil have supported themselves at the expense of the forest. They have also, however, inadvertently destroyed much of the forest without deriving any economic benefit from it whatsoever. A billion or more cubic meters of araucaria have been burned instead of used because conjunctures of timber and agricultural prices made it profitable. And now Brazil is short of softwood. Precious little has been done about this great waste.

Forest Policy

In Brazil land not privately owned belongs to the states, not to the federal government, so the responsibility for forest management falls largely on the state—largely, but not entirely, because since 1941 national parastatal agencies have existed to regulate forest exploitation and timber production. Federal authorities also influence the fate of forest land by sponsoring colonies, which in the recent history of southern Brazil have been failures—uneconomic settlements with military purposes, placed near the national borders. Both state and federal authorities issue titles and make sales of land, which has led to endless confusion and lawsuits in Paraná and elsewhere.[25]

Federal and state authorities have exhibited the same halfhearted concern for forest conservation in southern Brazil. Despite several pronouncements and laws ostensibly designed to prevent shortsighted exploitation of natural resources, state and national officials have stood by while the araucaria forest has all but disappeared. The Paraná state government has shown itself especially eager to sell rights of exploitation to development companies and colonization schemes, several of which, particularly before 1945, were foreign owned. British, American, Argentine, and Brazilian companies have found congenial the official attitude toward natural resources, which, in the words of one Brazilian geographer, has been "colonial."[26]

The first forest code in Brazil was passed in Rio Grande do Sul in 1889. Created to protect dwindling stands of useful timber, it had no such effect. In Paraná forest legislation dates from 1907 when the state began to limit the quantities of timber that might legally be felled. In 1941 the federal government set up the National Pine Institute (INP) to oversee the wood products industry in the araucaria zone. A 1965 forest code permitted only "rational exploitation" of araucaria. In the 1960s the INP was subsumed into the Brazilian Institute for Forest Development (IBDF), charged with making the best use of Brazil's forest resources. These laws and institutions, despite all their quotas and licenses, have failed to prevent araucaria stands from disappearing to the point that their economic contribution to the region is almost negligible.[27] The IBDF has a large job ahead of it in Amazonia, but in the araucaria zone it is too late: there is very little left to oversee.

Neither the state governments in southern Brazil nor the federal government in Brasília has devoted the necessary effort and resources to stem the tide of deforestation in the araucaria zone. They have not done so in part because they have more pressing concerns than the conservation of forests, in part because their attitude toward the bounty of nature has been "colonial," and in part because it has proved lucrative to permit land and lumber companies to operate without serious inhibition. In Paraná many officials of the state government (and their close relatives) were deeply involved in some of the companies developing the western part of the state in the 1950s and 1960s. On the whole, governmental authority has chosen to let private enterprise and market forces determine the fate of the araucaria forest.

Even had the state governments been more disposed toward a careful husbanding of timber resources, they would have had great difficulty preventing the disappearance of the araucaria forest. Much of the timber cutting has been illegal, taking place in isolated regions and bypassing the quotas of the INP. And much of the settlement and clearing has been undertaken by squatters, occupying state, company, or private lands. These spontaneous colonists have withstood every effort by companies and private landholders to evict them. So even enthusiastic and vigilant state authorities would have found it a daunting task to stop logging and conversion of forest to agricultural land, since market incentives—and latifundia conditions in the rest of the country—ensured an army of eager pioneers. With the tempered enthusiasm of the Brazilian authorities, forest exploitation in southern Brazil has scarcely been impeded. Indeed, by encouraging development by land companies, which often engaged in no more than logging, the state governments (at least in Paraná) accelerated the process.

Rather belatedly, authorities have embarked upon a program of reforestation in southern Brazil as part of a major national effort dating from 1965 legislation. Through a complicated system the federal government approves areas for reforestation and awards fiscal incentives to private companies that agree to plant new trees. Embezzlement and other inefficiencies have hampered the program, especially in the early years, so that the actual reforested area usually amounts to about 60 to 70 percent of the approved area. Aggregate statistics exist only for approved area.[28]

In the three southern states the reforested area amounts to only a tiny fraction of the deforested area (see table 1.6). Considering all species, a total of 1.42 million hectares have been approved for reforestation, of which perhaps 1 million hectares have been planted. This comes to less than 5 percent of the original araucaria forest area, and a good deal less of what was the total forest area a century ago in the three southern states.

Since the araucaria grows slowly, other species, notably loblolly pine and eucalyptus, have been preferred for plantations (1.17 million hectares of the 1.42 million hectares approved have been for pine and eucalyptus). Since the araucaria regenerates poorly, other species dominate wherever second-growth forests have emerged. Total afforestation with araucaria amounts to about 50,000 hectares, or 0.3 percent of the araucaria area deforested since 1900. Araucaria reforestation, which began in the 1940s in a small way, peaked in the early 1970s, and in recent years has all but ceased. Since 1980 the total annual area approved for araucaria reforestation has never exceeded 700 hectares. So although the original araucaria zone has significant plantations of pine and eucalyptus, the government's reforestation efforts since 1967 have concentrated on other areas (particularly Minas Gerais) and other species. The disappearance of the araucaria forest is to be permanent.[29]

Table 1.6 Area Approved for Reforestation in Brazil by Species and by State, 1967–1985 (in hectares)

State	Species							Total
	Pine	Eucalyptus	Araucaria	Nativas	Fruit trees	Heart of palm	Other	
Acre					1160.00			1160.00
Alagoas		2800.00			7585.00		6640.00	17025.00
Amapá	25660.00	1500.00			450.00		1050.00	28660.00
Amazonas		800.00			26925.00	13632.00	500.00	41857.00
Bahia	279637.05	194351.72		1305.79	25494.62	5200.00	2760.00	508749.18
Ceará					81653.35		2881.00	84534.35
Distrito Federal	9091.42	15724.69		468.11	1248.91			26533.13
Espírito Santo	1213.37	142971.22	45.00	2125.94	686.00		125.76	147167.29
Goiás	29719.99	81251.10	1400.45	3067.62	14996.30		6007.00	136442.46
Maranhão	6300.00	14799.50		0.50	37399.50		6200.00	64699.50
Mato Grosso	961.08	15450.00		10.52	950.00			17371.60
Mato Grosso do Sul	64525.69	457927.61		2212.11	1196.00		1345.00	527206.41

Minas Gerais	188340.42	1506407.27	559.94	14693.00	15799.06		569.67	1726369.36
Pará	3500.00	4000.00			18556.87	42260.00	8420.00	76736.87
Paraíba		1000.00			9065.00		56360.00	66425.00
Paraná	541863.93	63629.57	48968.66	4772.77	12166.10	131294.50	1578.98	804274.51
Pernambuco		4400.00		3.00	11906.65		27310.00	43619.65
Piauí		32100.00			84324.58			116424.58
Rio de Janeiro	2551.07	18195.30	14.67	1038.77	1.45	55.00		21856.26
Rio Grande da Norte		3450.00			59531.60		24274.50	87256.10
Rio Grande do Sul	150679.21	46780.63	7921.24	1852.40	10265.56	1228.24	55653.31	274380.59
Roraima					720.00			720.00
Santa Catarina	280512.08	33548.43	14772.98	1229.92	7244.85	3759.02	798.15	341865.43
São Paulo	192618.09	414771.78	3204.07	7241.20	21783.48	37313.90	3502.58	680435.10
Sergipe		500.00			1199.00			1699.00
Total Brazil	1777173.40	3056358.82	76887.01	40021.65	452308.88	234742.66	205975.95	5843468.37

Source: IBDF unpublished data.

Note: The actual reforested area usually equals 60 to 70% of the approved area, often less in the 1960s and more in the 1980s.

Conclusion

The overall impact of the transformation of the araucaria zone is difficult to assess. Perhaps the economic effects are most clear. Logging and sawmilling, once the single largest source of employment in Paraná (12,100 workers officially in 1960), is greatly diminished. The genus araucaria has almost ceased to exist as an economic asset. Socially the processes that produced deforestation have in many ways been in conflict with one another. Violence between land or lumber company gunmen and settlers has been routine in Paraná, occasionally requiring the intervention of the Brazilian army. Overlapping land titles and differing ambitions concerning land use have at times created a lawlessness Brazilians compare to the American Wild West. This violence has waned lately, as settlements mount and logging declines.[30]

The ecological consequences of deforestation in the araucaria zone are rather hard to specify. Certainly the araucaria ecosystem is on the verge of extinction; what the costs inherent in that may be no one can confidently say. To some it is a tragedy, to others, a matter of indifference. Some clear-cut disadvantages have emerged, however. The drainage basin of the Iguaçu River, the border between Paraná and Santa Catarina, has been so disturbed that the water level no longer permits navigation. Periodic flooding from disrupted drainage and drought produced by local climatic change are both unhappy possibilities.[31] Without luck and careful management—neither of which is in the least bit assured—the recent agricultural prosperity of the araucaria zone might become a fleeting phase. Only time will tell, and probably not very clearly, what the ecological cost of the conversion of araucaria forest will be.

Even should the costs turn out to be minimal, the replacement of forest by agriculture and pasture has scarcely been a tale of unmitigated success. The waste involved has been stupendous; only a minority of the potential 1.5 billion cubic meters of sawnwood ever found a use. Many more people might have profited if the same process had been more rationally organized. But whatever the results, however the costs may compare to the benefits, the path has already been chosen, and quite irrevocably. Through the absence of policy, itself a matter of deliberate choice, market incentives have led to the human occupation of the araucaria zone and a thorough transmutation of the landscape and society of southern Brazil.

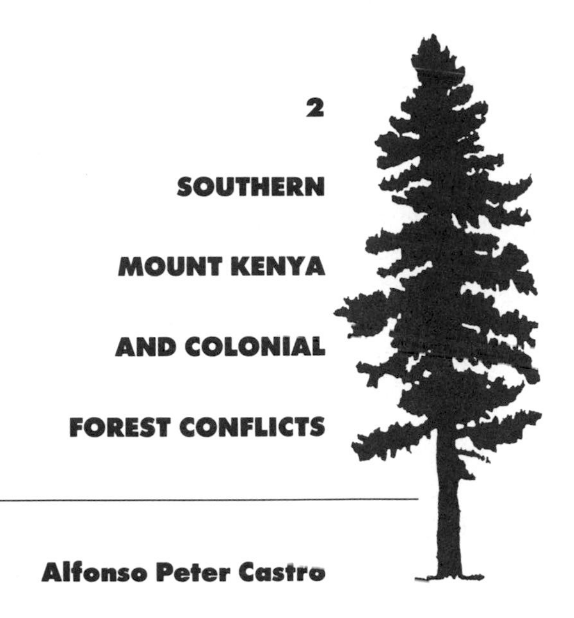

2 SOUTHERN MOUNT KENYA AND COLONIAL FOREST CONFLICTS

Alfonso Peter Castro

This essay presents a microhistory of the imposition of colonial government control over the forests of southern Mount Kenya. It examines the response of the Ndia and Gichugu Kikuyu peoples of Kirinyaga district to this intervention and how this response changed over time. The time frame covered spans the 1880s, when the whites first penetrated the interior of Kenya, to the 1950s and the Mau Mau rebellion. During this period Kirinyaga was at varying times part of Fort Hall, Nyeri, and Embu districts.

The transition from customary to state control and its impact on forest access, exploitation, and management are vital issues for understanding contemporary forest use patterns. Yet little information is currently available on these topics.[1] In such countries as Kenya, where the system of public administration was essentially inherited from colonial rule, knowledge of this past is indispensable for comprehending the traditions of formal forest management and local attitudes about it. Further, by becoming aware of the wealth of experience already available in terms of tropical forest management systems, we can better examine and consider the options and proposals that we have today.

From the perspective of Kenyan history, studies of colonial land alienation have usually centered on issues and conflicts generated by European settlement.[2] However, large tracts of land were also alienated by the colonial administration and set aside as "Crown Forests."[3] By the late 1930s over 13,000 square kilometers had been designated as government forest reserves.[4] The Kenyan Forest Department was responsible for their management. Much of the forest had been occupied or otherwise used by the indigenous people, and its loss often resulted in bitter and protracted controversies similar to those caused by land alienation for white settlement. The reservation of forests sometimes raised protests among white settlers as well, who asked that such resources be turned over to "free enterprise." Surprisingly, the socioeconomic history of the Kenyan forest sector has yet to receive extensive treatment.

The focal point of this study is the Mount Kenya forest bordering on and located within Kirinyaga district.[5] The Mount Kenya reserve, including national parkland, presently covers about 2,600 square kilometers, of which Kirinyaga has 355 square kilometers in a wedge-shaped portion radiating southward from the summit (see figure 2.1).[6] Mount Kenya, which rises to 5,199 meters in altitude, is characterized by a pronounced ecological zonation based on elevation, climate, and vegetation. The glaciers and screes at the summit soon yield to an afro-alpine moorland, which itself gives way at around 3,400 meters to a thick belt of bamboo (*Arundinaria alpina*). The bamboo is supplanted at about the 2,400-meter contour on the southern slopes by a moist montane forest with giant East African camphor (*Ocotea usambarensis*), a commercially valued hardwood, plus podo (*Podocarpus gracilor*), one of the few economically important indigenous softwoods. Other commercial timber species include *Surega procera, Cassipourea malosaria,* and *Macaranga kilimandscharica*. The camphor forest is located in the wettest part of Mount Kenya, with annual rainfall usually exceeding 2,000 millimeters per year. Below roughly 1,950 meters in elevation in Ndia and 1,800 meters in Gichugu the forest is replaced by farmland. At one time the primeval forest undoubtedly extended much farther down the slopes.[7]

The indigenous people of Ndia and Gichugu, like other Kikuyu, customarily regarded Mount Kenya as the resting place of Ngai—the supreme deity. When people prayed they faced the sacred peaks. On occasion groups of men also went into its forest "to pour out fat as a sacrifice to God that all things . . . might prosper, and diseases disappear."[8] However, the sacred mountain was important to the people of Kirinyaga in another way: it was a frontier for the expansion of their agricultural and pastoral economy. They also used the mountain and its resources for honey collection through the hanging of beehives or the finding of wild hives; for hunting, particularly to get ivory for trade

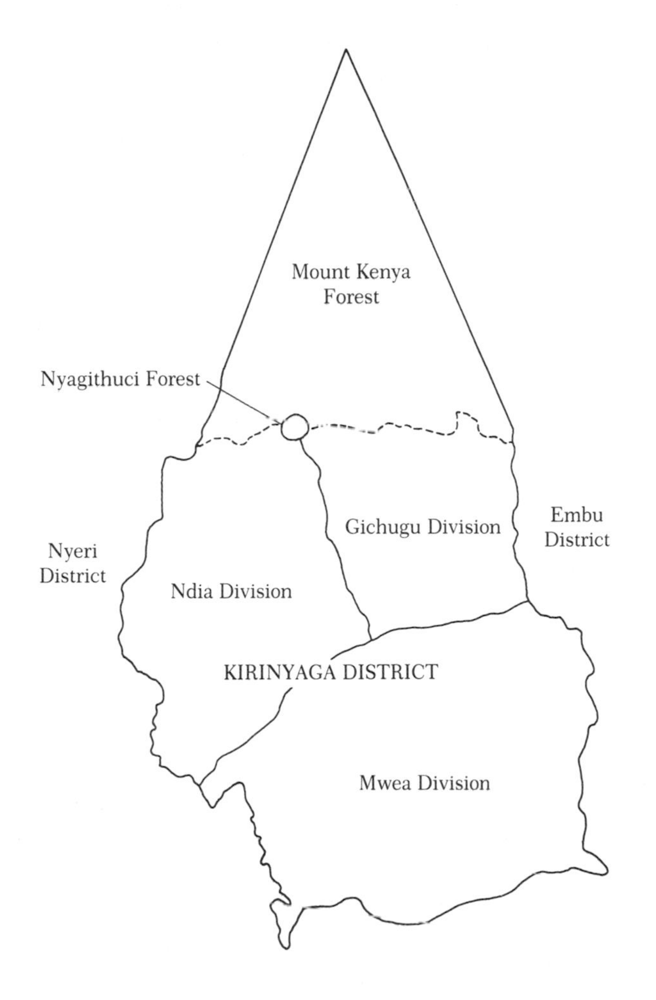

Figure 2.1 Southern Mount Kenya Forest, Kirinyaga District, Kenya. *Source:* Adapted from *Kirinyaga District Development Plan, 1977–1983* (Nairobi: Ministry of Economic Planning and Development, Government of Kenya, 1980).

with caravans from the coast; for gathering wild plants, not only for food, but also for medicines, dyes, and other products; and for the removal of wood for fuel, building, and other purposes. The Ndia and Gichugu Kikuyu had only lightly penetrated the upper reaches of the mountain by the turn of the century. A few medicinal plant collectors journeyed to the moorlands, and groups of men cut bamboo for trading and home use.[9] However, the issue of customary forest use became highly controversial in the colonial era when ownership rights to the forest were contested.

The essay is divided into seven sections. It begins by discussing the rise of European interest in Mount Kenya. Section two describes the actual transfer of the forest from local to government control. The third section reviews the forest management system introduced by the British and the African reaction to it. In section four Kikuyu claims presented to the 1932–1934 Kenya Land Commission are examined, and the following section relates the conflict that occurred with the settlement proposed by the commission. The sixth section considers conflicts associated with pitsawing during the 1940s and early 1950s and their relationship to the Mau Mau revolt. A brief conclusion follows.

The Rise of European Interests

Although Mount Kenya was observed from a distance by a European as early as 1849, over 30 years passed before white explorers visited the area. The Europeans were impressed by what they encountered: a land of pleasant climate where "magnificent" primeval forests were giving way to fertile and highly productive "gardens" cultivated by an "industrious" people.[10] One traveler in the countryside now comprising Kirinyaga district wrote that the land was "extraordinarily rich and fertile . . . and produces . . . a practically unlimited food supply."[11] He recorded that food was abundant despite a smallpox epidemic. Carl Peters stated that Kikuyuland was "beyond all question, the pearl of the English possessions, with the exception of Uganda."[12] The early European explorers appreciated the agricultural wealth of the Kikuyu because their caravans (and those of others involved in the ivory trade) depended on the availability of food surpluses for survival.

In 1895 the British declared Kenya a protectorate, and by 1900 a government station was established at Fort Hall in Murang'a, along the southwestern boundary of modern Kirinyaga.[13] Though the hostility of the Nyeri, Ndia, and Gichugu Kikuyu temporarily halted the expansion of colonial administration, "punitive" expeditions in 1904 and 1905 forced Kikuyu submission to British rule.[14] Administrative control was extended as far as the Rupingazi River, the present boundary between Kirinyaga and Embu.

The European evaluation of Kikuyuland and the Mount Kenya area, including the Africans' relationship to the forest, underwent considerable change in the early 1900s. The building of the Uganda railroad and the opening of Kenya to white settlement altered the potential value and use of land, natural resources, and the labor of the indigenous population.[15] Forests now became commercial assets, another resource to contribute to the development of the budding colonial economy. The colonial administration also began to see the indigenous population as a threat to the forests. This new attitude was heralded by Sir Charles Eliot, the first official advocate of making Kenya a "white man's country."[16] He thought highly of the southern Mount Kenya region and felt that it would attract European settlement. Unlike the earlier favorable accounts of Kikuyu agriculture, Eliot stated that their "wasteful methods" of cultivation destroyed forests and soils.[17]

The first move toward European use of the southern Mount Kenya forest came in the form of two applications for private timber concessions. Syndicates headed by the Earl of Warwick and Moreton Fruen each requested over 40,000 hectares in 1905. These applications as well as the alleged "lawlessness" of the Embu influenced the government's decision to attack them in 1906.[18] Provincial Sub-Commissioner Sidney Hinde believed that once the Embu were "pacified," the concession could be granted and European settlement would begin. However, the Colonial Office in London deferred its decision on the concessions until reports were submitted by Sir James Sadler, the protectorate's commissioner, and David E. Hutchins, a forester with experience in South Africa and India. The delay was motivated by the knowledge that considerable forest had already been alienated by white settlers without prior surveys as to its value.[19] Sadler's brief report proved inconclusive, but Hutchins's findings were more thorough and significant.

Hutchins stated that the timber, watershed protection, and climatic values of the forest were greater than previously thought.[20] He observed that the quality, composition, and value of the forest varied according to geographical position, with the southeastern slopes having the best forest in terms of commercial timber potential. Hutchins advised placing the Mount Kenya forest under government control rather than private European ownership to ensure proper management. His recommendation was accepted by the government.[21]

Hutchins was not against private exploitation of the forest. On the contrary, he outlined several schemes for developing its timber and agricultural potential, including plans for regulated private timber concessions that would use a railroad or the Tana River to transport logs to the coast for export.[22] His most provocative recommendation was to introduce European farms as a buffer zone between the forest and the Kikuyu. Hutchins proposed that settlers be offered land between roughly the 1,500- and 2,100-meter contour. According to Hutchins, the land use practices of the Kikuyu were simply too destructive to allow them to remain adjacent to the forest.[23] He accepted the view of geologist H. B. Muff that Kikuyu agriculture converted fertile soil into "barren rock." Hutchins also cited an estimate by Hinde that in the neighboring Nyandarua range the Kikuyu were causing the forest belt to recede a half mile a year.[24] Not only was deforestation causing soil erosion, but it was also said to disturb stream flow and disrupt rainfall.[25] Besides the environmental concerns, Hutchins suggested that leaving the Africans next to the forest posed a security risk. In a 1909 report Hutchins pointed out that "at each Kaffir war in South Africa, forests such as the Kenia forest were the refuge of the Kaffirs, and to finally dislodge them entailed an expenditure of millions and the loss of many lives."[26] Although land was opened for settlers on the sparsely populated western side of the mountain, the densely populated southern slopes remained closed to white settlers and land speculators.

The decision to deny the applications for concessions was made soon after Hutchins's report.[27] As late as 1912, however, some settlers complained that the government's intentions toward Mount Kenya were "shrouded in mystery."[28] But by 1919 it was clear that the administration wanted to institute conservation measures and systematic reforestation before allowing more forest to be distributed and destroyed. A conflict soon emerged between the Kenyan Forest Department and entrepreneurs interested in exploiting the country's timber resources.[29] The settlers saw the creation of government forest reserves as an attempt to "choke off" resources that "cry for development."[30] For example, Lord Cranworth accused the government of "hoarding" forests because it feared "private individuals or companies, or, worse still, a speculator," might "make a profit."[31] He also suggested that the foresters were overly protective of the camphor forest, noting, "These trees have been to the authorities even as an only child to its mother."[32] Cranworth concluded that the Forest Department had "successfully resisted" the "effort to establish private enterprise on Mount Kenya."[33]

The Demarcation of the Forest

While the fate of Mount Kenya was being debated in Nairobi, the Kirinyaga Kikuyu and their neighbors still occupied the forest. As late as 1910 no restrictions had been placed on their use of the forest. They continued as before, clearing fields, setting fires to extend or to improve pastures, cutting timber, collecting honey, and removing other forest produce.

Kikuyu forest exploitation became a subject of great controversy. It was said that "Pax Britannica"—the supposed end of "tribal warfare" brought on by the onset of colonial rule—caused deforestation to accelerate. According to Cranworth the Masai had been "the real forest conservator" because their raiding had forced the Kikuyu to maintain a belt of forest as a buffer zone.[34] Once the Masai threat was eliminated, it was argued, the Kikuyu initiated their "wholesale spoilation" of the forest.[35] The Routledges, who observed forest clearing near Nyeri and the Nyandarua range, protested the lack of preventive action by administrators.[36] McGregor Ross, director of public works for many years and a staunch defender of African land rights, testified at the Kenya Land Commission that massive deforestation occurred in the decade after 1900.[37] The surveys carried out by Hutchins confirmed the reports of wide-scale forest destruction.

These accounts contrasted with a description of southern Mount Kenya by Captain C. H. Stigand who visited Kirinyaga in 1907.[38] Like other writers, Stigand noted that the Kikuyu as a whole were "always extending and eating their way into the forest."[39] But along the southern edge of the Mount Kenya forest he encountered a different situation. Forest encroachment was halted, at least temporarily: "Much of the country round here is covered by scrub and bush which bushes up to the forest edge. It appears that this land used to be either good grazing ground or under cultivation. The people having lost many of their cattle, the undergrowth has quickly again claimed the land."[40] Unfortunately Stigand failed to mention the cause of this apparent depopulation. The likely reasons for this retrenchment were the costly punitive expeditions and smallpox epidemics. Whatever the source, Stigand was amazed by the lack of Kikuyu advance on the Mount Kenya forest.[41]

To some extent the accounts of Stigand and the other writers can be reconciled. Major G. St. J. Orde-Browne pointed out that deforestation on southeastern Mount Kenya was "retarded" compared to the damage sustained on the Nyandarua range.[42] Clearing may have increased in the period after Stigand visited Kirinyaga, since the district exported "a large quantity of grain" to Nairobi in 1908.[43] Misconceptions about Kikuyu agriculture may have also caused some writers to exaggerate or to overestimate its environmental impact. The issue of Kikuyu deforestation requires further historical attention, because it was used as the primary argument for establishing forest reserves.

The outcry about Kikuyu deforestation caused the administration to order the immediate demarcation of forest reserve boundaries on Mount Kenya and the Nyandarua range.[44] Despite numerous claims about disappearing forests, officials lacked accurate maps showing the extent of the montane woodland. Furthermore, there was no firm operational definition of what constituted "forest." Aware of this, the central government instructed provincial and forestry officials not to wait for precise yet time-consuming surveys before marking the borders. Years later the conservator of forests admitted that the boundaries were essentially "give-and-take" ones.[45] That is, in some places farmland and pasture were included within the forest line, while patches of woodland were occasionally excluded. To facilitate demarcation the Forest Department preferred to draw long, straight boundaries. These "lines of convenience" added an element of arbitrariness to the selection of "forest."[46] The provincial commissioner was delegated responsibility for settling any possible boundary disputes.

Demarcation commenced on the Nyandarua range, the area perceived as experiencing the worst forest damage.[47] By 1910 work was underway on the Nyeri side of Mount Kenya, reaching Kirinyaga roughly a year later. Using local "compulsory and unpaid labour,"[48] a wide strip was cleared between forest claimed by the government and land belonging to the Africans. The Kikuyus who lived north of the forest reserve line were evicted, their land rights being unrecognized. A line of eucalyptus trees was planted along the clearing to mark the exact boundary. Demarcation was completed by 1913. The administration now controlled Mount Kenya.

The Regulated Forest

From its inception in 1902 the Kenyan Forest Department stressed two policies: protection of forests for environmental and economic reasons and limited exploitation of woodland based on the principle of sustained yield.[49] Ordinances passed in 1902 and 1911 formed the framework for forest administration. The latter was modeled after the Indian Forest Act and the South African Cape Forest Ordinance.[50] These ordinances provided for the reservation or "gazettement" of government forests. They also restricted access to and use of the forest reserves, instituting a system of royalties and penalties for the removal of forest products. People were required to purchase a license to collect deadwood for fuel, poles, or other uses. In addition permits were required to cut timber, hang beehives, or graze cattle in the reserve. Farming, setting fires, and herding goats in the forest were prohibited. Restrictions on the removal of forest produce created problems for Kikuyus who collected medicinal plants and tree parts. For example, forestry officials rejected requests for permission to gather *ngaita* (*Rupanea rhododendroides*) seeds and *muthura* (*Ocotea kenyensis*) bark,[51] which were used in the treatment of respiratory ailments and other diseases. Violators of forest regulations faced fines or imprisonment.

The task of maintaining the forest boundary was held initially by the government-appointed Kikuyu headmen whose territory was adjacent to the border. Each headman was paid 3 rupees (approximately 4 shillings) per acre to keep the boundary cleared. This system was used until ordinances in 1915 and 1916 provided for Forest Guards.[52] In the early days the Guards, usually Africans, were often armed, but this became rare by the 1930s. The Forest Guards, along with Foresters, had the power to arrest suspected forest regulation offenders.

Unauthorized cultivation in the forest was prohibited. When the forest was demarcated all people residing within its boundaries were evicted (this is discussed in detail below). Some farming was allowed but only as part of the Forest Department's reforestation program. The department used the *taungya* system of combining tree planting with shifting agriculture.[53] First introduced in Burmese teak plantations in the 1860s, the taungya system has been one of the cheapest and most satisfactory methods of reforestation. In Kenya the method was called the "forest *Shamba*" system. Cultivators in the Nyeri division of the Mount Kenya forest, almost exclusively Kikuyu, made contracts with the Forest Department for three years during the late 1940s.[54] The laborers would be permitted to intercrop maize, beans, and other subsistence crops with newly planted tree seedlings. The laborers were responsible for weeding and maintaining the young trees. After three years the workers would be moved to another site for reforesting. Forest labor was similar to squatting on a European farm, in that the worker received temporary use of land in exchange for labor, with most of his remuneration coming from his garden plot instead of the small wage he was paid. Until the 1950s very few, if any, public services such as dispensaries and schools were provided to the workers.

Forest reservation, the new system of management, and staff vigilance caused an immediate decrease in damage by "natives and forest fires."[55] But honey hunters still remained a problem. Most honey collectors kept beehives made from the hollowed-out trunks of camphor or *Cordia abyssinica* trees, which required a permit to be hung in the forest. Some honey seekers at times found wild bees with their combs in a hollow camphor branch or trunk. In the process of smoking out the bees with a torch, fires would accidentally start.[56] Fires caused by honey hunters remained a problem throughout the colonial era. In 1934 an estimated 40 percent of all forest fires in Kenya were attributed to this cause. The Forest Department criticized judges for making the problem worse by giving offenders sentences that were too lenient.[57] In the 1940s honey hunters in the southern Mount Kenya forest destroyed many camphor trees. In some recorded cases near Castle forest station two men cut 20 camphor trees, one chopped 16, and another destroyed 9 trees.[58] The assistant conservator of forests stated that one large camphor, even a hollow one, could contain over five tons of timber, so that, given the camphor's high value, each destroyed tree represented a considerable financial loss. Livestock herders, attempting to improve grazing by burning off dry grass, were another cause of fires.

Although large-scale forest destruction was halted almost immediately, the collection of payment for removed produce proved more difficult. During World War I the Forest Department experienced a staff shortage throughout the Kikuyu province. As a consequence supervision over local forest use was minimal. One provincial report complained that "a large amount of revenue is lost by forest fees remaining uncollected . . . timber and firewood being stolen by native thieves."[59] The lack of enforcement probably softened some of the shock of forest alienation. The local Kikuyus found that they could still freely collect wood—even if officials did consider it "theft."

By the 1920s the Forest Department staff returned to its normal level. Not surprisingly, African objections to the loss of the forest became increasingly vocal during this decade. At the opening of the Embu Local Native Council in July 1925, and after a speech by Acting

Table 2.1 Compounded Forest Offenses, Southern Mount Kenya, August 1937 to March 1940

Date	Offense	Penalty (shillings)
Aug. 1937	fuel cutting	5
	stock grazing	10
Jan. 1938	honey hunting	10
	cutting poles	4
	fuel cutting	4 to 6
Apr. 1938	fuel cutting	2
July 1938	fuel cutting	1 to 3/50
Sept. 1938	goat grazing	3 to 6
	fuel cutting	1
Oct. 1938	fuel cutting	2
Nov. 1938	fuel cutting	2 to 4
Jan. 1939	fuel cutting	5
Apr. 1939	fuel cutting	2 to 6
Feb. 1940	fuel cutting	3
Mar. 1940	goat grazing	8
	sheep grazing	3/50

Source: Embu Archives, File FOR 13/3/1, "Forest Department, Offenses Compounded."
Note: Gaps between months indicate either no offenses were recorded or the page was missing.

Kenya Governor E. B. Denham, Chief Runyenjes spoke to complain about local beekeepers being denied entry into the forest. Runyenjes expressed fears of further European land alienation.[60] At the council's second meeting Runyenjes protested against the loss of "valuable assets," including salt licks, in the forest.[61] The following year another councillor asked "by what right Government had reserved the Forest."[62]

Women and honey collectors who volunteered their labor for such tasks as clearing boundaries were given firewood in exchange. In all other circumstances removed forest produce had to be paid for, a policy that was a frequent source of conflict. The Embu Local Native Council Minutes again provide evidence. Council members complained, or were reminded, about fees and permits required to remove timber and firewood and to place honey barrels.[63] Removing seeds, bark, and other tree or plant produce for medicinal purposes was also strictly regulated.[64]

Under the 1911 regulations the Forest Department established its own justice system. While serious offenses would be tried in court, minor violators could have their violations "compounded."[65] This meant that the offenders, in agreement with the conservator of forests, could pay money in compensation for the violation instead going to court and facing a possible prison term. Table 2.1 indicates the range in offenses and penalties paid for compounded offenses during the late 1930s and early 1940. Penalties tended to be meted out according to each particular case, but from the available data it would appear that honey-hunting and grazing violations received the heaviest fines, while illicit fuel cutting was the most common offense. The cost of disobeying regulations could be higher than following them. For example, in 1927 a permit to gather wood from the forest cost 2 shillings per month, with the compounded fine averaging 3 shillings per illicit load.[66] People living in Embu district who committed forest offenses needed to travel to Nyeri town to have their cases compounded until 1931, when women and children were allowed to have their cases compounded by the Embu district commissioner.[67]

The Kenya Land Commission

By the early 1930s the issue of European land alienation pervaded Kenyan politics.[68] Many white settlers and government officials feared that a Kikuyu insurrection was imminent.[69] In this atmosphere of crisis the secretary of state for the colonies convened the Kenya Land Commission in April 1932 to adjudicate African land grievances. Once these claims were evaluated, and future African land needs taken into account, the commission was supposed to draw permanent racial and tribal boundaries. The grievances of the Kiambu Kikuyus received the most attention; nevertheless, the commission also investigated requests by the Kirinyaga and Embu Kikuyus for the return of their forest land. Lambert wrote that in Embu the restoration of the forest was perceived as the most pressing land issue.[70]

The possibility that discontent could turn to violence was almost fulfilled in the southern Mount Kenya forest. Several incidents of people threatening the Forest Guard station were reported in September

1932.[71] In one case an African waving a *panga* (machete) told the Guard that the boundary no longer existed and the eucalyptus boundary markers would be chopped down. Less than a year later the local people were still said to be "very unruly," disputing the Guard's authority to prevent them from grazing stock or removing forest produce.[72] The assistant conservator of forests asked for "moral support," as well as any available tribal policemen.

Africans all along the southern Mount Kenya forest boundary presented land claims to the commission. Kikuyus from Nyeri, which then included the Kirinyaga and Embu districts, argued that their land rights had been ignored and violated during demarcation. Juhena Ngungi, from Ndia, testified that he showed district officials the "remains of huts and deserted places" in the forest.[73] "This land originally belonged to us and we used to get our best cultivation there and best wood," he attested. According to Ngungi his people were evicted between 1911 and 1912. Chiefs Njega and Gutu "were told to burn down all these huts on this land and they did it. There were over 100 and less than 1,000 huts."[74] Chief Njega himself testified, confirming Ngungi's account:

> After this boundary was marked out we could not get into the forest and we had our important trees there and we cannot go inside there again and get sufficient wood. As I was the Chief, Government came and told me that I should go into the forest and turn out all the people who were living there and burn down their huts, which I did. The huts were uncountable, very, very many. Even now you can see the remains of the huts. They were the huts of people who lived there before Government came to this country.[75]

The final sentence quoted above, about the people living in the forest before the coming of "Government," was of critical importance to the commission. Because the British believed that Kikuyu administration actually resulted in accelerated deforestation, the commission considered Kikuyu forest occupation as legitimate only if it existed prior to 1895.[76]

The Embu branch of the Kikuyu Central Association (KCA) presented the commission with a memorandum protesting the forest demarcation. The KCA stressed that the forest was set aside without local consent. They insisted that when the line was made the district commissioner told them it was a road to link Nyeri and Meru so that the Kikuyus could buy sheep and travel without difficulty. Once the road was completed and the eucalyptus trees planted, the district commissioner informed the people that they "would be fined or sent to prison" if they "touched these trees."[77] The Embu were "surprised" by the sudden restriction "because the forest was ours before the white man came to the country." Thus, the Embu KCA charged that the government seized the forest through trickery and deceit. They pressed for the return of their original forest land, including salt licks and sacrificial sites, plus adequate space for current farming and wood requirements.

The denouncing of the forest boundary was supported not only by those people considered "agitators" by the administration but by more "moderate" or "cooperative" Africans as well. H. E. Lambert, the Embu district commissioner, emphasized that the local forest grievance was being forwarded by "responsible persons."[78] In South Nyeri, Chief Njega's support of the claim was important as he had many ties to the government and was also involved in the process of demarcation. Another relatively pro-administration group that contested the forest boundary was the South Nyeri Progressive party. This group, which had supported the missionaries during the female circumcision controversy,[79] asserted that "we were not told, nor did we know, that the boundary lines were to be like walls of a cattle byre."[80] They had been told that the line was to prevent burning; but no other activities, such as grazing, cutting timber, or fetching firewood, were mentioned as being affected. The Kikuyus "did not know" that their forest would be taken away "forever."[81] The party stated that at least 104 people lost land in Ndia and Kichugu. "These are all of those whom we know, and there may be others who are not known who have died."[82]

The district commissioners testified at the commission, and both the South Nyeri and Embu officials generally supported the African claims. As the head administrative officer at the district level, the district commissioners were in a key position to gather and to evaluate the local land grievances. Before the commission convened, the district commissioners held special public meetings and used the local native councils as means of collecting information. The commission sought their advice as to the legitimacy of the claims.

John W. Pease, the South Nyeri district commissioner, said that "large numbers of *shambas* (farms) had been located within the present forest boundary."[83] He estimated that "600 to 800" people were removed from the forest in the "Kerugoya area." Pease suggested that "perhaps two-thirds" of the forest shambas existed in the "preadministrative" era, with some having been "occupied for a considerable period." When the forest was demarcated the Kikuyus lacked the political organization and education to present their views. Pease explained that the Forest Department's interest was in saving the forest from destruction rather than with "administration questions and the needs and interests of the local Kikuyu."[84]

Embu District Commissioner H. E. Lambert's investigation revealed that the local people had extensive claims to the forest. Lambert wrote that the "land was actually owned (in the usual native sense) within the present boundary to a varying depth up to about two miles."[85] The original plot owners could still be named and landholding boundaries pointed out. According to Lambert, land use was strictly regulated: "Natives from the lower areas who wished to cultivate in the higher land had to get the permission of the individual owners of pieces of land already demarcated and recognized as the individual's *ngambas* (plots); they could not merely choose any convenient sites higher up; the upper boundaries of definitely owned *ngambas* were the upper limits of cultivation."[86] Within the area that the Embu had "demarcated," cultivation involved only "clearing bush in glades and comparatively open spaces." Large forest trees were regarded as "tribal property" and could not be cut or burned by individual right-holders. Land above the cultivated zone, if it "could be said to be owned at all," belonged to the "tribe." The forest, "as far as destruction went," was "inviolable."[87] Although Lambert believed that Embu forest use "was a good deal less destructive than that usually ascribed to the Kikuyu," he suggested that Embu forest use had been in a "transition stage" and would have become more predatory on the forest once demand for land increased.[88]

Lambert agreed with the Embu claim that they did not understand nor were they told the purpose of forest demarcation. He pointed out that the work of making the forest line was carried out by "the compulsory and unpaid labour of the people themselves."[89] After its completion the Embu were allowed to harvest standing crops and ordered not to cultivate again. In Lambert's opinion the Embu had "definitely suffered a hardship in the loss of land and salt-lick."[90] But neither Lambert nor Pease recommended that the forest be returned directly to the local people. Lambert suggested that part of the forest be designated as African Reserve Forest, where limited cultivation could occur and the local native council would earn revenue from the removed forest produce fees. Pease proposed a sort of profit-sharing taungya system, with the Kikuyus leasing land from the Forest Department to plant food crops and trees and the revenue obtained from the later sale of the trees being divided between the Africans and the department. The commission did not use either recommendation.

The Nyagithuci Hill Controversy

The Kenya Land Commission decided that the Kirinyaga Kikuyu had suffered some legitimate loss of land during the forest demarcation.[91] The commission believed that placing the original boundary of Kikuyu occupation one mile north of the existing line was probably a "very conservative estimate." The commission rejected the Embu claims entirely. An important reason for this probably was testimony by the conservator of forests, who said that the Embu losses had not been significant.[92] The forest conservator insisted that there had been no complaints about the boundary since it was cut over 20 years before.

In evaluating the Kikuyu land claims, the commission considered all land grievances in terms of the "tribe" as a whole. Although individuals, lineages (*mabari*), and clans had presented specific claims, the commission felt that the Kikuyu themselves actually viewed land losses as a "tribal" affair.[93] The commission, like most Europeans, believed that the tribe was the fundamental sociopolitical unit in African society. This was a misconception. The Kikuyus certainly formed a cultural entity, in terms of a group of people who shared cultural and linguistic attributes. But the Kikuyus never formed the socially and politically unified entity of "tribe" as conceived by Europeans. Instead each Kikuyu group's interest was limited territorially. The gulf between European ideal and African reality became evident when the administration attempted to carry out the commission's recommended settlement of the Kikuyu land grievances.

The commission decided to have 8,500 hectares of land excised from the various government forests throughout Kikuyuland to compensate the "tribe's" losses. The conservator of forests selected "scrub-covered" instead of heavily wooded land.[94] One 809-hectare block of land was located in southern Mount Kenya adjacent to Ndia and Gichugu. The place was listed as "Narkothi Hill," and became known as "Neakochi," "Nyakithuci," and "Nyagithuci" hill in later correspondence and reports. The colonial secretary instructed district officials not to take action on the land commission's recommendations when they were announced in July 1934.[95] Kirinyaga had been transferred to Embu district by this time (since August 1933), and the Embu district commissioner reported less than a month later that no action had been taken other than informing the local native council of the decision.[96] By the beginning of 1935 the local people were complaining that the 809 hectares had not been turned over and that the assigned land was treeless.[97]

The decision to hand back the land to the local people was finally made in December 1935.[98] Two months later the Embu district commissioner reported that the 809 hectares had been "surveyed and handed over for occupation."[99] This decision was abruptly changed only months later when the new provincial commissioner, S. L. La Fontaine, suggested that the Kiambu Kikuyus receive preference in obtaining the added lands.[100] La Fontaine argued in correspondence that the Kiambu Kikuyu had lost six-sevenths of the Kikuyu land but only half of the added land was adjacent to their area. He pointed out that the Kiambu were upset with the commission's settlement, considering it inadequate, and they were angry that so much land was granted to outlying districts such as Nyeri and Embu. Instead of benefiting the entire tribe as the commission intended, the compensated land would most likely become the property of the Kikuyus residing in the other districts. Unless the Kiambu Kikuyus' grievances were addressed the situation "might easily burst into flame and have irrevocable results."[101] La Fontaine proposed that at least 1,500 of the 2,000 acres be reserved for the Kiambu group.

The policy switch created a controversy that lasted over three years. The high point was reached in November 1938, when Kenya Governor Sir Robert Brooke-Popham visited Nyagithuci. He encountered local Kikuyus cultivating and some evidence of deforestation. This set off charges and counter-charges among district, provincial, and central government officials as to who was to blame for the supposed local encroachment on the land reserved for the Kiambu.[102] The provincial commissioner was ordered to prepare the local inhabitants of Nyagithuci for "evacuation" pending an investigation by the district commissioner.[103] The provincial commissioner refused to order the local people to prepare for evacuation, saying it would be "most impolitic."[104] He did instruct the Kikuyus not to break any further ground and not to cut forest.

The provincial commissioner also met with local chiefs to obtain their version of the controversy.[105] The chiefs denied that any orders preventing the local people from resettling the area were ever issued. Chief Mugera stated that land disputes were absent in the resettled zone because "everyone admitted the right of the old rightful owners. Mbaris (lineages) did it all on their own." Chief Justin bluntly pointed out that the local people "did not want strangers. Landless Kiambu came to find an area. We said ours is small and our people are suing one another below (the) old (forest) line as land hunger." The local Kikuyus told the Kiambu they could come, but only as tenants at-will to the locally recognized right-holders. The district commissioner's report confirmed that four-fifths of the 344 occupiers of Nyagithuci held their land as prior right-holders, with another 55 occupiers (16 percent) possessing land as tenants of the original right-holders.[106] Almost half (163, or 47 percent) of the cultivators belonged to the same clan, the Unjiru. The district commissioner's survey showed that almost 800 acres were under cultivation, and he estimated that only 45 acres had been deforested. This amount was said to be "reasonable" given the area's timber and firewood shortage. The Embu district commissioner concluded that the government's prestige would be seriously damaged if the land were reclaimed for Kiambu resettlement.

After receiving the Embu district commissioner's report, the provincial commissioner advised the commissioner of lands and settlement to allow the local Kikuyus to remain in Nyagithuci.[107] The provincial commissioner wrote that "this land was . . . not lost to the Kikuyu as a tribe but lost to those Kikuyu natives in that vicinity." Two months later the Embu district commissioner received word that the local people could stay, but he was requested to set aside 120 hectares to accommodate any possible Kiambu Kikuyus.[108] Apparently very little Kiambu resettlement occurred, the only recorded case being four "families" that received 40 hectares in 1940.[109] The lack of Kiambu support for the resettlement scheme was a missing ingredient in the government plans. Throughout the years of controversy it was suggested that Nyagithuci was undesirable to the Kiambu because of "the distance from their relatives and the frequent depredations of game," the "cold reception" they might receive from the local people, or because the local Kikuyus and Embu would expect the newcomers to be "reborn" into a local clan.[110]

A much more important reason for the lack of Kiambu interest in Nyagithuci was the availability of land in the lower, warmer zone of the Kerugoya subdistrict and Embu. Known as the *Weru,* and located between the densely populated rolling countryside and the hot, barren

Mwea plains, this area attracted many Kiambu Kikuyus in the 1930s. Thus, many Kiambu Kikuyus were willing to settle far from relatives, risk attack from game, and even become blood brothers ("reborn") or tenants of clans to obtain land. The crucial difference was that in the relatively warm Weru two crops of maize per year could be grown, and the region's flatness and sparse population were ideal for a new farm innovation—the plow.[111] In contrast, the cool, damp Nyagithuci area could only grow one crop of maize, which was a major cash crop, per year. The Kiambu outsiders and the local Kikuyus could arrange all necessary transactions without much colonial interference. Instead of accepting Nyagithuci and having to battle the local Kikuyus, the Kiambu carried out their own settlement movement into the Weru while denouncing the land commission's settlement as unjust and insufficient.

From Sawyers to Mau Mau

The resolution of the Nyagithuci controversy brought a certain degree of calm to the issue of local versus colonial government control over the forest. But after World War II forest-related conflicts reappeared, revolving around the issue of pitsawing. Hutchins had originally recommended the introduction of European sawyers from South Africa.[112] However, the first sawyers in Kenya arrived from India.[113] Kikuyus learned the pitsawing trade and "developed high standards of workmanship."[114] An apprenticeship system was devised, with the apprentice working under a master sawyer.

World War II stimulated timber demand, causing a rapid expansion in the pitsawing trade.[115] This resulted in a swelling of the pitsawyer ranks, a breakdown in the apprenticeship system, and a decline in the quality of production. Along southern Mount Kenya the issue was not declining standards, but the granting of licenses. In the early 1940s pitsawing was of little local interest. When pitsawing concessions were offered at Chehe, on the Nyeri–Ndia border in the Mount Kenya forest, no one from Ndia applied.[116] In 1945 the Embu district commissioner observed that pitsawing was largely done by "a number of strangers, persons living in the district as tenants at-will free of any kind of payment."[117]

Local interest in pitsawing began in the mid-1940s. Both the Embu district commissioner and the local native council felt that local Africans should be given preference in licensing for pitsawing. The council decided to grant no more concessions to outsiders, though those already possessing them would be permitted to retain them.[118] Plans for expanding pitsawing were thwarted the following year by the absence of a forestry officer to supervise production.[119] No new concessions were granted while the district waited for a new officer to be posted. By 1947, even though the Ragati forest officer from neighboring Nyeri would occasionally visit for one or two days to mark trees for pitsawing, the Embu district commissioner reported that the Forest Department was becoming "unpopular with the local inhabitants."[120] For the local population the level of production permitted by the department was too low to meet their needs.

By 1949 a European forest officer had been appointed to the district and pitsawing had increased.[121] The district commissioner complained that there was no policy of cutting and replanting in the Mount Kenya forest.[122] In the next year the Forest Department refused to increase the number of pitsawing licenses, which angered local people because the trade was still largely in the hands of outsiders.[123] The people of Embu district perceived the Forest Department as "repressive rather than helpful."[124]

Up to 1951 most of the timber cut by pitsawyers in the Embu-Kirinyaga area was probably sold for use within the district. Pitsawyers were denied permission to export camphor in 1946.[125] The local native council recommended that pitsawyers offer their wood for sale first in Ndia and Kichugu "where there was thought to be big demand."[126] By the early 1950s, however, pitsawyers were allowed to export timber outside the district.[127] An important change in timber production was proposed in 1951, when the colonial authorities decided to allow A. Bond to establish a sawmill at Sagana in western Ndia.[128] Bond intended to convert timber cut by pitsawyers into planks and other wood products for export.[129] The Forest Department planned to have the mill produce up to 300 tons of timber per month.[130] At a special meeting of the local native council the conservator of forests said the number of pitsawyers might be increased slowly, and sawyers from Embu district would be given preference.[131] To facilitate production the government intended to organize the sawyers into a cooperative.[132]

The Embu council supported the sawmill proposal but it did not want the Bond sawmill to have exclusive control over timber marketing.[133] The forest conservator told the council that in "years to come" the people of Embu district would be able to have their own sawmill.[134] Another request was for licenses to be granted only to those local Africans who were recommended by a council subcommittee. The council's aim was to curtail the continued involvement of "lawless types" in pitsawing. Sawyers were to obtain a letter from their local chief certifying their "good character."[135]

The plans for the new mill, the pitsawyer cooperative, and related matters were disrupted by the Emergency declaration and subsequent events. At first the October 1952 Emergency order had little effect on pitsawing. The divisional forest officer, D. G. B. Leakey, discussed the possibility of suspending pitsawing in November 1952.[136] Leakey instructed the assistant forest conservator to "preserve the status quo" and to check up on the pitsawyers "as we don't want the forest to acquire a bad name." He urged the assistant conservator: "Don't hesitate to eject yourself any undesirable pitsawyers or pitsaw employees." Leakey felt that repatriating forest laborers, though, was "most undesirable and highly unlikely."

Events in the forest were already beyond the control of the assistant forest conservator or any other local official by late 1952. The southern Mount Kenya forest was first designated a "Special Area," with access being restricted to those who obtained permits allowing them to enter the forest. By mid-1953 most of the forest, except for the forest stations, was declared a "Prohibited Area" with all access closed.[137] The exempted areas included plantation forests, cultivated land, forest nurseries, and station buildings. This excision from the "Prohibited Area" was short-lived, because intensive Mau Mau activity forced the total closing of the southern Mount Kenya forest in October 1953.[138] For over two years Forest Department activities came to a halt. Access to the forest by local people remained restricted for several years afterward.[139]

Not surprisingly the sawyers were linked to the Mau Mau by colonial officials. The Embu district commissioner wrote in 1952 that "the outbreaks of oath giving were more common in the areas of the district near the forest boundaries and there is no doubt at all that considerable part was played by pitsawyers and their employees in the Crown Forest many of whom were Kikuyus. Soon after the beginning of this outbreak, information began to come in and there was a very considerable activity in months of October and November."[140]

That same year the African Affairs Department's *Annual Report* recorded that pitsawyers enlisted regular forest staff and labor into the

Mau Mau, "no doubt with a view to conducting their operations from the cover of forest."[141] The report noted that "this evil infiltration did not become overt till November." Louis Leakey underscored the fact that Mau Mau organizers would find a sympathetic reception from at least part of the Forest Department's staff, the forest laborers.[142] He observed that among the laborers were young men "who are restless because they know there is no security, not any prospect of owning part of the land themselves."

Though the pitsawyers' involvement in Mau Mau could be expected, the nature and character of their participation in Mau Mau remains unclear. There were numerous grievances associated with pitsawing, such as granting licenses and restrictions on exploitation because of the lack of Forest Department staff. But assumptions about pitsawyers as a homogeneous group can only go so far. Distinctions concerning outside and local sawyers were made, as well as between "lawless" types and those presumed to be "law-abiding." Not all sawyers supported Mau Mau; some later records mentioned "loyalist" pitsawyers.[143] Further research is required on the problem of whether distinctions based on the size (depending on number of employees) and scale of a pitsawyer's operation, employee or employer status, the origin of the sawyer (local or outside), or other relevant variables influenced sawyer participation in Mau Mau. What is evident is that pitsawyers as a whole, with their intimate knowledge of at least parts of the forest, were in a strategic position for aiding and participating in opposition to colonial rule.

Conclusion

Between 1914 and the late 1950s the southern Mount Kenya forest was the object of contention between the local people and the colonial state. Conservation of resources was a motive in the government's takeover of the forest, but the main intent was to set aside the area for colonial economic development. European settlers and later the colonial state itself were to be the chief beneficiaries of government forest management.

Local responses to government forest administration were varied and changed over time. At different points in time grievances were filed with the government, and local leaders complained during meetings with colonial officials. Some people chose to ignore the regulations, though this action often resulted in jail terms, fines, and other punishments. In later years the forest conflict merged with the larger anticolonial movement, and the southern Mount Kenya forest became a center of fighting between Kikuyus and the government. However, throughout the colonial era many people accepted government regulation of the forest, though they also may have supported the various efforts to regain control over it. According to available colonial records, violations of forest regulations did not appear to have been an acute problem.

Since independence the system of forest management used in colonial times has been continued, including a policy of establishing tree plantations of exotic species for timber and fuel purposes. For the most part local people have continued to respect the integrity of the forest boundary and regulations.[144] Disagreements about use of the forest are no longer defined in racial terms (for example, with the policy of Africanization all of the foresters currently posted to Kirinyaga are Africans). Nowadays conflicts over forest use are between the central government and the Kirinyaga County Council, and these conflicts are largely based upon administrative and territorial interests. With rapid population growth and increasing land scarcity, though, both central and local government may come under pressure in the future to excise large portions of forest in order to create settlement for the landless.[145]

3

THE IMPACT OF GERMAN COLONIAL RULE ON THE FORESTS OF TOGO

Candice L. Goucher

The modern nation of Togo is relatively small in land size, with a total area of 56,800 kilometers. However, it represents literally a slice of West African colonial history, having experienced German and French policy directly and British policy indirectly. Nearly the full range of West African vegetational zones, which occur in successive bands from west to east, are contained within Togo's borders. The several centuries of European presence in neighboring Gold Coast (Ghana), the activities of missions in southern Togo that frequently transgressed the commercial realm after 1822, and European trading ventures in the coastal region from about 1840 provided inroads for the subsequent, more intensive impact.

The colonial period did not formally begin in Togo until the establishment of a German protectorate in 1884. The political boundaries of the German colonial period were established by the Berlin West Africa Conference of 1884–1885. However, the period of actual German administrative rule in Togo was short-lived. In 1915, following the occupation of Togo by French and British forces and the subsequent Convention of Lomé, the French inherited most of Togo and with it the legacy of German policies on railroads, land concessions, plantations, and forests. The western territories, British Togo, came to join the Gold Coast, while the larger French-mandated colony became an independent republic in 1960.

While the colonial period comprised less than a single century, it was preceded by many more centuries of human activities that had significantly altered the precolonial environment. Evidence from archaeology, oral history, and early travelers' written accounts confirms the extensive loss of forest land prior to the twentieth century. Thus it is imperative that the impact of colonial rule also be assessed within the larger historical context of precolonial patterns of land use and forest exploitation.

Much of Togo falls within the "Dahomey Gap," a region marking a break in the forest canopy of Ghana and Nigeria that is characterized by a forest-savanna mosaic. Ferruginous soils, in addition to the ferralitic varieties of the contiguous forests, are found in this zone.[1] Patches of moist forest surrounded by savanna are found along the region's streams, hills, and plateau sites. In addition to its distinctive soil types, the zone is notable in receiving less rainfall than do the neighboring forests. Given this climatic predisposition toward reduced forest cover, Togo is situated in a region of greater ecological vulnerability.

Savanna vegetation dominates Togo today. The southwestern regions that receive heavier rainfall are covered with tropical forests. In addition some river valleys are forested. The narrow coastal zone consists of a series of inland lagoons that support mangrove and reed swamps. The predominance of areas of mixed forest and savanna creates a complex, problematic historical setting in which to evaluate precisely the loss of forests.

The Precolonial Period

Archaeological evidence points to the presence of large populations in areas of Togo that formerly must have sustained more abundant forests or woody vegetation.[2] During the sixteenth to the eighteenth centuries the climate of the Guinea Coast area (Sierra Leone to Dahomey) was probably drier than today, with numerous severe droughts.[3] This was followed by the establishment in the nineteenth century of a climate resembling that of today. Linguistic evidence also suggests the extent of environmental change. The former presence of forests north of the ninth parallel is documented by toponyms that survive the loss of forest vegetation: Lama and Lanmba ("those of the forest"), Lan ("forest"), and Lawda ("in the forest").[4]

The beginning date for forest exploitation and occupation remains a matter of speculation. It is now thought that penetration of forest zones in West Africa may have predated the development of iron technology.[5] There is indirect evidence that disturbance of forests took place in Togo with the settlement of large human populations and the beginnings of urban development. Certainly the existence by the seventeenth century of such precolonial centers as Notse, which covered at least 10 hectares and was enclosed by a large wall more than 10 kilometers in circumference, is one indication of the capacity of large populations to significantly affect the forest environment.[6] With dense populations (often needed to provide huge labor inputs), West African urban centers required the support of intensive agricultural bases and the availability of forest resources. Such products as wax and wood made possible the specialized (and fuel-consuming) crafts that these centers harbored: pottery making, brass casting, and blacksmithing. Recent studies of rural energy systems in Africa estimate a high wood consumption per household.[7] The potential impact of daily domestic activities on vegetation could constitute a threat to forest ecology.

Two major spheres of human activity had a particularly dramatic impact on the state of precolonial vegetation: agriculture and metallurgy. Agricultural intensification, which in turn supported increased populations, led to the widespread conversion of woodland to grass savanna.[8] Once converted these fields were more easily worked with iron tools. The early colonial forest experts' alarm over the progressive deforestation was largely triggered by their observation of traditional agricultural practices.[9] These comprised a variety of methods, dominated by a rotational system of shifting cultivation. Each portion of land was used once in each cycle spanning 10 years, sometimes in combination with various other crops. Land left fallow was invaded by secondary growth and by tall savanna grasses that could now be cleared satisfactorily by the traditional hoe accompanied by repeated burnings. Major crops included rice, yams, maize, sorghum, palm oil, millet, and plantain.[10] Regional trade in surplus foodstuffs was a characteristic feature of the precolonial economy. A declining environment quickly adopted crops like cassava (arriving sometime in the eighteenth century), which did well in poorer soils. As a further reflection of the declining productivity of the soils, the size of land traditionally required by individual farmers later became an issue during the colonial period.

Even more destructive of vegetation cover than precolonial agricultural practices were the fuel requirements of large-scale industrial complexes engaged in iron production. Former areas of iron smelting can be located by the presence of large slag mounds and the remains of the tall, induced-draft furnaces.[11] Between the fourteenth and nineteenth centuries, central Togo in the region of Bassar was the center for what may have been the largest complex for iron and steel production in West Africa. Even at the close of the nineteenth century, when oral traditions claim the industry was declining, some 500 furnaces were seen operating at a single smelting site.[12] Numerous smaller centers for iron production have been recorded from southern to northern Togo.[13] High quality iron produced in central Togo was traded at distances of hundreds of kilometers.

According to former iron smelters, without the proper charcoal iron could not be made.[14] Possibly the most crucial element in the chemistry of the smelting process at Bassar, as elsewhere, was the fuel. Charcoal produced for blacksmiths and smelters differed greatly from the fuel supplied for general domestic purposes. The woods needed by the iron industry were dense hardwoods that were long burning, often with a high silica and alkali content. Of great ecological consequence is the voracious appetite of the smelting furnace and the fact that the species preferred by iron smelters also tend to be slow-growing.

Knowledge of sources of charcoal was proprietary and as such was guarded by members of the industry. When in 1897 Friedrich Hupfeld asked the Bassar people about the charcoal used in smelting, he could obtain virtually no information.[15] Hupfeld speculated that the wood was traded from great distances. In 1908 another colonial official, Kersting, after residence in the Bassar region for more than 15 years elicited information about the woods used for fuel only by sending samples of the charcoal to Berlin for identification and analysis.[16] Preliminary metallurgical studies of slags from the smelting process indicate that a decisive change in charcoal fuel occurred between the fourteenth century and the eighteenth century.[17] It is likely that such technological change was an adaptation to deforestation brought about by the operation of the iron industry over centuries. Inability to compete with cheap European iron exports finally resulted in the decline of the traditional industry.

Fuel requirements probably continued to influence the pace of West African urban development throughout the second millennium. Wilks has shown that the economic transformation of the neighboring Asante nation was accompanied by large-scale clearance of Gold Coast forests.[18] One response to the deforestation that accompanied such development was the establishment of long-distance trade in firewood and charcoal. Forest products including timber were recorded by German officials in the trade from the Ho region of southern Togo to Asante during the nineteenth century.[19] Oral traditions also record the trade of iron from the Bassar region to Asante and the trade of charcoal from the same direction. Thus, it is possible that the period of expansion of the Asante state not only provided a major market for the iron of central Togo, but also literally fueled the final century of Bassar iron production.

Besides the direct cutting of trees for charcoal to fuel the smelting and smithing processes, there were long-term deleterious effects of pollution from smelting on vegetation and soils. Nineteenth-century German travelers in the Bassar region noted that the concentrated, low-lying smoke and fumes of smelting furnaces were visible at a considerable distance.[20] Evidence for the environmental effects of heavy particulates, oxides, fumes, ash, dust, and gaseous pollutants is now being studied in changes in soil profiles recording the last five centuries of industrial activity.

The effects of precolonial deforestation are immediately visible in the nineteenth-century photographs of smelting sites barren of almost all vegetation. Complaints of denuded hillsides and severe erosion in mining and smelting regions are found in the literature of the early colonial period and in local oral traditions. One official, Willi Koert, observed that the shallowness of some iron deposits was the result of the denuded landscape. The effects of erosion and weathering were noted with alarm.[21]

Early German studies of the territory's vegetation implicated annual grassburning and other traditional agricultural practices in the destruction of Togo's forests.[22] In the 1880s German expeditions into the central and northern regions encountered areas that were clearly being deforested. The colonial forester, O. F. Metzger, asserted that the vegetation of southern Togo had once been a dense, closed-canopy forest like that found in German Cameroon. He argued that only 1.5 percent of the original forest remained at the turn of the century, the rest having been converted to savanna.[23] This process of modification resulting in the loss of forests was observed to be ongoing, and the Germans considered the effects of deforestation on soil, climate, and rainfall to be a threat to the success of the colonial venture.[24] While German assessments may represent an exaggerated opinion of the nature of the original vegetation, it would be equally simplistic to continue to assume that precolonial forests had remained unaffected by African technology.

The Colonial Period

The German presence in southern Togo was established with the activities of Bremen and Basel missions in the early part of the nineteenth century. Production of palm oil for export brought Togo's population into early commercial contact with merchants from Bremen and Hamburg.[25] The period of colonial administration was initiated very cautiously, founded on treaties of protection that were negotiated in territories near the coast. Bismarck's diplomacy in Paris and Berlin (1884–1885) officially established the borders of the German colony.

Between 1884 and 1894 the newly formed administration, with financial support reluctantly supplied by the Reichstag, sought to expand its influence northward from the coast through trade and exploration. In economic terms the colony remained of marginal importance to Germany during the final two decades of the colonial period. Despite the brevity of German rule, significant increases in exports of key commodities resulted in considerable environmental change.

Initial ventures in colonial agriculture in the Togo colony were founded on an exuberant ignorance of local conditions. As early as the 1880s, and with virtually no scientific information on soils, climate, or ecology, the German colonial administrators presented to the German metropolitan public a strong case for the creation of large plantations. In 1888 Ernest Henrici, an explorer and planter, purchased a plot of land from the chief of Agome-Palimé.[26] He maintained that in Togo one did not have to deal with closed-canopy forests as in Cameroon but rather could find wooded savanna and rich soils.[27] Henrici concluded that Togo was suitable for all tropical cultivation, including coffee, cocoa, oil palm, and cotton. Despite his enthusiastic study of the region, Henrici's own coffee-planting ventures failed, as did the schemes of many others during the early years of the colonial period.[28]

German efforts at expanding agricultural exports focused more on the cultivated plants (cotton, coffee, cocoa) than on the collected products that were exported (palm oil, rubber, timber). In this respect the German administrators differed from their British counterparts across the Volta River. In these efforts the government figured prominently.

The development of plantations was seen as a means of mitigating the high transport costs encountered by merchants and their representatives. The fact that vegetation and climate varied with elevation and proximity to the coast and that much of the territory was dominated by a complicated savanna-forest mosaic increased the environmental complexity and thus the potential vulnerability of colonial endeavors.

Cotton received the most attention from the Germans.[29] Crops were planted by Wolf at the Bismarckburg station in 1888 and the following year in coastal regions near Anecho and Lomé. Particularly in the Misahohe and Atakpame districts, its cultivation spread rapidly. The volume of cotton exports rose from 14,453 kilograms in 1902 to about 502,000 kilograms in 1913, reflecting a much more significant growth than that of any other export sector.[30]

The real impetus toward agricultural production came about through the intervention of the Economic Committee of the Colonial Society.[31] During the first decade of the twentieth century agricultural studies were implemented; experts from the United States and Samoa were brought to Togo in an attempt to increase productivity. Plows were introduced and cotton-ginning stations were built. The committee also distributed improved seeds to farmers and provided price supports of 12 to 30 pfennigs per pound for both ginned and raw cotton.[32]

German attempts to increase productivity are reflected in the founding in 1904 of an agricultural training school and cotton experimental station at Nuatja. This represented in turn a measure of colonial faith in agricultural research and its potential contribution to productivity growth. Germany had taken a decisive lead in the establishment of agricultural experiment stations in Europe during the second half of the nineteenth century. Given the German lead in soil chemistry and institution building, it is perhaps less surprising that they were confident that improved agricultural technologies could readily be transferred.[33] By the early twentieth century most government stations in Togo, including those at Kete-Kraye, Yendi, Bismarckburg, Bassari, and Bafilo, had their own experimental garden plots or plantations. However, ignorance of the complexities of tropical (as opposed to temperate zone) agriculture presented a formidable barrier to the successful transfer of technology.[34]

The 1880s and 1890s also witnessed the expansion of indigenous entrepreneurship with the rise in importance of raw rubber exports. In the adjacent Gold Coast territories the volume of rubber exports in 1898 was 25 times what it had been in 1884.[35] By the time Henrici journeyed through the western Togo regions of Kewe, Agotime, and Agome in 1887, rubber trees were being exploited.[36] Officers Wolf and Ludwig Kling at the Bismarckburg station, a two-week march from the coast, noted the meteoric rise of the export rubber trade in 1888–1889.[37] As collecting for the rubber trade became more profitable, subsistence farming was often abandoned. In 1890 Clerk and Hall on their journey to Adele found the cost of foodstuffs to be very high because of the "daily comings and goings of the rubber buyers" and also partially because of poor soils.[38] Von Doering also showed concern over the dearth of yams as a result of the lucrative export trade.[39] Initially the rubber trade was frenetic; methods of collection were often wasteful and extremely destructive of trees. An additional environmental impact of the export trade was a result of the necessary transport of products down the Volta River to coastal markets. Between 1896 and 1922 virtually all of the silk cotton trees near the banks of the river were cut down for the construction of canoes.

Colonial rule marked the beginnings of a gradual reorientation of local economies and patterns of land use. Precolonial agriculture had never been purely subsistence. There continued to be considerable interregional trade in foodstuffs throughout the colonial period. The profits of the export market brought Africans to participate in the intensive collecting of forest products of systematic cultivation for export. In neighboring Asante almost all the forests were converted to agricultural lands during the colonial period as subsistence farming was abandoned in favor of cocoa cultivation.[40] As sales of land and contracts of purchase in the form of treaties consumed increasing amounts of land, the traditional farming practices that necessitated the use of even larger plots of land (much of which lay fallow) were increasingly threatened. In 1900 Dr. Hans Gruner, colonial administrator in Misahohe, claimed that two hectares of land per person could support a family. On closer investigation Dr. Rudolf Asmis, who chaired a land commission investigating the concessions of the Deutsche Togo-Gesellschaft, concluded that a total of 14 hectares per person was necessary.

The persistent problem of questionable land acquisitions figured in the nature and extent of colonial exploitation. As early as 1891 the colonial administration proposed provisions for the granting of subsidized concessions to cotton companies. For the time being, however, cotton planting was limited to small leaseholds around Anecho and Bagida. The following year Dr. Gruner initiated a series of government-supported experimental cocoa plantings at the Misahohe station. Widespread clearing of lands for other crops soon followed. The formation of the Economic Committee of the Colonial Society in 1896 greatly stimulated the transfer of lands for tropical agriculture in Togo. By 1900 a total of more than 1,400 acres was recorded as being under the control of the few major plantations operating.[41]

While Togo was never administered by chartered companies, the weight of government support affected the progress of agricultural clearing. The first colonial land legislation of 1904 provided that no land could be sold to foreigners without the governor's consent. This decision followed on two closely related developments: the increasing number of scientific studies and appraisals of Togo's potential for agricultural export crops and a series of large land appropriations. In the course of his 1897 journey, Ferdinand Wohltmann evaluated the soil and climate for possible cultivation of coffee and cocoa along the coast and in the mountains. As a result of his favorable findings, Sholto Douglas initiated the extensive Agu plantings. Similarly, Captain Mellin, the district officer at Yendi, cleared 500 hectares for planting after having been informed by the colony's forester of its agricultural potential.

By far the most controversial and consequential of land concessions was attempted by Friedrich Hupfeld. In 1897–1898, his "scientific mission" financed by the Deutsche Kolonialgesellschaft explored the potential resources of the hinterland. In fact, Hupfeld surveyed large areas of land for their suitability for the establishment of privately owned plantations and other planned commercial ventures. Subsequently, with less than a 500-mark investment, Hupfeld negotiated for 85,000 hectares in Agu and Buem. Alarmed by the size of this concession and made suspicious by rumors of improprieties in its acquisition, the governor of the colony delayed ratification of the deed. A furor ensued both in the colony and back in Germany. Hupfeld's land concessions had brought the threat of possible monopoly of agricultural production to the forefront of public debate. Merchants like J. K. Vietor relied on the independent producer, especially the small African farmer, and opposed Hupfeld. The opposition formed an organization of cotton producers, the Togo-Baumwollgesellschaft, which stalled but was unable to halt completely the growth of large planting concerns. Ultimately this failure spelled disaster for many of the remaining forests.

Tensions between commercial concerns and agricultural interests were also expressed in disputes over the colonial policy of land transfers. The statutory recognition of African rights in matters of land ownership helped to ensure that in 1904, of 718 registered properties, 635 belonged to Africans.[42] Not all agricultural lands were owned by their operators. The largest plantation, Togo Pflanzungs-Aktiengesellschaft, chartered in 1911, paid ground rent on the 7,500 hectares planted.

Persistent droughts adversely affected agricultural exports and were blamed for the failure of various government plantings, including Henrici's earliest plantation.[43] Of the 106,000 registered coffee trees that had been planted on the coast by 1896–1897, many subsequently succumbed to the drought. As Klose observed, "The trees lost leaves and completely withered in their upper branches, so that by May, 1897, the plantation of Vietor, which I visited at that time, presented a sorrowful sight."[44] By 1902 fewer than half the original number of coffee trees remained.[45] Despite these setbacks, an intensification of exports occurred in the decade between 1899 and 1909, almost tripling in value (with an increase of close to 4.8 million marks) while nearly doubling in volume.[46]

Overall, the variety of crops planted by the Germans was remarkable, with as many failed crops as successful ventures. For example, rubber trees were planted in the Bassar area under Officer Kersting's supervision, but they did not thrive in the region's dry, lateritic soils.[47] The firm of Boedecker & Meyer lost more than 18,000 marks on a poorly managed cotton venture that also succumbed to the recurrent droughts of the early twentieth century.[48] However, many enterprises prospered in the long run. Coconuts, sisal, cocoa, cotton kapok, corn, rubber, tobacco, coffee, and oil palm products were grown profitably for export.

Successful intensification of export agriculture and expansion of cleared lands resulted in the further loss of colonial forests. By 1914 the total amount of land cleared was estimated at 12,000 hectares. An additional 3,000 acres reportedly had been cleared and planted with timber seedlings at the various government stations by 1912.[49]

Colonial Trade and Transport

During the earliest decades of German trade along the Togo coast, wood was prominent among the items imported, not exported. Wood was brought by Bremen ships to the West African coast in 1875. In addition to iron goods, other forest products such as wax were also imported.[50] By the first decade of the twentieth century an unprecedented 500,000 marks' worth of construction timber was imported by the German colony in one three-year period.[51] The German officials justified the importation by blaming a local shortage of wood on the unregulated cutting prior to the period of German administrative rule. In truth, large quantities of wood were needed for such colonial projects as the construction of the wharf and railroad lines designed to link the coastal ports with the hinterland.

The first of the railroad lines, the Lomé-Anecho "Kustenbahn," was completed in 1904. It was designed to bring oil palm and cocoa to waiting ships on the coast. The Lomé-Palimé line, finished in 1907, was known as the "Cocoa Line," although cotton and rubber also traveled its length. The last line to be constructed under German administration was the "Cotton Line" from Lomé to Atakpame.[52]

Although the Germans never exploited the high-quality iron ores of the Bassar region, they clearly had that intent. The mining potential fueled the interest in a planned extension of the railroad to Banjeli.[53] In addition, trees planted later in the Mo-Kamaa region as part of the colony's forest program were seen as an investment in future mining and railway activities.

The construction of the railroads and the motor roads had an immediate environmental impact. Trees were cut and land was cleared for 325 kilometers of railway and for 1,215 kilometers of roads. Further agricultural exploration and land clearing was encouraged along these routes as the transportation links provided inexpensive access to specific export products of the interior. Developments in the commercial realm thus had a twofold impact: environmental exploitation resulted from the procurement of products and from their transportation.

Togo's environmental as well as economic well-being was closely linked to British activities and policies in the Gold Coast. The products of the German Togo hinterland (including goods originating in the Sudanic region) traditionally traveled an east–west axis and then followed the Volta River into British territory, ending the journey in the Gold Coast colony's ports, such as Keta. Northern regions of the Gold Coast suffered similarly for their connections with ecologically impoverished territories across the border. During the eighteenth and nineteenth centuries, and probably earlier, much of the charcoal for the Bassar industry was traded from the Asante and the northern hinterland. As a result, these northern regions continued to suffer shortages for decades. As one observer wrote in the 1930s, "There are now no trees left for charcoal to be made. . . . The timber situation here is certainly bad and speaking generally only the shea trees are left."[54]

Correspondence from German colonial officials describes the acute shortages of timber supplies facing the various regions by the early twentieth century. The administrator at the Sokodé-Bassar station pleaded with the Colonial Office in 1911 to aid in the provision of wood for the construction of bridges; local wood was reportedly poor and furthermore had to be hauled from as far away as 30 kilometers.[55] Such shortages made the acquisition of timber a top priority in numerous colonial districts throughout the period of German rule.

West African forests both attracted and repelled European settlement. The potential value of forest exploitation was immense, but most Europeans succumbed to illness after extended residence or journeys into the hinterland. The experiences of other German colonies alerted officials in Togo to the profitability of forest clearance for the establishment of plantations for export crops. As in other parts of West Africa, areas cleared of trees became grasslands and thus more susceptible to the tsetse fly and spread of disease.[56] Precarious financial and administrative circumstances in the short-lived German protectorate and the extent of precolonial deforestation combined to prevent the large-scale production of export timber during the colonial period.

Deforestation and Forestry Management

Official recognition of widespread deforestation and soil erosion in Togo initiated a concerted effort toward developing a colonial forest policy. Experimental plantings were undertaken in most of the districts. Implemented in 1906, Governor von Zech's policy was twofold: reliance on reforestation of severely deforested areas for colonial needs and the establishment in 1912 of regulated forest reserves in which forest clearance by woodcutting or burning was prohibited.

The Germans chose three principal areas in which to carry out their program of reforestation. These targeted regions were: (1) the confluence of the Haho and Baloe Rivers (28,000 hectares); (2) the area between the Mo and Kamaa Rivers (38,400 hectares), in anticipation of the projected railroad and iron-mining projects; and (3) a smaller area in the proximity of Sansanne-Mango (6,000 hectares).[57] German policy also concentrated on replanting some of the most severely affected areas around the district stations at Sokodé-Bassar and Haho.

At Yendi the district officer, Captain Mellin, planted 500 hectares over a two-year period (1909 to 1910).

In addition to the response warranted by a deteriorating ecology, there was clearly a commercial incentive for the creation of forest reserves. In Germany the price of wood had reportedly risen from 8.50 marks per 100 kilograms to 30 marks per 100 kilograms in 1910.[58] Interest in woods and forestry experiments to investigate commercial potential began much earlier.[59] *Daniella thurifera,* which had also been used by the traditional African iron industry, was intensively investigated as an export from the colony that could supply raw materials for the German paper industry.[60] It was hoped that other plantings would satisfy the colony's own timber needs. Teak, in particular, was planted in 1903 along colonial roads.[61]

August Full claimed that 80,000 hectares of savanna had been replanted by the Germans before the beginning of World War I.[62] This figure would seem to represent the ambitious 25-year projection rather than actual plantings accomplished during the program. When the British forester Unwin visited Togo late in the year 1911, only 1,650 hectares had been successfully reforested. By the end of German rule, the total area affected by the program was probably still under 3,000 hectares.[63] Reforestation was undertaken by the Germans as an investment, and its returns benefited the French.

Nearly 20 years after the departure of the Germans, many of the later French foresters, including A. Aubreville, attempted to study and summarize the German forestry efforts.[64] Jean Meniaud, head administrator of the French colonial Service des Bois, described Togo as "lagging behind" the three great West African exporters, yet not without potential. He wrote that "the forest regime of Togo is in an intermediate position between those great forested colonies from the first group (Ivory Coast, Cameroon, and Gabon) and those of the other West African colonies."[65]

Continued recognition of the intensive deforestation in precolonial times made protection of remaining colonial resources the prime forestry objective in the French colony. Little had been accomplished in the years since the departure of the Germans. The principal regulation (1920) was essentially a reproduction of the 1896 German ordinance that prohibited the cutting of colony trees without official permission, under penalties of fines and imprisonment.[66] The remaining forests of western Togo, under British authority, were protected as forest reserves.[67]

Conclusion

Given that the entire central administration of the German colony consisted of no more than 20 persons in 1914, colonial support of a forester and assistants was a significant investment. Deforestation was an important theme in colonial governmental and scientific reports. Despite this it should be emphasized that actual German accomplishments were slight. Because of expectations of a natural vegetation cover originally resembling other densely forested German colonies like Cameroon, the Germans were more critically aware of the environmental impact of precolonial activities than of their own. Simultaneously, while measures undertaken by colonial administrators changed traditional land use patterns, the substitution of cheap, often inferior manufactured goods from Europe for products of local industries increased. Consequently the decline of such local industries as traditional iron smelting reduced the most exorbitant levels of forest exploitation in Togo's history. Plantation and export agriculture dominated colonial policy as it affected the forests of Togo. The opposing interests of traders, indigenous cultivators, and forestry management spared the colony the full implementation of a system of plantation agriculture. Nonetheless, colonial Togo experienced the environmental consequences of profound economic transformation under German rule.

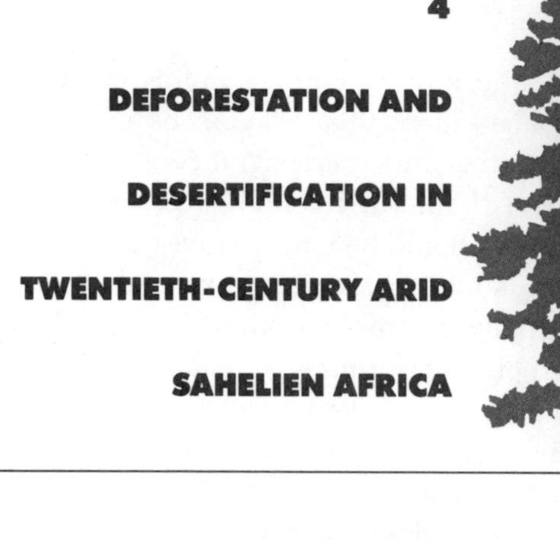

4

DEFORESTATION AND DESERTIFICATION IN TWENTIETH-CENTURY ARID SAHELIEN AFRICA

James T. Thomson

In the world's arid lands, particularly along the edges of major deserts, deforestation has posed a problem of growing magnitude in the twentieth century. Stripping ground cover from fragile soils exposes sands and clays to wind and water erosion and to laterization. Environmental degradation often quickly follows, as it has in the West African Sahel, the lands fringing the southern Sahara, upon which this essay focuses.

Deforestation and desertification have advanced in tandem in the Sahel throughout much of the twentieth century. Desertification is the consequence and deforestation the intervening variable. The sufficient (if sometimes unnecessary) causes of deforestation include population growth, low-efficiency stoves, and sometimes inefficient land use techniques (extensive rather than intensive farming practices). Land and tree tenure rules that often discourage investment in future supplies must also be counted among the causes of deforestation.

This essay examines relationships among these various causes and their consequences within the three central west Sahelien countries of Niger, Mali, and Upper Volta (Burkina Faso since 1984). For quantitative data it draws heavily on documents produced recently for the Club du Sahel.[1] Case study material from the Yatenga region of northern Upper Volta illustrates land use dynamics.

The essay argues as follows: during the half millenium preceding the twentieth century, deforestation posed few problems in the Sahel. Populations rarely reached the point where they threatened natural regeneration of renewable resources—trees, pastures, soil, and water.[2] Droughts of varying intensity struck the region frequently and unpredictably, temporarily impairing the productivity of the resource base. But when rains returned, trees, pastures, and soils yielded adequate amounts of wood, fodder, and crops, and renewable resource use returned to "normal" interdrought levels.

With the advent of the twentieth century and the culmination of French colonial conquests in the Sahel,[3] colonial activities began impinging on Sahelien economies and ecologies. This impact occurred indirectly, so far as woodstock exploitation went. With the exception of *Khaya senegalensis* (African mahogony), few commercially valuable stands existed. Insufficient rainfall and prohibitive transport costs to European markets barred Sahelien wood from the ranks of potential export commodities.[4]

Rapidly spreading rain-fed agriculture has spelled disaster for many Sahelien woodstocks. This in turn bodes no good for long-term equilibria of Sahelien environments or for the living standards of resident populations. The shift from complex intensive to cruder extensive agricultural techniques occurred by fits and starts throughout the Sahel as different areas responded directly or indirectly to colonial demands for cash, crop production, and migrant labor. This process began in the first decade of the twentieth century and grinds on today with unabated fury.

Given large-scale labor migration of the most able-bodied from the Sahel to richer coastal regions, or from Sahelien rural areas to the many burgeoning cities in West African arid areas, population growth has not automatically forced intensification of agricultural techniques.[5] Instead individuals have elected those methods that allow them to cultivate the most land in the least amount of time at the beginning of the rainy season (June through September in northern Sahel and May through October in the more humid southern portion of the region). Collapse of extended family institutions, which in earlier times usually imposed rational controls on resource exploitation, has ruptured old patterns and rhythms.[6] Most individuals or nuclear families now seek, as best they can, to cover their food needs by extensive farming. Labor bottlenecks in the rainy season, occasioned by labor migration and changing kinship structures, curtail crop yields. In drought years harvests often fall below the break-even point for these nuclear units. Since 1960 soil fertility has eroded dramatically in many areas, leading peasants to suppress traditionally long-duration fallows and to put ever more marginal soils to the hoe. Woodstock destruction follows in bush areas and fallows, formerly sites of substantial natural regeneration.

Inappropriate state policies concerning land tenure and tree property rights, especially as enforced (or not) in local communities, have pushed people to farm lands they might otherwise have left fallow. When farmers continuously crop their fields, little woodstock regeneration occurs. Seedlings that escape mutilation by the farmer's hoe often succumb to browsing local and transhumant livestock. Animals forage over most Sahelien village lands during the dry season. Inevitably they kill many seedlings or stunt growth to the point where trees never adequately recover.

Finally, cooking "stoves" in the Sahel function poorly. Clay pots balanced on three stones over a fire simply do not use available energy sources efficiently.[7]

We will now examine these points in five sections: traditional woodstock exploitation in the Sahel; impacts of colonial presence and economic policies on village and family authority structures, labor forces, landholding patterns, and land use practices; Sahelien conservation service policies and land tenure rules; quantitative data from three Sahelien states concerning woodstock evolution; and future prospects that wood reserves will reconstitute themselves or new sources be deliberately created.

Traditional Woodstock Use in the Sahel

Until the twentieth century, Sahelien woodstocks mainly protected the environment and regenerated soil fertility. They also furnished fuels, building poles, fruits, nuts, medicines, and sundry other products for consumptive uses. Man-land ratios remained low from the precolonial era (to 1900) through the first quarter of the twentieth century. Fallowing or entirely abandoning lands after several years' cultivation kept soil fertility near the potential peak determined by soils, characteristics, and prevailing rainfall levels. Few Sahelien regions boast rich soils. They typically succumb to water and wind erosion once ground cover has been removed.[8] But so long as bush remained abundant and most land carried some kind of vegetation, soil degradation posed no serious problem.

That situation has now changed drastically. Saheliens have cleared larger and larger areas surrounding agricultural villages and have pushed the farming frontier further and further out from the old settled communities. This process threatens to deforest some areas totally. In others, where peasant farmers have deliberately or through imposed legal constraints kept trees on their fields, agroforestry exists as a still-functioning though threatened system of environmental management. As suggested below, agroforestry appears to offer the best and perhaps the only hope of long-term environmental preservation in many parts of the Sahel.

Impacts of French Colonial Presence and Policies

In an underpopulated region poor in natural resources—the Sahel in 1900—two broad strategies offered themselves to colonial officials responsible for organizing the Sahelien colonies and ensuring they paid their way within the French empire.[9] They could export surplus population from heavily inhabited areas, for example, the Mossi plateau in central Upper Volta and the Hausa regions of south-central Niger. Or, they could develop the meager resource base by pushing cash crops (mainly peanuts and cotton). Administrators used monetary head taxes to get the leverage over local populations needed to pursue both strategies.[10] Initially they accepted payment in kind (grain or livestock), as had most traditional Sahelien states. However, administrators soon insisted on payment in cash.[11] This pressured Sahelien peasants because French specie was rare in the colonies.[12] Younger men emigrated in the dry season (November through May in most of the Sahel) to richer coastal countries, Ghana and Nigeria in the early period (1900 to 1940), Nigeria and Ivory Coast later on.[13] Sterling flowed temporarily to the franc zone from adjacent British colonies, but the developing French West African empire risked dependence on the economically more advanced British coastal colonies. French colonials tried to reorient labor flows from English to French territory, but without notable success until the post–World War II Ivorien economic boom.

Migrants, especially Mossi from Upper Volta, sought refuge from French colonial taxation and from recruitment into the military for forced labor or for "paid" jobs on plantations in British colonies.[14] They also sought to share in the real benefits offered by the booming, labor-hungry Gold Coast (Ghanaian) economy of the period 1910 to 1940.[15] Tensions within indigenous societies, pitting older against younger men in a competition for political and economic power, often fostered and then consolidated migration patterns once exit possibilities became established.[16]

Further north in the arid Sahelien regions, French colonials also leveraged peasants with the second strategy, pushing export crops with increasing vigor from the 1920s on to the end of the colonial era. Independence governments inherited and continued the policy to the beginning of the drought of the 1970s.[17] Peasants faced administrators bent on forcing them to plant, harvest, and sell peanuts and cotton to French colonial trading companies extending their networks from the coastal colonies of Senegal and the Ivory Coast into the Sahelien interior.[18]

Some French West African administrators, apparently supported by local chiefs, advocated the individualistic solution: individual or family fields devoted to producing cash crops suitable for export.[19] But in Mossi areas, because of resistance to cotton farming when food supplies remained short, collective fields organized by village chiefs appear to have been the standard approach adopted to ensure cash crop production.[20] Eventually, however, cash cropping became largely an individual phenomenon.[21]

The importance of these two strategies lies in the margins of the personal liberty they introduced into those sedentary Sahelien agricultural societies previously organized in fairly tight gerontocracies. In many situations this new leeway enabled younger males particularly to request, demand, or take their land inheritance well before the elder male head of the household unit died or, failing that opportunity, to establish a new autonomous existence beyond his authority. Up until the beginning of the drought in the 1970s, many younger men throughout the Sahel could pay their taxes and meet other cash expenses with earnings from migrant labor or cash cropping.

Nuances of this process of extended family breakdown may be fairly uniform within a single ethnic group; they may also differ. They certainly vary from one ethnic group to another. Numerous factors influence the exact manner in which kinship power unravels. An important one is often religious change. Throughout most of the Sahel, the salient conflict has pitted Islam, in this context an egalitarian force, against various indigenous religions, most of which involve some form of ancestor worship.[22] For all practical purposes, Islam has carried the day with the exception of a rear-guard action in the Mossi plateau and other Voltaic regions, and in parts of southern Mali. The organization of lineage ties, underlying economic relations (influenced both by local agricultural techniques and environmental characteristics, and by world market conditions), legal systems effective at the local level, and political factors have all also shaped the process of individualization of land control, retarding it here, advancing it there. As rural people have slowly gained access to better medical care, population growth rates jumped up to the range of 2 to 3 percent per year.[23]

Gradual disruption of controls over lineage members (in some cases not yet complete) has fragmented institutional structures that formerly upheld the organized active and passive systems of environmental management in the Sahel.[24] These have now largely collapsed into severely fragmented systems of land use. These in turn have threatened, and often led to, the complete breakdown of sustained-yield use of renewable resources.

Numerous observers argue that these developments have undercut the formerly prevalent local social security mechanisms. In hard times, when drought, diseases, and insect attacks reduced food supplies, joint exploitation of lineage or extended family fields and controlled sharing of produce meant that those who failed to cover subsistence needs in a given year could be carried by the rest of the group, on the tacit understanding that once restored to health they would extend the same security to other lineage members.[25] Now, spreading individualism frequently causes each man to be viewed as the master of his own fortune, whatever his situation and however little his capability to assume the responsibility. The individual is now also often viewed as the sole victim of his own misfortunes. The security net has largely unraveled.[26] Each person is now compelled to provide for himself and his immediate nuclear family. If personal security can be enhanced at others' expense, that is the going price, which many are willing—or see themselves forced—to pay.

As Marchal has lucidly and exhaustively argued of Mossi peasants in Upper Volta's Yatenga region, each individual who now operates a nuclear family enterprise of restricted size seeks to plant the maximum possible field area. In a calculated gamble to win the annual struggle for subsistence by spreading his bets, he sows various crops on tiny, widely dispersed fields.[27] This maximizes his chances of reaping at least some harvest. Quick and easy farming methods designed to spread the farm unit's scarce labor resources to the absolute breaking point in cultivating the greatest possible surface per worker have unfortunately replaced traditional techniques.[28] The latter, environmentally sound, techniques included manuring practices keyed to various soil types, intensive sowing and weeding of fields at critical times during the growing season, systematic fallowing, and water- and soil-conserving methods of cultivation.[29]

Inevitably crop yields on extensively cultivated fields are poorer in drought years than those on intensively worked lands because the former do not retain surface moisture to the same extent as the latter. Soil fertility declines over a period of years because lands are neither worked as intensively nor fallowed as in former times, and the subsistence gamble becomes ever harder to win.

These developments have been accompanied by—and flowed in part from—improvements in public health levels, which have slowly lifted population growth rates in all Sahelien states well above 2 percent a year. Populations now double in the Sahel every 20 to 30 years. This sort of demographic pressure easily provokes a land rush, as peasants move to stake claims to remaining common bush areas. Perspicacious individuals do their best to farm even marginal lands to establish usufructuary rights and so stave off as long as possible the subsistence crisis in their own families.

Migration has offered a safety valve in these circumstances.[30] Some people have moved their families elsewhere within their home countries to uncultivated, unclaimed lands. Others have become expatriates, living in neighboring states (e.g., Nigeria, the Ivory Coast, or formerly, Ghana) and earning their living there.[31] But events of recent years suggest that this recourse may be throttled off. Within the last decade Nigeria, Ghana, and Ivory Coast have all responded to internal political pressures by forcing "illegal" immigrants to leave.[32] If Saheliens are unable to migrate reintensification of agricultural practices becomes the only feasible solution, aside from direct efforts to cut population growth rates to more manageable levels.

Sahelien Conservation Policies and Land and Tree Tenure Rules

Contemporary Sahelien land tenure rules reflect both these pressures: the drive to secure fields adequate to support the owner's dependents and state policies calculated to encourage food production ("the land to he who tills it" is the commonest denominator of these state programs). But those who no longer live in extended family settings rarely see intensification as a practical alternative to extensive agricultural techniques. Thus the system pressures each individual to claim as many fields as possible and to farm them to solidify the claims. In Niger such activities have been common for at least three decades.[33] In a 1975 edict President Seyni Kountche sought to end the legal jockeying that many individuals instigated to increase their own land holdings. This has made land owners wary of loaning fields to land-poor individuals, for fear they will be unable to reclaim them. It has also forced those with "surplus" lands, some of which are fallowed to regenerate soils, to bring them back into cultivation for fear of losing them to an unauthorized "settler."[34] This reduces ground cover through rule-induced destruction of fallow.

In the northern Mossi region of Upper Volta, a comparable development followed the colonials' codification of the legal working rules of land tenure practiced in Yatenga during the late 1950s.[35] There is ample evidence that similar results have occurred elsewhere in the Sahel, after similar legal changes.[36]

Such systems mean land borrowers must now realistically be pessimistic about chances of converting a temporary field loan into permanent possession. Owners tend to reclaim fields fairly quickly, even if they then loan the same individuals other pieces of land. Thus borrowers have no reason to make long-term improvements on land they cultivate, even if they are so inclined. At best, seeking short-term gains, they might apply chemical fertilizers for a few seasons, or green manure the borrowed land if materials are available and application requires little additional effort. But investments in terracing, or in tree planting or nurturing—to regenerate soil fertility or reduce wind or water erosion—seem uneconomic for borrowers, if not irrational.

Little has been done in Mossi areas of southern Yatenga or Hausa regions of south-central Niger to develop new land loan forms requiring borrowers to apply soil conservation measures on loan fields. Such loan conditions may later be imposed as individuals see more clearly the stakes involved in preserving renewable resources. So far nothing like this kind of institutional innovation has surfaced.[37] Where field loans make up a sizeable portion of arable land, noninnovation in this area poses a long-term threat of considerable magnitude.[38]

Tree Tenure

Tree tenure rules pose formidable difficulties in many Sahelien African settings; in other areas, the issues exist in germ, but do not yet affect behavior. In essence the problem is this: Sahelien state forestry services generally claim residual, and often primary, authority over exploitation of all woodstocks within national borders. Woodstocks can be divided into two major types: state forests and related preserves, and trees growing on "private" lands (i.e., on villagers' fields) or in bush common areas. Unfortunately, since the founding of the colonial agency in 1934 Francophone West African forestry services have never had enough men, money, or equipment to translate these authority clauses into effective control.[39]

Area foresters who visit often enough to assert authority over woodstock use without, however, being able to enforce the law systematically,[40] suffer most from this situation. Villagers, concluding use control is the foresters' problem, have for the most part allowed others to harvest wood on their land without trying much to prevent it.[41] Foresters have been unable even to reach other areas regularly, so their legal claims to control wood use are irrelevant there. In such places, woodstock deterioration has also occurred, often because no one happened by to prevent uncontrolled cutting. But in such situations, at least, the possibility of individuals asserting authority to control use of their own woodstocks exists.

Locally made rules, locally enforced, can probably modify use behavior simply because recourse costs are very low. Recourse to the few roving foresters in charge of policing wood use in areas they "frequent" (visits often do not exceed two or three a year) is much more expensive because they so rarely appear when needed.

Forestry service personnel have been spread too thinly, given the generalized mandate to "control woodstock use" enshrined in the forestry codes of most Sahelien states. In particular, foresters often fail to stop illegal cutting in state forests. If local communities could themselves regulate the use of trees on village fields, some would rapidly deforest their lands but others would develop systems of renewable resource controls. Freed from local resource management tasks, foresters could better patrol state domain lands. They could also do extension work in communities where people want to rebuild the woodstock, or exploit state forests for sustained yield on a participatory basis, or all three.[42] At present, they are spread too thinly to do any of these tasks effectively.

Deforestation in the Sahel: Qualitative and Quantitative Data

For the first half of the twentieth century, little quantitative data exists about Sahelien deforestation. Nobody collected figures. The forest service was then, as it remains today, the last and least of Sahelien development agencies.[43] Indeed, it took French West African colonial administrators 35 years after creating the West African empire to set up a conservation service for all constituent colonies.[44]

In French West Africa at mid-century, the combined forest services of the eight colonies in the block counted only 36 forest officers, all products of the French national forestry school at Nancy; 44 *controleurs,* with two years of training at an agricultural college followed by a year at the forestry school at Barre, France; and 36 African assistants.[45] The bulk of this personnel undoubtedly operated in the Ivory Coast, with its important commercial logging operations; in consequence, only a handful of foresters controlled wood use in the vast Sahelien expanses.

Qualitative material about deforestation does exist. It appears first in scattered remarks by French colonial administrators and then increasingly in expert foresters' observations, insights, and arguments. As foresters set up state preserves in the colonies they measured and described in detail expropriated areas. But the vast regions of virgin and second-growth bush (*terres vacantes et sans maitre,* or unclaimed lands, as foresters and legislators blithely styled other peoples' fallow reserves) remained largely uninventoried. However, the latter far outweigh state forest in terms of sheer bulk of ligneous material.

It took the fierce drought of the 1970s to catapult the Sahelien woodstock into public prominence as a critical element in any future effort to stem desertification on the arid West African steppes. Quantitative data begin to appear after 1975. By the early 1980s estimates became available concerning gross land areas and the types of wood cover associated with them.

In this section we will first review experts' observations concerning Niger in the 1930s, briefer comments on the situation in northern Upper Volta in the same decade, and comments on Senegal at midcentury. Then we will present data and conclusions prepared by multidisciplinary teams for the Club du Sahel[46] about woodstock conditions, forestry services, and forestry policy in the early 1980s in Mali, Upper Volta, and Niger.[47]

From 1930 to 1960

Niger. In Niger an experienced tropical forester, A. Aubreville, saw clear evidence in 1936 of deforestation and desertification in several regions. He carefully distinguished naturally occurring from man-induced desertification. The former he related to various macronatural causes: (1) long-term consequences of Saharan dessication from the Pleistocene or Glacial period to the present; (2) contemporary year-to-year variations in the intertropical convergence zone;[48] and (3) naturally occurring wind erosion.

Saharan dessication created live dunes.[49] It also fostered water erosion by reducing vegetable cover. Erosion in turn slowly smothered bottomland vegetation in high steppe valleys by depositing an impermeable layer of clay soil on valley floors.[50]

The second cause, unpredictable short-term weather patterns, produces shifts in controlling wind patterns over West Africa, and in the harmattan winds, which sweep out of East Africa over the Sahara and into the Sahel. These give rise to the *brousse tigrée* formations (strips of bush vegetation separated by intervening denuded spaces) typical of much of the Sahel. Droughts weaken existing (and often aging climax) stands of bush. Wind erosion, depositing a layer of smothering fine clay soils, completes the process of desertification in each tiny local environment.[51]

Aubreville distinguished these natural causes from the second type of desertification, which results mainly from land clearing.[52] In 1853 the German explorer Heinrich Barth passed somewhat further south through the same region along the contemporary Niger/Nigeria border.[53] He considered it relatively forested.[54] Eighty years later, Aubreville identified the now common phenomenon of deforestation in concentric circles around major urban centers.[55] At the time of Aubreville's visit in 1935, fully 50 kilometers to the east of Maradi toward Tessaoua had been completely cleared for millet cultivation. In fact, stable dunes had been so stripped of wood north of Maradi that Stebbing, an English colonial on tour up into Niger from northern Nigeria, reported the Sahara had advanced to within three miles of Maradi.[56] (The region directly east of Maradi today counts relatively well-developed stands of *Acacia albida*, including large numbers of specimens under 10 years of age.) Further east in the area between the urban centers of Maradi and Tessaoua, the woodstock remained much the same as at Barth's passage. It included "the indigenous vegetation, such as I described it for the sandy, subsahelien areas. Trees are fairly spread out, but some fine specimens exist. The grasses are not very thick, so bush fires don't do much damage. The whole is relatively wooded."[57]

Aubreville found deforestation a problem in many of the more populated regions of the colony. The Niger River valley and two ancient dry water courses, the Dallol Bosso and the Dallol Maouri, had been largely cleared. Denuded areas in south-central Niger included the Adar Doutchi-Maggia valley, Govir (especially between Birnin Konni and Madaoua, on the southern border of the colony), parts of Maradi (discussed above), and parts of Damagaram (Zinder Department).[58] In many of these areas, the only trees remaining in 1935 were those peasants had deliberately left on their fields, particularly *Acacia albida*.[59]

Upper Volta. During the same period in Upper Volta, the process was not yet so far advanced, but it was certainly discernible. In the more arid northern Yatenga region around Ouahigouya, a *commandant de cercle* noted as early as 1924 that "the question of reforestation is of primary interest for this region. As yet, neither plantations nor nurseries have been established. The vast size of the *cercle*, and the administration's minimal means of action have made it impossible to do anything in this regard. However, the desperate war against bush fires and unreasonable felling of trees, which the *cercle* has undertaken, appears to be producing the desired consequences."[60]

As the French geographer Jean-Yves Marchal notes in his commentary on this passage: "The problem of 'deforestation' (firewood, building poles, and browse during the dry season) and its effects (erosion) have already attracted attention, but without anyone thinking seriously about trying to find a remedy. Deforestation would only get worse and worse."[61]

Senegal. By the 1950s in Senegal, the head of the colonial forestry service, P. Foury, detected all symptoms of the situation now endemic throughout much of the Sahel: a growing imbalance between supply and demand for trees, wood, and wood products. Furthermore, the accelerating loss of soil fertility throughout the peanut growing regions of Senegal, and the narrowly "agricultural" solutions proposed as remedies for this problem, disturbed him.

Agronomists advocated organic or chemical fertilizers and plow cultivation. Foury calculated that local herds were too few to produce enough manure, and chemical fertilizers were too expensive for most peasant farmers. Plowing even gently sloping land aggravated soil destruction by breaking up the ground surface and fostering sheet and gully erosion.[62]

Foury instead saw salvation for the land in more effective silvicultural practices. These would build on a long tradition of agroforestry, at least among the important Serer ethnic group. *Acacia albida,* long a favorite Serer field tree, both fixes nitrogen and responds well to pruning and lopping. Thus while the tree restores soil fertility, it can also supply local fuel needs.[63] Foury outlined ways *A. albida* might easily be planted by local people, along with *Borassus aethiopum,* the ronier or fan palm. Serer mat weavers, dependent on the *Borassus* for raw materials, had by the 1950s already begun planting new stands against the day when existing ones would be totally destroyed.[64] The tree would also provide strong, termite-resistant construction wood.

In this same context, Foury suggested the forestry code be modified to permit pruning *A. albida,* as well as other protected species, both because they resist lopping well and because they afford, in many places, the only available wood supply. This comment implies unprotected species had already been decimated in such areas.

Finally, the urban wood supply problem drew his attention. He declared himself extremely skeptical of plantations, both on cost calculations and because land of suitable quality could rarely be obtained close to urban centers.[65]

In sum then, Foury ticked off all major difficulties of deforestation in the Sahel, as these are currently conceived, and proposed solutions. The solutions relied almost entirely on indigenous species and dovetailed fully with existing agricultural practices. The fact that Foury, as chief forester of the colony, had begun to think seriously in these terms suggests the looming gravity of deforestation in Senegal at the time.

Overview

To sum up the preceding discussion, Lord Hailey's mid-century estimates provide a sort of benchmark, though subject to the usual reservations concerning the reliability of figures: "French authorities estimate the total woodland area of French West Africa at 656,000 square miles (out of a total land area of 1,798,374 square miles), of which 192,000 are described as productive or partially productive [or 10.7 percent of the total land surface, and 36.5 percent of the wooded terrain]."[66]

These global figures concern a block of colonies spanning the gamut of ecological types from tropical rain forest to empty Saharan spaces. They suggest forests accounted for a major proportion of territory in populated regions (from the coast to the Sahel). Wooded areas must have exceeded half of the total subdesert area (the Saharan expanses of Mali, Niger, and Mauretania, practically devoid of trees, constituted more than half of then-French West Africa).

The proportion of wooded land in the Sahel would have been much lower, given the more arid conditions, than in coastal colonies (Dahomey, Ivory Coast, and Guinea), which contained important rain forest belts. Upper Volta, in the intermediate savanna zone, would have occupied a median position.

Contemporary Woodstock Use and Future Prospects

Most recent figures on deforestation and its causes in the Sahelien states suggest strongly that agricultural land clearing and cutting for urban fuel supply cause most of the woodstock destruction currently observed in the region. "Commercial" lumbering plays no role at all, at least as far as any export market is concerned. Demand for building poles accounts for no more than 10 percent of consumptive use of the woodstock. Another important factor promoting deforestation has been the failure of Sahelien forestry services to devise and implement policies encouraging individual, family, or group investment in future supplies. These points will now be explored in turn, using figures available for Niger, Upper Volta, and Mali.

Niger.[67] The 1981 CILSS/Club du Sahel report on forestry and ecological problems presents estimates of deforestation in the country based on the often contradictory figures available in contemporary documents. Nonetheless, the report permits calculations of trends and orders of magnitude in land clearing and wood consumption. These highlight wood supply difficulties that Niger will face over the next two decades. They suggest as well growing competition between agricultural and forestry land uses.

Niger faces a looming wood supply deficit and an associated ecological crisis of major proportions as rocketing demands for wood and for agricultural land slice deeper and deeper into existing wooded areas, reducing woodstock capital (see table 4.1).

Estimates of land area currently under cultivation conflict. In general, LANDSAT image-based estimates run 50 percent higher than Nigerien official calculations. Cereal grain production per hectare, depending on rainfall and terrain, may thus be well below official estimates of 250 to 450 kilograms.[68] If so, substantially greater land areas will be required to feed the more than 3.2 million individuals projected to swell total population over the next two decades (unless cultivation techniques change markedly). Inevitably lands now fallowed will be drawn into cultivation, and existing forest reserves will be severely diminished if agricultural productivity does not rapidly improve. Intensive farming systems must supplant the currently prevalent extensive forms.

While farming and wood use practices may change, nothing guarantees adoption of appropriate agroforestry techniques or sustained-yield management of existing woodstocks.

These calculations are supported by the results of a recent case study in a typical farming village of Maradi department. Of a total village area of 4,800 hectares, 23 percent (1,100 hectares) was cultivated in 1957. By 1975, 48 percent (2,300 hectares) had been cleared for farming. Over the intervening eighteen years field areas had been extended at the rate of 4 percent annually. If this trend continues unchanged, by 1990 all but 15 percent (720 hectares) of the village lands will be under cultivation. Not only will fallowing have been practically eliminated as a soil regeneration technique, but the woodstock will have been largely destroyed.[69]

Upper Volta.[70] Calculations for the 20 years 1981 to 2000 appear to suggest Voltaic wood production will still cover the country's global

Table 4.1 Woodstock Perspectives, Niger

	1981	2000	
Population (yearly population growth rate = 2.77%)	5,697,000	8,887,000	
Wooded areas (fallow included)			
Low estimate	9 million sq. ha. (34,000 sq. mi.)	9 million sq. ha. (34,000 sq. mi.)[a]	
High estimate	14 million sq. ha. (54,000 sq. mi.)	14 million sq. ha. (54,000 sq. mi.)[a]	
Estimated wood production overall average (for 350–800 mm rainfall zone)		Management unchanged[b]	Maximum management[c]
Average yearly production rate	.33 cu. m./ha.	.33 cu. m./ha.	1.00 cu. m./ha.
Total supply			
Low area estimate	2.97 million cu. m./yr.[d]	2.97 million cu. m./yr.[d]	9.00 million cu. m./yr.[e]
High area estimate	4.62 million cu. m./yr.[d]	4.62 million cu. m./yr.[d]	14.00 million cu. m./yr.[e]
Estimated wood consumption per capita			
Average yearly consumption rate	1.00 cu. m./yr.	.65 cu. m./yr.[f]	
Estimated total demand	5.697 cu. m./yr.	5.777 million cu. m./yr.	

a. Assumes total wooded area remains constant.
b. Assumes 1981 management techniques continue unchanged.
c. Assumes maximum sustained-yield management techniques.
d. Estimated hectares × average yearly production rate of .33 cu. m./ha./yr.
e. Estimated hectares × average yearly production rate of 1.00 cu. m./ha./yr.
f. Assumes new conservation measures (fuel-efficient wood stoves, etc.) and shifts to alternative fuels cut per capita wood consumption by one-third.

demand (95 percent for fuel, 5 percent for construction) to the end of the century. However, as the Club du Sahel report authors stress, the situation is rather more precarious than this summary statement suggests. The following figures provide a basis for trend calculations. Subsequent discussion will point out further complications.

These projections take no account of wood production, either from new plantations or from lands again retired to fallow. But neither do they take account of regional differences within the country, in terms of distribution of natural forests and lands of sufficient fertility to permit agriculture—and therefore tree growing. In fact, as the Club du Sahel report makes clear, the northern half of the country and the Mossi plateau (which cuts a swath from fairly close to the southern border up through the center of Upper Volta to its northern frontier) already face severe land and wood shortages. At present the southern half of the country (excluding the Mossi plateau) is much more favorably placed in both respects.

The report authors project a land rush, beginning with northern Mossi peoples and spreading southward down throughout the plateau, to the underpopulated parts of the country. Land disputes can be foreseen. In the event more intensive agricultural techniques and more intensive reforestation programs are not widely adopted, by the early twenty-first century the country's renewable resource base will be severely, if not irreparably, damaged (see table 4.2).

Confirmation of these assertions so far as the northern Mossi plateau—Yatenga—is concerned can be found in Marchal's work on desertification in the Department du Nord. Extrapolations from aerial photographs taken of the Yatenga region during 1952 to 1956 and 1971 to 1973 reveal progressive deforestation around Ouahigouya, the regional capital. Table 4.3 summarizes these changes in a rough circle of 35 kilometers, centered on Ouahigouya, the regional capital.[71] Once soils have been eroded to the point of exposing the underlying laterite, such areas are more or less definitively excluded from the capital of productive land available to residents, whether it be for food, meat, or wood production.[72]

For this same area, Marchal calculates that 81,883 steres (32,753 metric tons) of wood were produced naturally on the total wooded area of 22,849 hectares (good condition) plus 8,891 hectares (degraded), or 31,740 hectares altogether. Of this he reckons half, or 16,400 metric tons, could be harvested without curbing natural growth by mutilating trees or bushes. At the same time wood demand for the population

Table 4.2 Woodstock Perspectives, Upper Volta

	1980	2000
Population (yearly population growth rate = 2.00%)	6,250,000	9,285,000
Wooded areas (fallow included)	10,520,000 ha.[a]	5,020,000 ha.[b]
Estimated wood production overall average (for 350–1300 mm rainfall zone)	1.5 cu. m./ha./yr.	1.5 cu. m./ha./yr.
Estimated average wood volume over all wooded land types	21.6 cu. m./ha./yr.	21.6 cu. m./ha./yr.
Estimated total wood supply	15,780,000 cu. m./yr. (10,520,000 ha. × 1.5 cu. m./yr.)	7,530,000 cu. m./yr. (5,020,000 ha. × 1.5 cu. m./yr.)
Estimated wood consumption per capita	6,250,000[d] (6,250,000 × 1.0 cu. m./yr.)	9,285,000[e] (9,285,000 × 1.0 cu. m./yr.[f])

a. Total natural forest area, including 4.5 million ha. of fallowed lands.
b. Reductions of supply of wooded land on annual basis: 50,000 ha. of natural forest cleared, or 1 million ha. less by 2000; 250,000 ha. of fallowed lands cleared, or entire 4,500,000 ha. of 1980 fallowed lands cleared by 1998.
c. Assumes (inaccurately for first decade) sustained-yield management techniques based on systematic cuts will be used on all wooded lands to generate maximum yield.
d. This would require wood output from 4,166,650 ha. of wooded lands managed for maximum sustained yield; possible.
e. This would require wood output from 6,550,000 ha. of wooded lands managed for maximum sustained yield; impossible by these calculations because only 5,020,000 ha. remain, for a shortfall of 1,520,000 ha.
f. Assumes increasing wood shortages will not induce new conservation measures (fuel-efficient wood stoves, etc.).

Table 4.3 Deforestation around Ouahigouya, Upper Volta

Land categories	1952 (%)	1973 (%)
Wooded, good condition	61,356 ha. (35.4)	22,846 ha. (13.2)
Wooded, degraded state	7,396 ha. (4.3)	8,891 ha. (5.1)
Fields	98,266 ha. (56.7)	115,168 ha. (66.5)
Eroded areas	5,790 ha. (3.3)	18,795 ha. (10.9)

Source: Marchal, *Société, espace et desertification*, 222.

resident in the area amounted to 156,452 steres or 62,600 metric tons, calculated as follows:[73]

Ouahigouya: 20,000 pop. × 1.93 steres/capita =	38,600 steres
Villages and camps: 103,500 pop. × 1.08 steres/capita =	111,780 steres
Blacksmith quarters: 33 × 184 steres/annum/quarter =	6,072 steres
Total =	156,452 steres

As Marchal concludes, continuation of this trend can only occur at the price of extensive mining of woodstock capital and over-exploitation of wood supplies in villages located at greater and greater distances from the regional capital.[74] Of 121 villages in the region studied, he calculates only 52 (43 percent) cover or exceed their wood demand from sources located on village lands. The remaining 69 (57 percent) now face growing deficits that have to be met either by increased imports or reductions in consumption. He judges that the region as a whole slipped from equilibrium to deficit supply during the 1950s.[75]

Mali.[76] Mali enjoys a relatively satisfactory situation, if current estimates of wood supply conditions can be taken as roughly indicative of deforestation rates. Of five relatively populated regions in the southern part of the country, two—Mopti and Segou—have already outstripped available supplies. Two others—Koulikoro and Sikasso—continue to cover local demand with a comfortable margin of surplus. Finally, Kayes still produces twice its wood supply requirements through unaided natural regeneration.

However, within each region disparities already exist. The more humid savanna zones to the south dispose of substantially greater wood supplies than more northerly, arid areas. Another indicator of the relatively adequate wood supply is the Club du Sahel estimate that rural

Table 4.4 Woodstock Perspectives, Mali

	1976	2000	
Population (yearly population growth rate = 2.60%)	6,309,000	11,681,461	
Wooded areas[a] (fallow excluded)	31,674,300 ha.	22,996,700 ha.[b]	
Estimated wood production overall average (for 400–1400 mm rainfall zone)	0.24 cu. m./ha./yr.[c]	0.24 cu. m./ha./yr.[c]	
Estimated total wood supply	6,915,000 cu. m.[d]	5,035,000 cu. m.[d]	
Estimated wood consumption per capita	6,309,000 cu. m./yr. (6,309,000 × 1.0 cu. m./yr.)	5,840,730 cu. m./yr. (11,681,461 × 0.5 cu. m./yr.[e])	11,681,461 cu. m./yr. (11,681,461 × 1.0 cu. m./yr.[f])

a. Wooded areas in this case do *not* include fallows, which would increase estimated production somewhat in all zones.
b. Assumes 8,677,600 ha. will have been converted to agricultural (field or fallow) purposes from wooded areas.
c. This is the estimated average productivity of wooded areas over all zones indicated. In the 400 mm–600 mm area, productivity is estimated to be .13 cu. m./ha.; in the 1200+ mm area, it is estimated to be .35 cu. m./ha.
d. These production figures were calculated by multiplying estimated (1976) and projected (2000) surfaces by rainfall zone from 400–1400 mm by the estimated productivity for each zone. These totals do *not* include wood produced in the arid VI and VII regions.
e. Assumes consumption per capita drops as supplies tighten, either through conservation measures or through shifts to alternative fuels.
f. Assumes consumption remains unchanged despite increasing scarcity.

residents currently farm only one-quarter of the Malien arable zone (400 millometers—1,400 millometers isohyets). But projections from current trends suggest the situation will gradually worsen. Demand for arable land, in particular, menaces the existing woodstock.

Figures indicate that even if wood scarcity and rising prices throttle back consumption to 0.5 cubic meters per capita per annum, global supply will not meet global demand by year 2000 (see table 4.4). If demand continues unabated, the situation will become much worse. It is worth noting in this context that Mali is projected to retain a very substantial "nonagricultural" area to the end of the twentieth century. This figure, 229,967 square kilometers (46.7 percent), is then expected to be devoted to fields and fallow lands.

Trend projections thus suggest deforestation and wood supply difficulties will increase, though more slowly than in many other parts of the Sahel. Transportation difficulties make it unlikely that large amounts of wood will be shifted over long distance (more than 160 kilometers or 100 miles). Arid rural areas, where wood production is sharply limited by natural conditions, are ill-equipped to meet a concentrated demand for fuelwood and building poles. Thus those near urban centers face a particularly bleak future, unless changes occur.

Conclusion

The trend projections sketched out above assume some changes in human use of wood, particularly reduction in consumption as supplies dwindle. Fragmentary evidence concerning interest in and adoption of more efficient wood stoves lends credibility to this assumption.[77]

More problematic are two other assumptions: (1) natural stands of bush, as well as the woodstock on farmers' fields, can be managed much more aggressively for sustained-yield use; and (2) farmers will adopt more intensive land use practices, for example, cover crops to augment soil fertility on short-term fallows, and traditional agroforestry practices coupled with better cultivation of smaller land surfaces.[78] All are technically feasible to some degree. Whether Saheliens —foresters, peasants, and herders as well—can in fact realize these improvements is questionable, though by no means improbable. The real question, however, is whether individuals will be sufficiently motivated to work out ways in which better environmental management practices can be achieved and retained into the twenty-first century.[79]

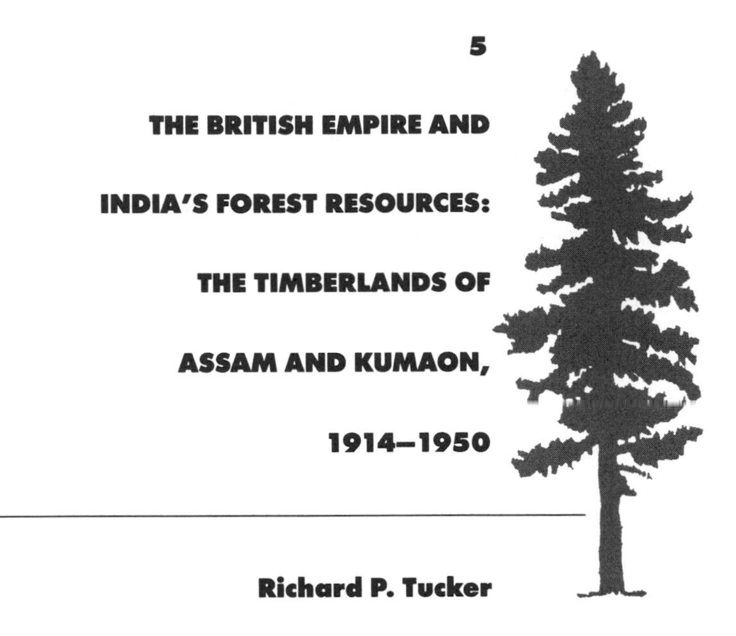

5

THE BRITISH EMPIRE AND INDIA'S FOREST RESOURCES: THE TIMBERLANDS OF ASSAM AND KUMAON, 1914–1950

Richard P. Tucker

In each forested region of India the first era of massive deforestation occurred shortly after it was absorbed into the British Empire, at a time of transition in the area's political economy. Whatever the time for each area, from roughly 1770 to 1860, it was a period of new commercial possibilities for timber sales or alternative cash crops, with no public agency yet in place to enforce any systematic timber management or balanced land use policy. Then gradually, as the imperial bureaucracy became established on the land, greater planning and efficiency of forest use evolved, especially after the Indian Forest Service was established in 1865. The nineteenth century saw massive land clearances for agricultural expansion, but in the later decades there was also careful demarcation and detailed assessment of major timber stands, first in the commercially valuable forests accessible to transport lines and urban centers, and gradually farther into the periphery of the subcontinent in the formerly inaccessible Himalayan mountain regions of the Northwest and Northeast.[1]

A second great wave of deforestation came in the 1940s with the demands of World War II and the transition to independence for India and Pakistan in 1947. This essay will focus on the years leading to and encompassing the second era, in which timberlands reflected the series of global cataclysms that began in 1914. It will assess the impact on the extent and composition of two vulnerable forest regions: the conifer and broad-leafed forests of the western Himalayas and the subtropical mixed forest lands of Assam in the northeast. Both are crucial for the watersheds of major north Indian rivers: northwestern Uttar Pradesh on the upper Ganges and Assam astride the Brahmaputra. Both are adjacent to dense lowland agricultural populations with their demands for forest products. Both were subject to the bureaucratic and economic forces of the British Empire, and therefore to global events. Both by now have suffered severe ecological degradation.

But the two areas' specific patterns of resource exploitation have been very different. The deforestation of Assam has resulted from the complex of political and cultural factors that led inexorably to the explosions of 1983; its political and ecological downward spiral promises only to continue in the future. The western Himalayas, in contrast, have seen a longer and slower decline of their forest cover, and their recent social and political history has been much less turbulent. There the mountains have been washing away with a slower steadiness, and local peasant protests specifically focusing against commercial timber operations have become a familiar aspect of the scene since the early 1920s. By tracing trends in these two contrasting but equally fragile environments, this essay will highlight the dynamics of land policy, economic development, demographic and ethnic movements, and settings of landscape and vegetation in Uttar Pradesh and Assam.

The Timber Belt of Kumaon

West of the Nepal border lie the Himalayan mountain districts of Uttar Pradesh, or U.P., as the state is known by English and Hindi speakers alike. (Under the British Raj before 1947, it was called the United Provinces; hence U.P. then as now.) The state as a whole dominates the upper plains of the Ganges basin. With a population now passing 100 million, it faces intense pressure on its fertile alluvial soils. From the early nineteenth century both subsistence and commercial agriculture expanded into the forested belt below the outermost Himalayan hills. But in contrast with the vast expansion of tea plantations in Assam, capitalist export agriculture never became an important element of the Himalayan hills in the northwest. Again in sharp contrast to Assam, there was little shifting agriculture to wear away at the vitality of the forest. And there has been no great peasant migration from the fertile though crowded U.P. plains to the thin soils of the hills. Instead peasant agriculture in the hills has seen the steady multiplication of individual farm plots, often terraced; the intensification of livestock grazing; and a considerable outmigration to the plains. The changing extent and composition of these hill forests has been largely a reflection of commercial forestry management under the terms of the 1878 Indian Forest Law, plus subsistence pressures related to settled agriculture and transhumant grazing.

In the U.P. forest region of Kumaon[2] three distinct zones of topography and vegetation must be considered for any assessment of ecological change during this period. The belt that daunted settlers throughout the colonial era is the *tarai,* a malarial swamp that lies just below the outermost hills and accumulates heavy rainfall during the monsoon months of June through October. The tarai effectively discouraged plains people from expanding northward until the antimalaria campaigns of the 1950s, thereby serving as a buffer zone for the hills. Timber cutting was a far more profitable venture, for the Himalayan foothills from somewhat west of U.P. and for 2,500 kilometers southeastward as far as Assam produce vast quantities of *sal* (*Shorea robust*) timber, a versatile broad-leafed hardwood that has always been the chief object of north Indian forestry management. Under British rule the sal forests of the north began to be cut in the 1770s, as soon as Calcutta grew large enough to require building timber and fuelwood from farther reaches of the Ganges basin.

Even a thousand miles inland from the Bay of Bengal the elevation in the tarai is hardly above 300 meters. Immediately beyond the jungle belt rise the Siwalik hills, the outermost range of the Himalayas; in Kumaon their steep slopes reach to summits above 2,000 meters. From there other ranges of roughly similar height roll roughly 100 miles northward, cut by the gorges of several Ganges tributaries, until beyond the Ganges headwaters the High Himalaya towers to over 6,000 meters. The dominant forest community in this region, with elevations between 500 and 2,300 meters, centers on the *chir* pine (*Pinus longifolia*), especially on the dry south-facing slopes where reproduction of other species is extremely difficult in the long dry months before the monsoon rains.

Above the chir belt—at elevations of 1,800 to 3,600 meters, which begin to approach alpine conditions—several other conifers dominate: small stands of the prized *deodar* cedar (*Cedrus deodara*) in the northwest, the *kail* or blue pine (*Pinus excelsa* or *wallichiana*) both of which were widely lumbered by 1900, and the softer woods of the West Himalayan spruce (*Picea smithiana*) at 2,100—3,300 meters and silver fir (*Abies pindrow*) at 2,200–3,300 meters. These stands have been important for general timber supplies for the regional market, though the spruce and fir were not cut on a large scale until the era of aggressive economic development that began after independence.

The years before 1914 were a period of consolidation of forestry management in northern U.P. based on the 1878 forest law, which with amendments has been the basis of Indian forestry ever since. Until then bureaucratic management of the land was entirely in the hands of the revenue officials, centering in the district collector's office. Neither before, nor at any time since then, did the revenue authorities have the staff, training, or interest to manage forest or grazing lands in any detail. Their interest lay with agriculture: revenue from crop production was the fiscal mainstay of the districts. Even after 1878 the collector retained the power to grant small parcels of forest land to peasants who proved able to plough it; the Forest Department was rarely able to prevent the slow depletion of its acreage. The Revenue Department's principal interest remained the expansion of agricultural production, not by intensification of production on existing tilled fields but by cutting into pasture and forest as the hill populace gradually expanded and families produced additional labor.

The 1878 law did empower the Forest Department to survey and set aside reserved forests where timber stands could be commercially profitable and managed for a perpetual crop of trees or where the forest was so remote that it had no potential for crops and where, as on steep, fragile mountainsides, its preservation was necessary in order to stabilize watersheds. The process of establishing the reserves was painstakingly slow, occupying much of the foresters' time in the following years. By the early 1890s large reserves were established in the lowland chir belt, and timber harvests primarily for railway sleepers (crossties) provided the main source of the department's budget. Higher into the Siwaliks a few thousand hectares were reserved to assure an adequate fuelwood supply for the growing European towns in the hills. The chir pine belt in the Siwaliks and beyond was left to the slow processes of livestock grazing and encroachment for new crop terracing until after 1900, for chir had no commercial uses yet.

Shortly after 1900 one of the most important innovations in the history of Kumaon appeared: the beginning of commercial resin tapping, for which the chir pine was ideal. Methods of distilling chir resin were crude at first, and transport of the product from the factory at Bhowali at an altitude of 1500 meters was difficult. Large-scale resin production became a key element of the forest economy only after 1918, but its promise spurred the foresters on to establish large new reserves in the middle hills, out of forests that had remained since 1878 in the interim category of Protected Forest, under the minimal management of the Revenue Department. The New Reserves, as the 1911 to 1916 areas were called, promised to expand the Forest Department's scale of operations and profits dramatically. But local, national, and world events conspired to disrupt these plans, with severe and permanent degradation of the Kumaon forests as the ultimate consequence.

Stresses on the broader economy of the hills during World War I were the key to the forests' postwar crisis. The war produced severe pressures on India's productivity, but the direct effect on the forests was fairly limited. In the early stages of the war little timber was mobilized for the struggle in Europe. But by 1917 the wartime Indian Munitions Board's timber branch ordered structural timbers for bridges, piers, buildings, and so forth from the Forest Department, primarily for shipment to Egypt and Mesopotamia. New port facilities had to be created for these shipments in Bombay, Karachi, and Rangoon. Railway building, mostly also in Egypt and Mesopotamia, used almost 2,700 kilometers of wooden sleepers, at approximately 3,000 sleepers per kilometer.[3]

Little of this timber was taken from north India; forests nearer the coasts were adequate to meet the sudden demand. The indirect impact of the war on the U.P. hills was more significant. Throughout the war the Forest Department's budgets were curtailed and younger staff members left for military service until the end of 1918. Consequently, planned harvests in the forest reserves did not meet the targets for each year that had been set in the forests' Working Plans. The planned rotations (of as much as 150 years for the slow-growing deodar) were designed to cull diseased and "overaged" trees, thereby gradually improving the health and productivity of the forests. In other words, in terms of the production of commercially valuable species the war was a setback in the management of the reserves: by the silvicultural standards of the time too few trees were cut and too little revenue was generated.[4]

The wartime labor shortage extended to another sphere as well, the daily labor of timber cutting and transport. Thousands of men from the Kumaon and Garhwal hills left for military service; the exploits of several of their regiments are still remembered vividly in the traditions of the British army. Other Pahari men (*pahar* is a Hindi word for hill or mountain) migrated to the north Indian plains for wage labor there, accelerating a flow that has continued to the present and that has provided the single largest source of cash income for farm families of the hills.

This would have had little significance for the vegetation cover of the Himalayan hill region—the first aerial photographs in the 1930s would not have registered any thinning tree cover—except for the political tensions that resulted from the labor shortage and added an entirely new dimension to the ecological affairs of the hills. For forest labor, including the work of resin tapping, men could earn a small daily wage. Indeed, the Forest Department perennially pointed to this cash flow as a vital source of subsistence income for the Pahari economy. In contrast the traditional porter labor throughout the hills and even in many plains areas of north India, *begar* as it was usually called, was unpaid. India's traditional form of corvee labor, it was justified by both rajas and the British Raj as the equivalent of land taxes.

But Mohandas Gandhi had returned permanently to India in 1915, and others like C. F. Andrews joined him in declaring that *begar* was the empire's method of extracting free labor from unwilling landowners, humiliating them by this "feudal" exploitation. By 1916 the first active workers of the Indian National Congress in the Kumaon hills organized a new scheme of wage payments for coolie labor. Many hill peasants joined in threatening to boycott all further begar demands. Because this coolie labor was the sinew of local administration (the region had no motorable roads and few tracks wide enough even for bullock carts until after 1920), the Kumaon administration under Commissioner Wyndham moved speedily after the Armistice in Europe. By late 1920 a new system of entirely paid porterage in the Kumaon districts was ready to be introduced.[5]

But it was too late to avoid the spread of political tension to the broader arena of Gandhi's first Non-Cooperation Movement, the 1920–1922 nationwide confrontation with the British Raj. The movement featured civil disobedience against laws that Congress leaders believed unjust, and no injustice was deeper than that which deprived the masses of their inherent right to subsistence. For the people of Kumaon nationalist protests were unprecedented and galvanizing. To them the forest law was the pure expression of the immorality of British law, for it had given the Forest Department the right to exclude villagers and their livestock from reserves when the department considered it necessary for the growth of new seedlings. The New Reserves of 1911 to 1916 were the visible symbol of oppression.

One of the most difficult moments in relations between foresters and peasants, anywhere in India and later in other parts of the Empire as well, was the moment when a new reserve was declared and fenced. The new restrictions could be made palatable to nearby villagers only by careful communication between officials and villagers during the transition period. The war years in Kumaon left little time for announcing the new reserves or making the small adjustments that might satisfy the village people.

Moreover, the first months of 1921 were the hottest and driest in many years. By early March, the season when villagers traditionally burned the dry grasses of the hills to produce a faster, more lush new growth in the monsoon rains, the forests of the western Himalayas were in extreme fire danger. In a few weeks that March and April many thousand acres of resinous chir in the New Reserves and large young sal plantations below flared out of control. The U.P. government had already come to fear mass peasant uprisings in the plains districts, and the hill fires, suddenly the worst in modern India's history, led the government to urgent action. A special commission under Commissioner Wyndham surveyed the entire range of Pahari grievances about the forest law, and by the dry season of 1922 the government adopted its recommendation: the New Reserves were to be canceled and their administration returned to the Revenue Department.[6]

The Forest Department's outraged protests were set aside, despite their warning that forests left to the revenue officials would be given no silvicultural treatment, no reforestation, and no measures against soil erosion. But the damage was now twice done. Many Civil Forests, as they were henceforth called, had hardly a tree on them by the 1960s when they were turned over either to village councils or once again to the Forest Department. The intricate economic and political forces triggered by World War I had taken their ecological toll.

The war had another major consequence of a very different sort: it precipitated intensive efforts to find industrial uses for "India's forest wealth," as an influential book of the time called it.[7] Consequently a variety of tree species that had previously been considered of no commercial value began to be logged, and new uses for familiar species were developed. India's forests were entering the industrial age, and their exploitation was beginning to be much more complex.

World War I brought Britain to the sober realization that the crown jewel of her empire, India, could no longer be assured of the safe supply lines to Britain that had determined imperial policy for a century. With military threats from other powers compounded by rising transport costs of European goods the Indian Empire would henceforth be advised to look to its own industrial resources more seriously than in the past. Moreover, Indian industrialists had begun to challenge their British counterparts, while Gandhi's Congress was boycotting imported cloth. In an ideological parallel, Adam Smith would soon give way to John Maynard Keynes, and the government of British India would begin to take a more positive role in India's industrial economy.[8]

The clearest indication of more active governmental involvement in the timber economy concerned the chir pine. Since the turn of the century foresters in U.P. had been eager to exploit the revenue potential of resin manufacture. Immediately after the war the Forest Department constructed India's most sophisticated resin processing plant outside the city of Bareilly, which lay south of the tarai beyond Kumaon's southeast corner.[9] The Forest Department defended the expenditure of its funds on a large-scale industrial plant by assuring its critics that the factory would bring a new era of productivity to the Siwalik chir belt and prosperity to the local villagers, and also by insisting that no private entrepreneur could be found with adequate capital to set up a plant large enough to be profitable. But was the Bareilly plant immediately profitable? As soon as it began operating, it produced the dominant share of India's total production of resin products and in addition began exporting to Europe. Several small-scale, privately financed resin-processing plants farther northwest toward Kashmir, where the chir pine belt ends, could only supply local markets in the growing towns of Punjab.

But profitability was another matter. As an experiment in innovative technology and somewhat speculative financing, the Bareilly plant soon proved to be inefficient in some operations, and costs remained obstinately high into the mid-1920s. The advocates of the old principle that government should not intervene in the industrial economy were ready to counterattack. In the U.P. legislative assembly during the budget debate over the Forest Department in 1923, the defenders of laissez-faire industrialization challenged the foresters' move into industrial marketing.[10] The debate ended inconclusively: the Bareilly factory remained in operation and proved so successful that it still dominates India's resin industry. But the U.P. Forest Department adopted great caution in launching any further expensive forest products industries.

Yet the resin venture did exemplify what the Indian Industrial Commission had recommended at the end of the war: industrial self-sufficiency for the country. Research on the industrial uses of timber had begun in India in the early 1870s with the effort to creosote railway sleepers made from soft conifer such as chir so that they would withstand the attacks of white ants and the competition from European wood and iron sleepers. The effort bore little success for many years. But by the turn of the century the need for other experimentation on forest products was a major impetus to establish the Forest Research Institute (FRI) in Dehra Dun, at the westernmost corner of the U.P. Siwaliks. In short order the FRI became the colonial world's premier research station and a model for later centers in Britain's tropics. But the resistance to governmental interference in commerce and industry meant that prior to 1914 its work was primarily silvicultural: developing the systematic forest botany on which twentieth-century forestry management is built. Only after the war did industrial experimentation flourish at the FRI.[11] When its vast modern headquarters were built in the early 1920s, its experimental laboratories were allotted a large acreage. Research on resin products remained at Bareilly, but during the 1920s research began at the FRI on a wide variety of possible new tree products. The exploitation of many new tree species became a major possibility especially from forests such as mixed broadleaf tracts in the tarai that previously had been ignored except for their fuelwood potential. U.P. and other provinces appointed a new officer, the forest utilization officer, whose days were spent not so much in surveying the reserves as in studying the chemical and physical properties of each timber, and beyond that, in discussing potential market demands. Forestry was moving into marketing and publicity, and this too took an administrative budget that the old guard in the government attacked. If the annual balance sheet of the Bareilly resin plant was open to challenge, the profitability of the forest utilization officers and their ventures into creating future markets in India and abroad was at first negative. This might be acceptable in corporate finance, the conservatives argued, but it was not an acceptable use of the public treasury.

Neither side clearly won this debate on the industrialization of India's forests in the 1920s, for this was only an early stage of the transformation of industrial policy and the role of the public sector. As for the government's role in exploiting natural resources, the results by

1929 were modest. Had the years of relative prosperity and a rewarding marketplace continued longer the "progressive" foresters might well have scored greater successes. But when the bottom dropped out of the timber market by early 1930, and a decade of depression was followed by another world war and the subsequent transition to an independent government, nearly all expansion of the forest products industry was halted for 20 years and more. The ecological consequence was that lowland mixed broadleaf forests, and similarly the spruce and fir stands of the higher Himalayas, were given a long reprieve from the pressures of commercial forestry. Only the 1950s, which are beyond the scope of this study, transformed them.

The decade since 1918 had seen the full professionalization of the Indian Forest Service. Specialties of great importance for the long-run character of commercial forests were reaching their maturity. One of the most important examples was the vexing problem of how to regenerate the great sal forests, which had become the core of timber supplies for India's railways when northern deodar supplies were depleted in the late nineteenth century. It had proved very difficult to find the optimal pattern of light and moisture in which sal seedlings would thrive; more articles were written in professional journals and more study tours of the tarai forests were organized on this than on any other subject. Regeneration success rates were considerably higher by the late 1920s, assuring the continued supply of sal sleepers for maintenance of India's great network of rails when imports of European sleepers faded away. The sal forests from northwest of U.P. all the way to the heart of Assam in the Brahmaputra valley were becoming a showpiece of modern forestry practice in India.

But railway supplies were a controlled market; ever since the 1860s both quantities and prices had been determined not by fluctuations of price levels on an open market but by contracts negotiated between the Railway Department and the Forest Department. Other timber harvesting, for the urban fuelwood and construction markets of north India, had always been determined far more by open market demand and the perpetually unstable price levels of a volatile market. Research on price levels and the timber market before the 1950s has barely begun for Third World countries. In India available data are massive but poorly coordinated. Only 2 rough pictures for the 1920s are available now.[12] The immediate post-1918 period for about three years saw a slump in timber demand and large unsold inventories of timber in Forest Department storage. This in turn meant lower operating budgets for forestry. The next decade on the whole saw higher price levels and expanding demand for timber. Despite this demand, with generally unstable markets in the colonial economy over the decade,[13] timber marketing remained a volatile, high-risk investment for private entrepreneurs. Forest Department reports frequently indicate the difficulty the department faced in attracting private bidders at its timber auctions. Only a few Indian timber merchants succeeded in building personal fortunes before 1929, in stark contrast to the timber fortunes that became commonplace throughout the Himalayas in the 1940s and 1950s.

Management of the sal and chir forests had become professional and routinized. The U.P. forest service was the most prestigious in the country. After the Kumaon debacle of the early 1920s acreage under reserves changed little until after independence. Private forests were a relatively small percentage of the tree cover. And except for inaccessible conifer stands of the remote high forests of the north, nearly all remaining woodlands of any quality were under regular management. (This contrasts dramatically, as we will see, with the forests of Assam, whose commercial exploitation was still in its early stages.)

Annual statistics of total timber production in U.P. in the 1920s show a remarkably steady output, expanding somewhat but not fluctuating widely. During the war production had been steady at 8 to 9 million cubic feet annually. After the slight postwar dip, production reached a decade high of 11.3 million cubic feet in 1924, and remained close to that until 1929.[14]

Comparing the prosperous years of the 1920s with the Depression years that followed gives some indication of the effect of national and international market conditions on north Indian regional timber production. To set the context it is worth noting that international timber exports from the British Indian Empire suffered sharply in the early Depression years, as did most raw materials from most primary producing countries. Within India price levels generally fell by over 40 percent between September 1929 and March 1933.[15] Agricultural exports, which constituted the bulk of India's export earnings, were particularly hard hit. Total export earnings fell from $610 million in 1929 to $181 million three years later. In the case of tea, a plantation crop that had replaced subtropical forest in Assam and elsewhere in the northeast (discussed below) both production and export price levels fell dramatically for several years.

As for timber, exports from British India were entirely restricted to Burmese teak and mixed tropical hardwoods from the Andaman Islands in the Bay of Bengal. Andaman timber sales had roughly doubled during the 1920s, supplying specialized European markets for furniture material. During the Depression years, income declined by over 30 percent, but since prices presumably fell, this seems to indicate relatively stable production volume. Burmese teak exports had for years been very volatile. Production had hit as much as 60,000 tons in 1912–1913; it then fluctuated between 14,000 and 29,000 tons by 1931. Timber export earnings were also hard hit: total figures fell from a high of Rs. 135 million in 1928 to a 1932 low of Rs. 26 million. In sum, the Depression reduced both harvests and profits from specific forests, but these were confined to peripheral corners of the imperial possession.[16]

The Depression's effects on the forests of U.P. were more indirect. From the 1924 timber production of 11.3 million cubic feet, a low was reached in 1931 with 8.2 million, but by 1937 figures recovered to 11.2 million.[17] All in all, this was a far less severe fluctuation than any food crop's exports suffered. The relation between quantity and price levels has not yet been carefully studied, but an oblique indicator, total departmental income, varied equally little except for moderate declines in the early 1930s. This suggests that timber prices on local and regional markets in the north, despite sharp short-term fluctuations that troubled small-scale investors, were about as stable as long-term demand.

In the U.P. in these years the major change was the largely unmeasured depletion of Civil Forests and private woodlots for fuelwood, and this in turn reflected the demographic trends of the time. India's population had risen at relatively low rates throughout the nineteenth century, even in the area of greatest concentration, the Ganges basin. But the last great check, the influenza epidemic of 1918 and 1919, was now behind. Decennial censuses beginning with 1921 show far faster population increases, even in the hills.[18] These in turn inexorably raised firewood consumption, as even our very approximate statistics demonstrate. By this time fuelwood consumption had outstripped commercial logging in total biomass use, as it was doing in most Third World countries. Thus a full analysis of the impact of global economic structures on Third World forest use must ultimately come to grips with the controversies over indirect casual connections between colonial political economies and the Third World demographic explosion. This paper can only skirt that debate, moving directly to effects on fuelwood and forest cover.

From a 1914–1915 figure of 7.7 million cubic feet of fuelwood, the province's Forest Department increased its cutting to 24 million in 1918–1919, and as high as 30.6 million in 1923. Thereafter it stabilized just below 25 million cubic feet throughout the Depression years. As one would expect, demand for fuelwood was inelastic; the impact of world depression on the single largest use of wood was minimal. And since these statistics represent only fuelwood provided through Forest Department channels, giving no indication of cutting from Civil Forests or private woodlots, the subsistence demand on U.P. forests must have been several times the commercial demand. This was registered only in the long slow decline in the density and regeneration of the Himalayan hill forests.

Within the Reserved Forests there was constant concern over maintaining long-range rotational cuttings of timber through the 1920s. Under the fiscal austerity of the government from the mid-1920s onward and with the constant efforts to retrench staff and administrative costs that characterized the entire last two decades of British India, no new British foresters were trained after 1925 and total departmental staff was cut by as much as one third after 1929. The ecological effect of all this is a matter of dispute. On the one hand, the quality of silvicultural work on the reserves unquestionably suffered, as fewer and fewer professionals were available for a still vast acreage of forest. On the other hand, because foresters were always reluctant to allow unsupervised logging by haphazard or unscrupulous contractors, some important reserves were harvested at far below their scheduled rates. Did this inhibit the long-run sustained-yield maintenance of sal and other stands, or did it provide a buffer supply for meeting the intense demand of the war years that followed? Most likely the most important factor was the last, the years of breathing space that the slack harvests of the 1930s gave the trees before they were sacrificed in vast numbers to the war machine.

The opening assertion of this essay was that the 1940s were the second era of massive cutting in India's forests. From the beginning of 1942 onward timber management was placed on an emergency basis, with supplies and prices from the Reserved Forests strictly controlled by the Wartime Mobilization Board and the Forest Department.[19] Timber for bridges, harbors, railways, buildings, and many other uses placed all foresters and their timber stands under severe strain. Schedules of rotational harvesting were accelerated greatly. But by early 1945, when the end of the struggle was in sight, the national forestry board concluded that harvests under its command had been orderly, and no long-range ecological damage had been done to the reserves. No one seems to have disputed their assessment.[20]

In the private sector the story was grimly different. There could be no serious attempt to control urban price levels for construction wood and particularly for fuelwood. Scarcities of wood in north Indian cities were severe throughout the war, and private contractors made the most of rapidly rising prices. Private forest owners eagerly sold their standing trees for easy windfall profits, and investors made fortunes in a single season in timber contracting.[21] In Nainital, the chief town of Kumaon, local citizens nearly rioted in 1942 against the contractors and so began a process of confrontation between peasants and contractors that is intensifying even today.

The Forest Department was appalled by the carnage.[22] By early 1945 it began pressing for nationalization of all private timber as the only hope of preserving the small woodlots that still dotted the landscape. The result was the U.P. Private Forests Act of 1948, but private owners had several years in which to turn their trees into rupees before the "confiscation," as they saw it, could be accomplished. By the time the new national forest law was passed in 1952, there were few trees left on the former private lands, and not many more in the Civil Forests, which had become scrub grazing grounds for scrub livestock. Only the reserves were still in reasonably healthy condition, and they would come under increasing pressure from all interests in the accelerated economic development of the Five Year Plans.

Forest Depletion in Assam, 1941–1947

On the easternmost fringe of the Himalayan ranges, where the Brahmaputra has made its great arc to the southwest and the Bay of Bengal, Assam's dense forest belt has been under siege since the early nineteenth century.[23] In contrast to U.P., whose total land area of 294,413 square kilometers in 1930 included only 13,512 square kilometers of government forest, Assam's 142,854 square kilometers included 53,950 square kilometers under government forest control, one of the highest percentages in India. A large portion of this was inaccessible from the outside world, and particularly from commercial and demographic centers until very recently. This summary will center on two of Assam's three major forest regions, both in the Brahmaputra basin.

In the lower reaches, on rich alluvium, stand forests of sal, the farthest eastward extension of the great sub-Himalayan sal belt. Because of the perennially warm, moist climate these are the finest quality of all sal stands. The down-river districts of Kamrup and Goalpara boast the most extensive stands; the adjacent low hills of Darrang and Goalpara also have large sal reserves. All are of great economic value. Farther northeast, as the gorges become steeper and the hills higher and less accessible, lie Lakhimpur and Sibsagar districts near the borders of Burma and China. These still-remote forests now grow mixed subtropical broadleaf evergreen trees. In earlier times they grew various potentially marketable species; during the late colonial era the foresters' principal attention focused on the towering *hollong* (*Dipterocarpus marocarpus*).[24]

Even more than in most parts of India, the history of Assam's forests has been intertwined with the intricate ethnic and cultural patterns of the state.[25] The remote high hills of Assam and adjacent regions are home to a wide variety of tribal groups whose subsistence has been based primarily on shifting agriculture, or *jhum,* as it is locally known. Until recently tribal populations were thin enough that they presented no fatal threat to the mixed forest, if left to themselves. But the Brahmaputra lowlands supported a much more dense, rapidly growing, and culturally different populace of Hindu rice farmers. In the twentieth century Assam has had the fastest-growing population of any state in India: from 3.3 million in 1901 to 15 million in 1971, nearly all of the growth before 1947 occurring in the lower areas of settled agriculture.[26] Further, the traditional settlers of the lowlands are Assamese speakers, while the tribals speak a totally separate set of languages. Most challenging of all, down-river in Bengal lies perhaps the densest rural population in the world; by the late nineteenth century Bengali peasants, most of them Muslim, began drifting up-river into the fertile Assamese forest fringe. The British Raj had only rudimentary administrative operations outside the lowlands of Assam, and transport, commerce, and industry were less developed than in other parts of India. Even before World War I, one cause of depletion of Assam's vegetation was the steady encroachment of immigrant peasants on the forest and jhum lands of lower Assam.[27]

The other transformation, one that quickly penetrated far into the hills, was the tea industry. After 1833, when the East India Company's new charter allowed foreigners to own rural land in India, European tea planters quickly bought Assam hill land. By 1871, 700,000 acres were owned by the planters, though as yet only 56,000 acres were actually producing tea. By 1900 there were 764 working tea estates in Assam, producing 145 million pounds of tea annually for export.[28] Most of the plantation workers were outsiders; by 1900, 400,000 alien workers had been brought into Assam under appalling conditions. The ethnic struggle for land in Assam, then, was a four-way conflict, and beyond this the tea planters for many years held the dominant financial and political leverage. No wonder that the Forest Department, though separated from the Bengal Forest Department in 1874, had succeeded in demarcating only a few thousand acres of lowland sal as Reserved Forest by 1914.

The tea industry of northeast India is a classic case of a foreign-dominated plantation economy that controlled a colony's land use patterns and was highly sensitive to markets in the industrialized world. When we consider the impact of World War I on Assam's forest lands, the most immediately evident factor is the wartime prosperity in the tea industry. Prices in Europe rose; dividends to the planters for their role in aiding the war effort rose correspondingly; and acreage under cropping extended rapidly. The first years after 1918 saw a brief oversupply in England's warehouses and a minor depression in the Assam hills, but this was succeeded by a decade of prosperity and further expansion of the acreage under tea.[29]

The Depression hit the tea industry heavily. The all-India wholesale price index for tea dropped by 53 percent in four years. By 1933 the tea industry responded with an international system regulating supplies, which assured a rebound of profitability by 1934. Economic strength continued to increase until the war, at which point, although Assam was threatened by great military danger, the tea industry entered another era of wartime prosperity. The British government controlled all tea production and consumption beginning in 1940, and from early 1942 the rising competition of Indonesian tea was ended by the Japanese occupation there. The tea industry in Assam and adjacent northern Bengal expanded another 20 percent by 1945.

What was the impact on Assam's forests? Tea planting continued to expand onto lands previously in government forest, village commons, or private ownership. The tea industry also slowly came to have a stake in the adjacent forests, because massive amounts of wood were required for tea chests. Throughout the nineteenth century the planters preferred birch chests imported from Scandinavia; plywood technology was not yet available in India. Later the Japanese began supplying tea chests for northeastern India.

World War I brought intensive efforts to substitute Indian wood for the whole range of forest product imports. The Forest Department experimented with Assamese species for sleepers on which to lay the lines of new railways that moved, painfully slowly, up the gorges toward Sibsagar and Lakhimpur. They had little success in this quarter, but the war spurred more successful efforts to begin a plywood industry in Assam. By the early 1920s several sawmills were opened along the Brahmaputra valley, experimenting with various subtropical woods from upper Assam. Here lay a significant contrast with the simultaneous efforts to build a resin industry in U.P. In Assam the Forest Department was still a fledgling; there was no possibility of finding either capital or personnel within the department for the necessary sawmill experiments. Attracting private capital from Calcutta was the only possible course. In the early 1920s several Calcutta firms (controlled by Marwari or Bengali businessmen, not Assamese) were given long-term leases on exceptionally favorable terms, in order to encourage the commercialization of the upper Assam forests.[30]

This indicated a fundamental, widely shared view of the balance between agriculture and forest lands in Assam, which was different in degree from attitudes toward land use in far more densely settled U.P. "Wasteland," a term generally designating land not under settled agriculture or forest reserve, was a great opportunity for settling immigrant peasants. In Assam the Revenue Department, for whom "nonproductive" land was truly a waste because it produced no taxes, consistently pressed for opening more land to the plow. The Forest Department acquiesced on the principle that peasants' need for land that could be terraced for wet rice and other grains should take first priority in land allocation.[31] The foresters were as oriented to development as their confreres in other agencies. They would not disagree with the government's 1938 report, which stressed "that indigenous people alone would be unable, without the aid of immigrant settlers, to develop the province's enormous wasteland resources within a reasonable period."[32] However, by 1920 the Forest Department also realized that in the long run this would threaten timber supplies and watershed stabilization. Hence the department began to accelerate the painstaking survey of vegetation and land-potential classification in previously unclassified government forests, especially so as to delineate sal forests which arguably should be kept from the plow.

This was also a pragmatic response to the growing up-river flood of Bengali Muslim peasants. In lowland Goalpara district the population had risen by 2 percent between the 1891 and 1901 censuses; in the next 10 years it rose by 3 percent. From there the settlers moved up-river, occupying Nowgong and Kamrup district wastes; by this time they were beginning to challenge tribal lands as well. They had on their side the general British assumption that settled agriculture was a far more productive use of land than jhum. In the 1920s the movement began to reach upper Burma, where lowland peasants had no experience of the climate and agricultural patterns. Despite these difficulties, land hunger and political pressures continued to force Bengali peasants eastward, and when East Pakistan was created in 1947 from eastern Bengal, a new wave of Hindu peasants moved from Muslim Pakistan into Assam. In the 20 years ending in 1950 some 1,508,000 acres were turned to settled agriculture by the immigrants.

How was all this reflected in the timber extraction of the government forests? Before the early 1920s the Assam Forest Department was a small cadre, a few officers whose work was limited to serving short-term commercial demands, mostly in the sal districts. One report, surveying the 75 years before 1925, complained that "the provision of staff and improvements in the forests have depended upon the profit of each year, and not one single budget presented by the Department has even emerged without large cuts by Finance."[33] The government forests of the 1920s which had long been awaiting survey and reservation, were so burdened with peasants' rights that purchasing those rights threatened to be prohibitively expensive for the government.

Nonetheless, it was a boom period for Assam's forests. During the highly profitable years from 1925 to 1929, the reserves increased by some 400 percent, largely in the sal belt. But for the following decade severe retrenchments in staff and budget left Assamese forestry virtually where it had been in 1918, except that shifting cultivation in the Unclassed State Forests of the hill regions was inexorably expanding.

World War II brought major changes in forest production. Between 1939 and 1945 timber production in the reserves more than doubled, while fuelwood cutting there more than tripled. The height of this pressure came after 1942 when heavy concentrations of troops moved up the Brahmaputra valley to stem Japanese forces in Burma. Profits rose far more dramatically still: total receipts for railway timber rose five times, and military timber use raised eight times more revenue. Much of the profits went into the hands of private contractors from Calcutta as well as the Assamese towns, just as happened in U.P. at the same time. The majority of the contractors were small operators who simply cut and exported logs using only hand tools. With the exception of the two larger sawmills established in the early 1920s, no timber processing was done within Assam. Thus the extreme inefficiency and wastefulness of lumbering continued well toward the present.

In one way the war made major changes in forestry technology, not only in Assam but throughout the Himalayan region. Especially in Assam the war effort required new motor roads as well as emergency rail lines. These roads were used by motor lorries and by jeeps, some of which are still in operation almost 40 years later. Foresters in U.P. agree that the motorization of forestry that happened during the war years was a turning point in the mobility of foresters, bureaucrats, and politicians alike. No longer thereafter did foresters spend nine months each year in their reserves; they became more familiar with desk work and more remote from their forests. The scale of timber exports could now be increased to meet the economic demands of independence. New roads could be used, as in many parts of the tropics since that time, by peasants looking for new land to till.

Assam's forests, having been put under sudden new pressures before 1945, underwent yet another major trauma in 1947, when the influx of Hindu refugees from East Pakistan moved westward into the Calcutta region and northeastward into Assam at the same time that transport lines for Assamese timber were severed at the Pakistani border. The sudden new pressure of the immigrants accompanied severe disruption of the state's administrative machinery and its forest management. The valuable sal forests of Sylhet in the south became part of Pakistan just when the Forest Department had to face severe political pressures to de-reserve existing forest tracts. From an ecological perspective, some commentators suggest that this was a period of temporary reprieve from commercial logging, especially in upper Assam. The full picture of the transition in Assam's land use during the immediate aftermath of independence is still not entirely clear.[34] But it can safely be concluded that the political and ethnic turmoil of that period led to similar, but more intensive, conflicts in recent years, on a steadily shrinking base of land and vegetation. The recent political turmoil there has made regular economic and administrative life very difficult; the forests are among the victims.

Conclusion

To what extent should we speak of these changes as being related to international economic forces? It may be more appropriate to refer to the whole complex of political, military, and economic structures of the dying years of colonialism. The two great wars and the devolution of political power in South Asia were dominant forces in the second and fifth decades of the twentieth century. Market forces were more prominent between 1918 and 1939, but in U.P. and Assam there was little international export of timber. The only export plantation crop of significance was tea, which slowly expanded its acreage, encroaching largely on forests; the boom years for tea had been the nineteenth century. But here Assam's tie-in to international markets directly affected timber use too. Though Assam grew little teak or exportable tropical timber, production of packing cases and the plywood industry began to emerge in the 1920s. The Assam government's development policy encouraged this, but it was soon caught in the Depression. The worldwide economic crisis had effects placing Assam in the mainstream of troubles facing the "primary producing" areas: low prices, low production, management cutbacks. Parallel to all this were powerful local factors; first of these was the growing pressure of peasant immigration, which in turn was related to new political pressures centering in the new Muslim League state government after the 1937 elections.

Ecologically, though, we may come to somewhat different conclusions. All the negative vocabulary of economics for the 1930s may be inappropriate for trends in vegetation and land use. Slack markets meant less pressure for production in plantations and forests, though reduced labor probably increased subsistence pressure on the Assamese hills from the industrial unemployed. Similarly in the U.P. hills, reserved forest targets were not met in the 1930s, and resin production fell somewhat; Forest Department profits declined and management of the forests was reduced. At least by planning standards, this provided a reserve stock of timber for meeting wartime demands.

The massive cuttings of 1942 through 1945 were under strict control in government forests. But in the private economy of civilian life, the usual profiteering for urban demand showed the severe price inflation caused by scarcity, which meant a financial boom and ecological disaster on private lands, continuing into the time around the 1949 Private Forests Act. The disruptions of independence in Assam were far greater, when East Pakistan took Sylhet and cut off markets for Assam's tea and timber.

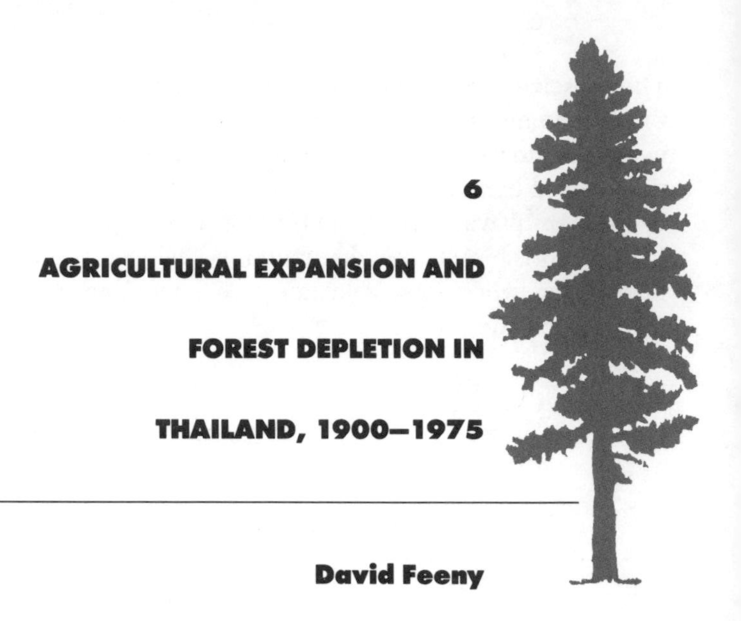

6

AGRICULTURAL EXPANSION AND FOREST DEPLETION IN THAILAND, 1900–1975

David Feeny

Deforestation has become an important issue in Thailand. It has been associated with soil erosion, siltation, flooding, and the depletion and in some cases extinction of wildlife species. Flooding and droughts in the lower Central Plain have focused attention on conditions in the northern watershed areas. Population pressure, shifting cultivation, the harvest of forest products for both domestic and international markets, opium production, and the expanding areas in upland crops in response to new export opportunities are all said to play an important role in accounting for the depletion of forests in Thailand.

While the issue of deforestation in Thailand is timely, it is not new. In order to provide some tentative explanations of the causes and consequences of deforestation in Thailand, this essay will exploit the quantitative records on the area under cultivation and forest product harvesting. In the process an imprecise record of the extent of deforestation over the twentieth century in Thailand will be established.

Increases in the area under cultivation have been the primary factor responsible for growth in agricultural output and export in Thailand. This generalization appears to be valid for the nineteenth century and first half of the twentieth century. Only recently have increases in productivity begun to play a more important role in accounting for output growth.

Output growth, thus, occurred largely through deforestation. In the first section of the essay rough estimates of the rate of expansion of the area under cultivation are derived. In the second section some of the direct evidence on forestry production is briefly discussed. Forestry policy and the evolution of concern about forest depletion in Thailand are examined in the third section, and conclusions follow.

Estimates of the Growth in the Area under Cultivation

Because most of the area currently under cultivation in Thailand was once covered by forests, especially outside the lower Central Plain, one approach to measuring the extent of forest cover in Thailand is to measure the rate of expansion of cropland. Thus the reliability and completeness of Thai agricultural statistics need to be assessed.

A significant underestimation of the area under paddy cultivation over the 1906 to 1955 period is very likely. It is also reasonably certain that underestimation of the area in other crops was even more substantial. This conclusion, when combined with the omission of area estimates for many crops, indicates that the major official crop area series figures significantly understate the area under cultivation in Thailand. The trend in the area under cultivation, however, is likely to have been captured in a reasonably accurate way. Though the degree of underreporting of the crops included probably decreased somewhat over time, the overall trend is basically reliable, even if the levels are not. Behrman reached much the same conclusion in his analysis of the Ministry of Agriculture production data for the 1937 to 1963 period. He states that the rice estimates are probably the most reliable and that "the estimates generally are much better indicators of trends than of absolute levels."[1]

Table 6.1 presents various estimates of the rate of growth of the area under cultivation for the 1906 to 1955 period and selected subperiods. Apparently the area under cultivation grew at roughly 3 percent per year over this period. Much of the expansion of the area under cultivation outside of the Chao Phraya River delta occurred through deforestation. As for the delta itself, while much of its vegetation was already secondary, the expansion of the area under cultivation still resulted in forest and shrub clearing.[2] Slash and burn cultivation was

Table 6.1 Average Annual Rates of Change of Cropped Area, Paddy Area, Paddy Production, Paddy Yield, and Population in Thailand, 1905–1955 (percentage)

Period	Average annual rate of change: Major crop area	Major crop area, including rubber	Paddy area planted	Paddy production	Paddy yield	Population, Skinner series
1905/6–1955	—	—	3.06	—	—	1.95
1906/7–1955	—	—	2.95	2.11	−0.82	1.96
1911/12–1955	3.11	—	3.00	2.11	−0.87	2.04
1913/14–1955	2.50	2.70	2.36	2.23	−0.13	2.06

Sources: See tables 6A-1, 6A-2, 6A-3.

often practiced by new settlers to provide initial subsistence and to clear the land in order to make it suitable for broadcast paddy cultivation or other crops.[3]

For the period since 1950 data are available on the cropped area for a number of products. Furthermore there are multiple sources of data including Ministry of Agriculture reports, the agricultural censuses, and the National Statistical Office crop-cutting surveys. More recently aerial photography and satellite imagery surveys have been conducted. A number of observers have examined the quality of the data and established adjustments to the official Ministry of Agriculture series that correct some of the major discrepancies.

While the quality of the data for the more recent period is improved, there is still substantial disagreement among the major sources. Ministry of Agriculture estimates of the area under cultivation are consistently below estimates derived from surveys of agricultural holdings, which in turn appear to underestimate the "true" area in use. Satellite imagery and aerial photography survey estimates consistently exceed those given by the other two methods but also tend to overestimate the area in use.[4]

Table 6.2 presents data on the total harvested area and harvested paddy area for the 1950 through 1975 period. The data, taken from Damrongsak, have the advantage of incorporating a number of reasonable adjustments designed to remove known inaccuracies in the official data. In his growth accounting study of Thai agricultural change

Table 6.2 Paddy Area Harvested and Total Cropped Area Harvested in Thailand, 1950–1975

Year	Paddy area harvested (thousands of ha.)	Total area harvested (thousands of ha.)
1950	5,824.00	8,191.84
1951	6,309.76	8,698.08
1952	5,643.20	8,124.64
1953	6,524.00	9,074.08
1954	4,976.16	7,839.20
1955	5,913.28	8,657.92
1956	6,338.24	9,214.72
1957	4,715.68	7,695.04
1958	5,685.92	8,772.32
1959	5,789.12	9,078.08
1960	6,207.52	9,847.84
1961	6,221.44	9,925.12
1962	6,499.84	10,279.36
1963	6,672.16	11,610.24
1964	6,567.52	10,839.04
1965	6,555.52	11,143.52
1966	7,701.60	12,712.32
1967	6,388.16	11,446.40
1968	6,937.92	12,146.40
1969	7,985.12	13,595.68
1970	7,539.68	13,569.44
1971	7,805.12	14,193.76
1972	7,457.92	14,196.64
1973	8,357.28	15,828.00
1974	8,066.56	15,687.36
1975	9,203.52	16,960.16

Source: Damrongsak, "Sources of Agricultural Output Growth," 146.

over this period Damrongsak concluded that technical progress accounted for between 26.8 and 38.2 percent of output growth over the 1950 through 1976 period, depending on the method of estimation used to derive the aggregate agricultural production function.[5] The technical progress variable captures technical progress as well as the effects of improvements in input quality, management, irri-

Table 6.3 Average Annual Rates of Change in Cropped Area in Thailand, 1911–1976 (percentage)

Period	Total area	Total area, including rubber[a]	Paddy area planted	Agricultural output in constant prices	Agricultural capital in constant prices	Agricultural labor force
1950–76	3.03	—	—	4.00	5.22	2.28
1950–75	2.95	—	1.91	—	—	—
1955–75	3.41	—	2.19	—	—	—
1911–75	3.75	—	2.75	—	—	—
1912–75	3.45	—	2.42	—	—	—
1913–75	3.36	3.34	2.31	—	—	—

Source: See tables 6A-1, 6A-2, and 6.2, and Damrongsak, "Sources of Agricultural Output Growth," 34, 113, 128, 146.

a. Rubber is included in the Damrongsak 1950–76 series. For the earlier period, estimates in table 6A-2 include rubber and were used in computing the average annual rate of change for 1913–75.

gation, education, extension, and the regional specialization and diversification trends that were important over this period. If technical progress accounted for 26.8 percent of output growth, then the growth of land, labor, and capital would have accounted for 25.0, 27.4, and 20.8 percent respectively. Similarly, if technical progress accounted for as much as 38.2 percent of output growth, then land, labor, and capital input growth accounted for 20.1, 21.4, and 20.4 percent respectively. Thus, unlike the earlier period, the growth of output is not primarily accounted for by the growth of land and labor inputs. Whereas in the earlier period (table 6.1) area grew faster than output, in the 1950 to 1975 period, the reverse was true.

The rate of growth of the total area under cultivation was, however, approximately as rapid as in the 1906 to 1955 period, around 3 percent per year (see table 6.3). Splicing together the relatively more complete coverage for the period since 1950 with the less complete earlier coverage, the total area under cultivation appears to have grown at roughly 3.4 percent per year. This figure tends to overestimate the area growth because the degree of coverage increased over time.[6] Thus it may be regarded as an upper-bound estimate. Perhaps the roughly 3 percent per year figure is more reliable for the whole period.

In the post–World War II period the dominance of paddy cultivation in Thai agriculture has declined and the relative importance of such new export crops as maize, kenaf, and cassava has grown. Much of the growth in the production of these upland crops has come at the direct expense of forest land. While a significant portion of the rapid expansion in the area under paddy cultivation in the pre–World War II period was concentrated in the lowland areas of the Central Plain that were less heavily forested, much of the expansion in the post–World War II period took place in formerly wooded areas.[7] As in the past, abundant land and favorable external markets have been the source of much of the growth in agricultural output. Growth by clearing continues into the present.

Forest depletion has, thus, largely occurred as a result of the clearing of land for cultivation in response to economic incentives for commercial agriculture and/or subsistence production. Favorable prices and population growth underwrote the rapid expansion of paddy production in the nineteenth century and first half of the twentieth century. The favorable markets for upland crops and even more rapid rates of population growth underwrote a continuation of the process of forest clearing for agricultural cultivation in the post–World War II period. Malaria eradication programs further contributed to forest depletion by making previously infested areas safer for cultivation and augmenting the rate of population growth, leading to more clearing for both commercial and subsistence production. Commercial forestry as well as harvesting for local use (discussed below) have further contributed to forest depletion.

Evidence on Forest Area and Production

The estimates of the expansion of the area under cultivation presented above are indirect evidence on the rate of forest clearing. Fragmentary data are also available on the forested area, timber cutting, timber exports, the production of major forest products, the export of forest products, and government timber royalty revenues. This essay will focus on the first three.

A very rough idea of the rate of forest clearing can also be obtained by piecing together various estimates of the percent of the total area of

Table 6.4 Estimates of the Area under Forests and Rates of Change in the Area under Forests in Thailand, 1913–1980

Year	Percent of total area in forest	Area in forest (thousands of ha.)	Source and comments
1913	75	38,514	Graham (1913, p. 347); includes forests, marsh, and jungle.
1930	70	35,946.4	Ministry of Commerce (1930, p. 35).
1947	63		Tsujii (1980, p. 29); taken from Ministry of Agriculture data.
1949	69	32,600	Donner (1978, p. 71); area in forests and pasture.
1955	63	32,129	Sukhum (1955, p. 8).
1956	58	30,288.3	Pendleton (1962, p. 134); area in forests and pasture.
1959	58	30,010	Charlermrath (1972, p. 20); official estimate.
1961	56	29,000	Donner (1978, p. 133); estimate from aerial photography survey.
1961	52		Charlermrath (1972, p. 24); estimate of forestry official.
1963	53	27,100	ADB (1969, p. 475); estimate based on FAO world forest inventory.
1965	53	27,300	Donner (1978, p. 22); author indicates that this estimate which is based on a land-use survey is probably an overestimate.

1965	<40		Charlermrath (1972, p. 24); estimate of forestry official.
1966	51	26,500	Krit (1966, p. 5).
1969/70	52	26,900	Land Development Department estimates based on aerial photography.
1970	39–49	20,000–25,000	Donner (1978, p. 134); author's estimate.
1970	30		Tsujii (1980, p. 29); estimate of forestry expert.
1974	37	19,040	NESDB (1977, p. 149); estimate based on satellite imagery.
1975	41	21,068	World Bank estimate based on satellite imagery.
1978	25	13,018	Wilson (1983, p. 133); estimate based on satellite imagery.
1980	<30		NESDB (1981, p. 7).

Period	Average Annual Rate of Change in Forest Area (percentage)
1930–74	−1.43
1930–75	−1.18

Note: The total area in Thailand is 51,352,000 ha.; see Donner, *Five Faces of Thailand*, 907.

modern Thailand covered by forests. Some of the estimates are based on land use surveys, some on more systematic aerial photography or satellite imagery surveys, and a number are just guesses made by well-informed observers (see table 6.4). It appears that in the early part of this century, forests covered around 70 percent of Thailand, but the proportion had fallen to 50 or 60 percent by the 1960s and to roughly 40 percent by the mid-1970s. Today less than 30 percent of the forested area remains. Depending on the benchmark estimate accepted, since 1974 or 1975 the area under forests has declined by 1.43 to 1.18 percent per year.

The changes in the area under forests are compared to the changes in the area under crops in table 6.5. It appears that increases in the area under cultivation account for an increasing proportion of the decline in the forested area over time. This result is consistent with the increasing importance of upland crops and concentration of the expansion in the area under cultivation in areas outside the lowlands of the Central Plain.

The longest time series data on timber cutting are for teak, which is produced in northern Thailand (see table 6.6). Incompleteness due to illegal cutting is a serious deficiency in the timber cutting data and one that is especially important for a valuable species like teak. A stump inventory conducted in six northern provinces covering the 1937 to 1956 period concluded that illegal cutting was 148.6 percent of the legal cut in those provinces.[8] Other observers indicate that the total teak cut is three times the legal cut.[9] Teak production data problems are further confounded by the fact that some of the production is floated to market via the Salween River, some via the Mekong River, and some via the Chao Phraya River, the most popular route and the one for which the data is the most accurate and complete.[10] Because the first two routes become much less important over time, the data presumably become more complete and accurate.

While the industry was initially export-oriented, over time production has increasingly been destined for domestic use. The quantity of teak exports peaked in 1905 to 1909. Since then teak exports have declined in absolute terms and as a share of the value of total exports. Teak also accounts for a declining fraction of the volume of recorded timber production, falling from roughly 34 percent in 1932 through 1936 to around 11 percent by 1970 through 1973 and less than 7 percent in 1974 through 1978.

Teak production grew rapidly with the entry of European companies

Table 6.5 Comparison of Changes in the Area under Forests and in the Cropped Area in Thailand, 1913–1975

Year	Area under forests (thousands of ha.)[a]	Change in area under forests (thousands of ha.)	Cropped area (from table 6A-1, in thousands of ha.)	(from table 6.2)	Change in cropped area (thousands of ha.)	Change in cropped area (percentage)
1913	38,514	—	2,195.70	—	—	—
		2,567.6		—	1,053.64	41
1930	35,946.4		3,249.34	—	—	—
		3,817.4			2,939.6	77
1955	32,129	—	6,188.94	8,657.92	—	—
		3,129			1,267.2	41
1961	29,000	—	—	9,925.12	—	—
		7,932	—		7,035.04	89
1975	21,068	—	—	16,960.16	—	—
1913–55		6,385			3,993.24	63
1913–75		17,446			14,764.46	85

a. Sources are given in table 6.4.

Table 6.6 Average Annual Production of Teak, Other Timber, Total Timber, Firewood, and Charcoal in Thailand, 1890–1978 (cubic meters)

Period	Teak	Other timber	Total timber	Firewood	Charcoal
Up to 1890[a]	62,500 to 69,500	—	—	—	—
1890s[a]	83,400 to 97,300	—	—	—	—
1898–1907[b]	130,268	—	—	—	—
1900–1930[c]	222,907	—	—	—	—
1932–36[d]	195,171	377,327	572,498	788,017	220,357
1937–41	141,019	485,432	626,451	861,934	337,587
1942–46	60,993	460,488	521,481	1,297,987	544,499
1947–51	209,380	929,434	1,138,814	1,557,705	629,030
1952–56	294,462	1,270,841	1,565,303	1,137,015	669,639
1957–61	158,362	1,248,812	1,407,174	1,196,795	598,890
1962–66	153,662	1,646,091	1,799,753	1,392,811	632,777
1967–71	261,013	1,984,995	2,246,008	1,395,455	495,143
1972–73	183,188	1,989,661	2,172,848	1,261,607	417,345
1974–78[e]	196,800	2,806,000	3,002,800	993,800	279,400

Source: Unless otherwise noted, data are taken from *Statistical Yearbook Thailand*, nos. 20–22, 24, 29–31.

a. Data are taken from Smyth, *Five Years in Siam*, 1:71.

b. Data are for teak floated via the Chao Phraya valley only, as recorded at the main duty station at Paknampoh, and therefore omit shipments via the Salween and Mekong as well as teak used in northern Thailand. Data are taken from Dickson, "The Teak Industry," 172.

c. Data are taken from Ministry of Commerce (1930), 130.

d. Data are taken from Sukhum, *Brief on Forestry Information*, 61–62.

e. Data are taken from Wilson, *Thailand*, 138.

after the closure of the upper Burma forests in 1885.[11] Timber stocks rapidly depleted. In 1896 the Royal Forest Department was created specifically to regulate teak cutting. Experiments with teak plantations were begun as early as 1906 and the taungya method was introduced in 1942. Starting in the immediate post–World War II period efforts were begun to replace foreign timber companies with indigenous firms and the government's own Forest Industry Organization. After 1952 no new leases were issued to foreign firms and by 1960 the foreign concessionaires had been eliminated. The continued depletion of teak stocks led to an increase from 10 percent to 40 percent in the export duty on teak and the lowering of the import duties on timber to 10 percent in the mid-1960s. A teak export ban in late 1977 followed alarming stock estimates based on satellite images.[12]

Because teak grows in isolated pockets in the forest, its production trends are not accurate proxy for deforestation resulting from overall forestry production. Fortunately, Royal Forest Department data on nonteak timber and other forest resource production are available for the period from 1932 to the present (table 6.6). The data are clearly incomplete, omitting a large amount of illegal cutting and the harvesting of timber and other forest products for home use. Use surveys indicate that overall forest production is probably three to four times the level indicated in the official estimates. Thus it is likely that, as in the case of teak, the official production estimates significantly understate the actual level.

Traditionally Thai farmers have collected timber from nearby forests for construction, firewood, and other uses. Production for home use is undercounted in the official estimates. The Royal Forest Department firewood and charcoal estimates shown in table 6.6 primarily reflect production for use by the major commercial consumers of these products, the railway and the tobacco curing and ceramics industries.

In addition to the production of teak, firewood, and charcoal, Thai forest production includes yang wood, yang oil, rosewood, boxwood, sapanwood, ebony, other woods, rattan canes, bamboo, cardamons, sticklac, resins, gambodge, attap palm, tanning bark, wood oils, and pulp.[13] The two major forest types are mixed deciduous and evergreen.[14]

The Evolution of Thai Forest Conservation Policy

Modern Thai forestry policy began with the commercial exploitation of the northern teak forests in the late nineteenth century. The creation of the Royal Forest Department in 1896 and passage of the Forest Protection Act of 1897 resulted from that commercialization. While the conservation of timber stocks spurred creation of the Forest Department, the introduction of controls also involved a transfer of the administration of the northern forests from the local Lao chiefs to the central government in Bangkok. This also enhanced the central government's revenue base.[15] The Royal Forest Department gradually introduced and tightened regulations on leases, felling cycles, minimum girth requirements, and replanting. Its focus was on the regulation of commercial exploitation. The approach was patterned after that taken by the Forest Department in India and Burma from which the Thai government recruited foreign experts to help in the creation of its forest service. The impact of British policy and practice was further enhanced because many Thai foresters in the pre–World War II period attended forestry colleges in India or Burma.

Under a 1913–1914 decree forest species were divided into reserved and unreserved.[16] Reserved species (such as teak and yang) could only be legally cut if a license had been obtained and a fee paid. The decree did, however, allow for free permits for cutting for home or charitable use. Under provisions of a 1948 act, households residing near forest areas were allowed to cut up to 26 cubic meters per person for use in construction of their homes. Not unexpectedly, many farmers in northern Thailand regularly built sturdy houses with extra-large house posts and then sold the house to timber merchants and started all over again after several years had passed. While the 1960 Forest Act repealed that provision, a loophole in the 1941 and 1960 forest acts allows households to possess legally up to 0.2 cubic meters of lumber destined for domestic use. Harvesting for domestic and even commercial purposes has continued through this provision of the law.

The importance of such home-use production is evident in data collected by the 1953 farm survey. For the whole kingdom, 89.9 percent of farm families cut wood for home use, 3.6 percent earned income through the sale of wood, and the value of fuelwood cut for home use represented 8.1 percent of the value of total farm production.[17]

While the Forest Protection Act of 1897 and the 1913–1914 decree focused on the regulation of commercial forestry, conservation issues were discussed within official circles as early as the 1910s. In 1916 the forestry department first proposed that national forest reserves be created in Thailand as they already had been in a number of other countries.[18] In the 1920s Graham argued that conservation was still not adequately appreciated and that forestry policy concentrated too heavily on exploitation.[19]

In the 1930s Thai forest legislation began to reflect a concern with the conservation of forests that extended beyond the regulation of teak cutting. The Forest Reservation Act of 1936–1937 allowed the government to designate reserved and protected forests.[20] Additional legislation in 1939, 1941, 1944, 1948, 1951, 1953, and 1954 further enhanced the ability of the government to preserve forest and watershed areas.[21] (These acts were later replaced by the National Reserved Forest Act of 1964.) Writing in 1941 Thompson concluded that "Siam is just beginning to appreciate the importance of permanent forests in relation to climate and rainfall."[22]

By the early 1950s a forest reserve goal of at least 50 percent of the area of Thailand had been set. While there was little discussion of exactly how the specific target was decided upon, the general motivation was the desire to preserve basic soil, water, and wildlife resources.[23] The target was incorporated into the 1959 World Bank report on Thailand and the First Five Year Plan (1961 to 1966).[24] Continued deforestation resulted in a downward revision of the target. The goal reported in the Fourth Plan (1977 to 1981) was 37 percent.[25]

Wildlife preservation legislation began with the Wildlife Elephant Preservation Act of 1900 and a subsequent 1921 law on wild elephants. In 1933 a wildlife preservation act was drafted but enactment was delayed for 27 years.[26] Decrees in 1942 and 1953 were designed to protect bird nesting sites. Finally in 1950 the Association for the Conservation of Wildlife was formed and in 1960 and 1964 major legislation on wildlife conservation was passed.[27]

Conservation provisions were also contained in the 1901 Mining Act and its 1967 replacement. Other relevant conservation provisions were contained in land settlement legislation. The National Park Act was passed in 1961.

In sum, the scope of Thai forest and related conservation legislation gradually evolved from an initial narrow focus on the regulation of commercial teak cutting to a broader concern with conservation issues that became evident over the period from the 1930s through the 1950s. By the early 1960s the earlier concerns culminated in the passage of more comprehensive legislation on forest, park, and wildlife preservation.

Under the forest legislation briefly described above virtually all forests in Thailand are publicly owned. Their exploitation is supposedly regulated by permits issued by the Royal Forest Department. In spite of provisions for obtaining permits for domestic use, there is considerable tension between the traditional Thai villager's view of forests and forest products as common property resources and official government policy.[28] Villagers resent the cumbersome procedures and expenses (both formal and informal) involved in obtaining permits to harvest timber legally and often circumvent them. Circumvention is especially common in officially reserved forest areas where villagers frequently harvest timber and clear land for cultivation. These activities are often in fact accommodated after the fact by the local administration officials of the Ministry of Interior who issue land registration documents and provide social services in the newly settled areas. Reportedly villagers tend to avoid encroachments on officially unreserved forest areas, feeling that they are the "property" of influential persons and therefore violations will be detected and violators punished.[29] Several observers have called for the creation of small forests managed by local governments for local use and the promotion of private woodlots.[30]

The conservation of forests and preservation of important watershed areas have been frequently articulated goals in Thailand over the last several decades. Observers argue that forests protect soil fertility, reduce erosion, regulate the water supply, retard salination, reduce silting and flooding, and preserve groundwater supplies.[31] Forest destruction of the last fifty years has resulted in a serious deterioration in basic soil, forest, and water resources as well as a depletion of wildlife.[32] Many observers have been critical of government efforts to restrict forest cutting and curtail swidden cultivation.[33] Strict enforcement of forestry department production quotas would lower production from over 2 million cubic meters per year to 1.6 million cubic meters.[34]

Increasing concern over the effects of swidden cultivation on soil, water, and forest resources has also been expressed. The problem is especially acute in northern Thailand, where hill tribesmen and lowland Thais compete for the use of hill slopes.[35] The Asian Development Bank estimates that roughly 1.5 million people in Thailand—especially in the north, west, and northeast—depend regularly on swidden cultivation, clearing 500,000 hectares each year.[36] Swidden cultivation results in a destruction of primary forests and interruption of secondary growth, often resulting in the growth of grasslands, soil erosion, flooding, and groundwater shortages. Watershed planning and bans on swidden cultivation at middle elevations and on virgin forest land have been proposed. A broad range of solutions including research on forestry and alternate upland crops, plantations and wage labor opportunities, and developments in the property rights systems are needed to bring swidden cultivation in Thailand under control and integrate it with sound national land use policy.[37]

The discussions of the importance of forest conservation are, however, not new. Discussions and recommendations made in the pre–World War II period are very similar to those made more recently. By the 1930s irrigation officials were already concerned over the effects of swidden cultivation in the hill districts on siltation, flooding, and the regulation of water flows.[38] As in the cases of irrigation and agricultural research policy debates in Thailand, there is a great deal of continuity between the interwar and post–World War II periods.[39] Concern over deforestation and other environmental problems has, however, become much more widespread in Thailand, especially in the last decade.

While there has been a marked change in Thai conservation policy and in the awareness of ecological and environmental problems, there appear to have been relatively few serious examinations of the socially optimal use of land in Thailand. In addition, while the major and minor causes of forest depletion may be enumerated and their consequences discussed, it is not possible on the basis of the shaky evidence available to partition the historical pattern of forest depletion among its causes or to decide how much of the depletion is due to each factor and the interactions among them.

Forest depletion in Thailand occurred because it was privately profitable. Private return to land clearing for commercial and/or subsistence production was sufficient to underwrite a large expansion in the area under cultivation. The expansion was, of course, in part fueled by rapid rates of population growth. Private returns drove clearing in both lowland and upland areas, both for settled and shifting cultivation. Growth in domestic and foreign demand (decreasingly important over time) made timber cutting and the harvesting of forest products privately profitable, especially when public forest resources could be exploited for only nominal fees (formal and/or informal). These processes were facilitated by law enforcement of existing legislation, the exemptions and provisions in the legislation itself, and the inefficiencies inherent in the administration of forest lands in Thailand.[40]

From a contemporary perspective it now appears that forest depletion in Thailand has probably gone too far or at least far enough. That conclusion is, however, impressionistic and is not based on comprehensive or rigorous evidence. But in the earlier periods during which Thailand was a land-abundant, forest-abundant, and labor-scarce country, growth through clearing was quite reasonable and contributed significantly to the impressive growth in per capita Thai incomes experienced by virtually all classes.

The crucial issue is the optimal rate of exploitation. It now appears that the current rate is too high. This problem occurs because there are two sets of divergences between the private profitability of land clearing and forest production and the social rate of return on those activities. First, there is a divergence caused by the open and free access resource nature of Thai public lands and their management. Second, there is a divergence arising because some of the costs of clearing and harvesting are not borne by the agents who reap the direct benefits, causing them to overexploit the resource from the socially optimal point of view.

While the legislation described above indicates that forests are public property subject to regulation by the Forest Department, the actual pattern of practices admits many exceptions, both legal and illegal. For example, under the 1954 Land Code all areas not claimed within 180 days of the passage of that act became property of the government. Provincial governors, however, have the power to allow villagers to continue to file claims for newly cleared areas and have routinely continued to do so. Thus local administration officials who are under the Ministry of Interior have allowed local practice to override national conservation policy. Timber cutting by villagers is another example: because villagers do not own the forest and cannot exclude others from using it, they have little incentive to conserve and every incentive to capture the gains from cutting before someone else does. In addition there is widespread evasion of forestry regulations. Both large- and small-scale commercial operators exploit publicly managed forest resources for private gain. In all of these cases deficiencies in the property rights system and its administration and enforcement mean that private parties have incentives to overuse the forest resource.

The second divergence arises through a set of negative externalities. Some of the costs of land clearing are borne downstream through erosion, silting, and flooding. As population density downstream has risen and the number of people affected by flooding has increased, the marginal social cost of the clearing of forest lands has risen over time. The costs of land clearing or forest production also include the depletion of the genetic pool as various flora and fauna become increasingly rare. An intergenerational externality may arise in that the current population may insufficiently take into account the interests of future generations. An externality also arises because forests are also valued for aesthetic and recreational uses as well as for income generation purposes.

In contemporary Thailand the erosion and flooding externality is probably the most important one, although we lack rigorous evidence on the magnitude of any of these externalities. In spite of the depletion of forests that has accompanied the increasing population density, it still may be true that the highest social value of land use in Thailand is crop cultivation rather than forest cover, except for the ridge tops and steepest slopes. The rise in population density increases the relative scarcity of both forest and crop lands.[41]

Four policy implications follow. First, institutional changes are needed to redesign the property rights system and its application to reduce (subject to transaction cost) the first divergence. Second, institutional changes are needed to alter private incentives so that private parties are faced with returns that better approximate the social returns. Third, because clearing occurs because it is privately profitable, the generation of more lucrative opportunities in intensified agriculture and industrial employment is crucial in reducing the rate of forest depletion. Both the carrot and the stick are needed. Finally, once legislation and institutions reflect a careful consideration of the factors that determine the socially optimal rate of forest use, more efficient and effective enforcement mechanisms will be needed. The probability of detecting violations will have to be high enough and penalties heavy enough to discourage socially suboptimal illegal uses. Charlermrath has suggested that one agency (a merger of various forestry, police, and other agencies concerned with land use), a Ministry of Natural Resources and Environmental Protection, be established to create the bureaucratic incentives for vigorous enforcement.[42]

Conclusion

The clearing of land for cultivation, the cutting of forests for timber, and the collection of forest products all contribute significantly to forest depletion in all regions in Thailand. The relative importance of the factors does, however, vary considerably across the regions. In the North legal and illegal forest cutting, the extension of the area under cultivation, and shifting cultivation all seriously threaten the forest ecology. In the Northeast the extension of the area under cultivation may play a relatively more important role with forest cutting also contributing significantly. In the South rubber cultivation and mining activities appear to play the leading roles. There is considerable regional diversity and examination of the trends and causes within each region is important to the research agenda.

In Thailand the conservation problems posed by the harvesting of forest products by rural dwellers for their own use are exacerbated by the even more serious deforestation problems posed by shifting cultivation, illegal commercial timber operations, and land clearing for the extension of the area under cultivation. These processes often interact and have been intensified through the expansion of the highway network. Land on which illegal timber operations have been conducted is often further cleared through the gathering of firewood and timber by the local population and finally completely cleared for cultivation. All of the following steps toward solving Thailand's increasingly serious conservation problems have been suggested: intensification of the cultivation of already cleared areas, the enhancement of economic incentives for more intensive cultivation, research on improved agricultural practices (especially for rain-fed areas), institutional changes to improve the efficiency of the irrigation system and further intensify cultivation, the promotion of replanting, the more rapid issuing of title deeds on legally cleared land, the privatization of some forest lands, restrictions of the growth in the capacity of the sawmilling industry, the promotion of taungya plantations in hill areas, the promotion of conservation education, tighter enforcement of existing legislation, the simplification of existing forestry regulations that apply to villagers, the curtailment of the cultivation of opium and swidden cultivation, more intensive forest management, and the development of a realistic comprehensive forestry policy.[43]

In sum, a primary source of the growth in Thai agricultural output has been an expansion of the area under cultivation. Growth by clearing was especially important in the period from 1900 to 1950 and continues to play a crucial role. As a result, large forest areas have been cleared and forest, soil, and water resources have been depleted. The problem is especially acute in northern Thailand.

Awareness of conservation problems has been apparent for over sixty years and has grown over time. Deficiencies in policy, management, and enforcement, however, have persisted and are likely to continue. Forests in Thailand are often treated as an open and free access resource; overexploitation has been the result.

APPENDIX The Reliability of Thai Agricultural Statistics and Rice Balance Sheet Estimates

As argued in the main text, it is important to assess the reliability of Thai agricultural statistics. The traditional system of data collection for area, production, and yield statistics for major crops in Thailand involves initial reporting by the village headman (who is not an employee of the central government). The reports are channeled through the commune leader to the district officer (who is a central government employee), who in turn passes them on to provincial officials who forward the results to Bangkok. In this system, there are few incentives for accurate or complete reporting. Thus in the early 1950s the Rice Department (Ministry of Agriculture) instituted procedures in compiling the reports that ensured at least minimal checks for internal consistency in the reports on paddy production, yields, and area.[44]

The 1963 agricultural census conducted by the National Statistical Office (NSO) generated production estimates for nonrice crops that differed sharply from those published by the Ministry of Agriculture. Starting in 1966 the NSO began conducting crop-cutting surveys. In general the estimates differed significantly from those made by the Rice Department. Because the NSO rice production estimates generally exceeded the Rice Department estimates, and rice balance sheet estimates (see below) also indicated that the Rice Department consistently underestimated rice production, a number of observers, including the national income accountants, have routinely adjusted the Rice Department figures upward accordingly.

Thus, even in the post–World War II period, the data for the most important crop, paddy, are far from totally reliable. The margin of error for other crops or at more disaggregated levels is probably greater.

Crop reporting by village headmen was the system relied upon in the pre–World War II period and thus there is every reason to be skeptical about the accuracy of the data. Official estimates of the major crop area, paddy area, paddy output, and paddy yields are shown in table 6A-1. For the period prior to 1927–1928 the area under major crops includes paddy, tobacco, pepper, maize, cotton, peas, and sesame; for the period from 1927–1928 through 1955 the area under coconuts is also included. Paddy area dominates, always accounting for over 90 percent of the total estimated cropped area.[45]

The official estimates undoubtedly overstate the dominance of rice.

Table 6A-1 Whole Kingdom Paddy Area, Paddy Output, Paddy Yield, and Major Crop Area for Thailand, 1905/6–1955

Year	Paddy area planted (thousands of ha.)	Paddy output (metric tons)	Paddy yield (kg./ha.)	Major crop area (thousands of ha.)
1905/6	1,281.23	n.a.	n.a.	n.a.
1906/7	1,388.42	2,614,472	1,883	n.a.
1907/8	1,385.23	2,616,637	1,889	n.a.
1908/9	1,253.15	2,387,267	1,905	n.a.
1909/10	1,752.86	3,068,314	1,750	n.a.
1910/11	1,454.75	2,976,743	2,046	n.a.
1911/12	1,571.69	2,903,648	1,847	1,606.96
1912/13	1,974.73	3,698,885	1,873	2,007.61
1913/14	2,162.32	2,876,064	1,330	2,195.70
1914/15	2,038.30	3,126,732	1,534	2,066.70
1915/16	2,072.24	3,046,692	1,470	2,104.48
1916/17	2,169.44	3,816,160	1,759	2,198.14
1917/18	2,222.50	3,013,143	1,356	2,247.41
1918/19	2,158.57	3,411,275	1,580	2,184.43
1919/20	2,478.80	2,288,574	922	2,504.31
1920/21	2,446.90	4,311,735	1,761	2,475.01
1921/22	2,595.87	4,266,166	1,644	2,621.68
1922/23	2,527.26	4,375,187	1,731	2,558.76
1923/24	2,686.60	4,434,923	1,652	2,710.34
1924/25	2,776.98	4,941,573	1,779	2,807.77
1925/26	2,736.54	4,193,102	1,532	2,766.33
1926/27	2,894.68	5,226,037	1,804	2,923.02
1927/28	2,927.75	4,564,096	1,560	3,008.04
1928/29	2,849.68	3,882,165	1,361	2,931.87
1929/30	3,035.86	3,874,834	1,275	3,114.56
1930/31	3,180.08	4,826,301	1,515	3,249.34
1931/32	3,090.79	4,068,530	1,316	3,167.14
1932/33	3,213.79	5,116,405	1,590	3,291.45
1933/34	3,245.30	5,007,734	1,541	3,329.43
1934/35	3,336.69	4,597,786	1,376	3,425.83
1935/36	3,377.70	4,726,983	1,399	3,458.96
1936/37	3,258.18	3,379,856	1,039	3,337.86
1937/38	3,369.98	4,555,706	1,350	3,461.54

Table 6A-1 Continued

Year	Paddy area planted (thousands of ha.)	Paddy output (metric tons)	Paddy yield (kg./ha.)	Major crop area (thousands of ha.)
1938/39	3,507.02	4,523,663	1,290	3,595.86
1939/40	3,463.88	4,560,463	1,316	3,552.46
1940	3,806.98	4,923,350	1,294	3,892.20
1941	3,969.25	5,120,097	1,290	4,077.25
1942	4,398.63	3,868,806	880	4,528.41
1943	4,314.72	5,702,005	1,322	4,474.39
1944	4,240.04	5,107,635	1,205	4,392.21
1945	3,942.39	3,699,322	938	4,066.61
1946	3,981.99	4,442,271	1,116	4,099.91
1947	4,825.00	5,453,110	1,130	4,979.66
1948	5,211.75	6,768,852	1,299	5,371.03
1949	5,268.22	6,618,908	1,256	5,557.77
1950	5,539.98	6,715,813	1,212	5,872.26
1951	5,959.26	7,254,318	1,217	6,323.01
1952	5,368.12	6,538,029	1,218	5,736.87
1953	6,171.93	8,159,456	1,322	6,556.26
1954	5,557.16	5,653,604	1,017	5,967.03
1955	5,769.57	7,262,453	1,259	6,188.94

Sources: All data were originally taken from official Thai government sources. The data for 1905/6 through 1941 are from Feeny, *Political Economy*, 138, 140. Data for 1942 through 1955 are from *Statistical Year Book of Thailand*, no. 21 (1939–40 to 1944), and no. 22 (1945 to 1955).

The degree of understatement of the area under cultivation was likely to have been greater for crops other than rice. Some increasingly important crops like rubber were omitted from the official estimates that combine the official cropped area estimates with separate estimates of the area in rubber are presented in table 6A-2. Paddy accounts for at least 85 percent of this total area estimate.

Information from a more comprehensive land use survey conducted in the interwar period indicates that paddy accounted for 60 percent of the total area utilized. Tobacco, maize, cotton, peas, sesame, and pepper together accounted for 0.6 percent. Homesteads, gardens, fruit, and other small crops accounted for the remaining 39.4 percent.[46]

Table 6A-2 Area Planted in Rubber Trees and Total Area under Major Crops in Thailand, 1913–1955

Year	Rubber area (thousands of ha.)	Total area in major crops, including rubber (thousands of ha.)
1913	1.60	2,197.30
1914	1.60	2,968.30
1915	1.92	2,106.40
1916	5.44	2,203.58
1917	10.72	2,258.13
1918	10.72	2,195.15
1919	14.72	2,519.03
1920	16.48	2,491.49
1921	16.48	2,638.16
1922	16.48	2,575.24
1923	16.48	2,726.82
1924	16.48	2,824.25
1925	16.48	2,782.81
1926	22.40	2,945.92
1927	107.36	3,039.23
1928	107.36	3,039.23
1929	107.36	3,221.92
1930	107.36	3,356.70
1932	140.48	3,431.93
1932	140.48	3,431.93
1933	140.48	3,469.91
1934	140.48	3,566.31
1935	140.48	3,599.44
1936	140.48	3,478.34
1937	140.48	3,602.02
1938	184.64	3,780.50
1939	228.80	3,781.26
1940	272.96	4,165.16
1941	290.56	4,367.81
1942	n.a.	n.a.
1943	n.a.	n.a.
1944	n.a.	n.a.
1945	n.a.	n.a.
1946	n.a.	n.a.

Table 6A-2 Continued

Year	Rubber area (thousands of ha.)	Total area in major crops, including rubber (thousands of ha.)
1947	n.a.	n.a.
1948	290.56	5,661.59
1949	346.56	5,904.32
1950	361.44	6,233.70
1951	383.84	6,706.85
1952	393.44	6,130.31
1953	441.92	6,998.18
1954	526.24	6,493.27
1955	530.88	6,719.82

Source: Data on the area planted in rubber are taken from Stifel, "Growth of the Rubber Economy," 130. His estimates are based on export and domestic usage estimates, taking into account the short-run price elasticity of supply and the effect of the age composition of the trees on average yields. The sources for the major crop area (without rubber) are given in table 6A-1.

The official major crop area data omit homesteads, gardens, and most fruits; even when they are included the dominance of paddy is clear.

Thus in order to assess the overall reliability of the cropped area estimates, we need to examine the paddy area and output series in some detail. The decline in paddy yields evident from 1920 through the mid-1950s (table 6A-1) was reflected in contemporary discussion of the need for investments in rice research and irrigation. The level of paddy yields reflected in the official estimates roughly corresponds to the scattered evidence available on the yields obtained by farmers in various regions.[47] Most of the observations available are for individual farms, villages, or districts in the Central Plain, a region in which yields generally exceed the national average. These observations tend to indicate that the official whole kingdom yield estimates may be a bit low, but the extent of the bias, if any, is not clear.

Under the assumption that the yield data are roughly accurate, the reliability of the area and output data can be assessed by constructing a rice balance sheet that takes into account exports, human consumption, seed, feed, and losses in milling and storage and compares estimated production based on usage to the official estimates. Each component of the rice balance sheet will be discussed.

Data on rice exports are readily available from the foreign trade series. Nearly all rice exports flowed through the port of Bangkok, the port for which the trade data are the most complete and reliable. Thus this component of the rice balance sheet estimate is probably accurately measured.[48]

Time series on per capita rice consumption are unavailable. We do, however, have a number of benchmark estimates and observations on consumption per person at the village level, and informed contemporary estimates will be constructed assuming two different constant levels of per capita consumption, 170 kilograms and 144 kilograms.[49] The former implies a daily intake from rice of 1,709 calories. If this intake is combined with the 1962 Household Expenditure Survey estimates, which revealed that rice accounted for 85 percent of the total caloric intake, this implies a total daily consumption of 2,011 calories.[50] Supanee and Wagner, using the FAO method, estimate a daily per capita caloric intake requirement of 1,932. Thus the 170 kilograms of rice per person per year consumption figure implies an essentially well-nourished population, which is consistent with both contemporary accounts and the rapid rate of natural increase experienced over the period. The 144 kilograms per person per year estimate was used to test the sensitivity of the results and also corresponds to the assumed consumption level used by Ministry of Agriculture officials in 1950 when the first rice balance sheet estimates were constructed. The appropriateness of assuming a constant rate of per capita consumption will be examined below.

Assuming a rate of per capita rice consumption, total consumption estimates can be constructed by multiplying that rate by estimates of the population. Unfortunately population estimates for the 1906 to 1955 period are not totally reliable and there is some controversy over which set of estimates is the most accurate. Skinner provides the most comprehensive and probably most reliable set of estimates (for 1825–1955). Sternstein disputes Skinner's gradual acceleration in the rate of growth of the Thai population over the nineteenth century and instead argues that the rate of growth of the Thai population was slower in the nineteenth century, but more rapid in the twentieth. Finally, Bourgeois-Pichat argues that Skinner underestimated the degree of underenumeration in the 1929 census and presents alternate estimates for the 1919 through 1956 period that largely agree with the pattern of change reflected in the Skinner series. Caldwell has reviewed the debate of the pre-World War II population estimate

Table 6A-3 Various Estimates of the Thai Population, 1906–1956 (in thousands)

Year	Estimate from Skinner	Estimate from Bourgeois-Pichat	Estimate from Sternstein	Census estimate
1906	7,896	8,364	6,200	n.a.
1911	8,432	8,947	7,204	8,266
1919	9,608	9,966	9,158	9,207
1929	12,059	12,433	11,606	11,506
1937	14,721	14,549	14,218	14,464
1947	17,643	17,647	18,117	17,443
1956	20,865[a]	20,776	23,286	20,095[b]

Sources: The Skinner series is taken from Skinner, *Chinese Society in Thailand*, 79, 183. Bourgeois-Pichat is taken from Bourgeois-Pichat, "Attempt to Appraise." The data for the 1906–19 period were interpolated from Bourgeois-Pichat's 1919 estimate assuming that the Thai population grew at 1.36 percent per year over that period, the rate of growth given by the census estimates for the 1911–19 period. The Sternstein estimates are from Sternstein, "Critique of Thai Population Data" and Cochrane, "Population of Thailand," 6. The figure for 1906 is interpolated from the graph presented in Sternstein. Census data are taken from Feeny, *Political Economy*, 147. Annual data were interpolated between these benchmarks for all three series for use in constructing the rice balance sheet estimates.

a. The 1956 Skinner estimate was interpolated from the 1955 estimate by adding 1.8813 percent.

b. The 1956 census estimate is based on a survey rather than a census.

and concludes that the evidence tends to favor Skinner rather than Sternstein.[51]

All three series will be used to construct the rice balance sheet estimates. Unfortunately the Bourgeois-Pichat and Sternstein series had to be interpolated over the 1906 to 1919 and 1906 to 1920 periods respectively.[52] Given that the Bourgeois-Pichat series is basically similar to the Skinner series; that Caldwell concluded that the evidence favored Skinner with respect to Sternstein; and that there is no need to interpolate the early twentieth-century population estimates for the Skinner series, that series should provide the most reliable population estimates for use in the rice balance sheet estimates (see tables 6A-3 and 6A-4).

Two paddy seed rates in kilograms per hectare were used, 125 and 100. Each of these figures is in broad agreement with contemporary observations on seed use and reflect the higher rate of seed use under

Table 6A-4 Average Annual Rates of Growth of Population in Percent Per Year for Thailand, 1906–1955

	Population estimates			
Period	Skinner	Bourgeois-Pichat	Sternstein	Census
1906–11	1.32	—	—	—
1911–19	1.65	—	3.05	1.36
1919–29	2.30	2.24	2.40	2.25
1929–37	2.52	1.98	2.57	2.90
1937–47	1.83	1.95	2.45	1.89
1947–56	1.88	1.82	2.83	1.59
1906–55	1.96	1.84	2.67	—

Source: See table 6A-3.

broadcast cultivation, which accounted for the majority of the area under paddy cultivation in this period.[53] A slight overestimate of seed use may result from the use of these rates.

The Thai national income accountants estimate that 3 percent of the paddy crop is accounted for as animal feed and losses. The milling rate (weight of milled rice as a percent of the weight of paddy) was assumed to be 60 percent, a lower-bound figure reflecting performance rates under hand-milling and small upcountry mills. The results are not particularly sensitive to either of these assumed parameter values.

The rice balance sheet estimates were constructed according to the equation:

$$[(XI.POP_t + EX_{t+1})/0.6 + X2.AREA_t]1.03093 = OUTPUT_t$$

where

XI = per capita rice consumption, 170 or 144 kg,

POP_t = population in year t,

EX_{t1} = rice exports in the following year, t + 1

$X2$ = the seeding rate in kg per hectare, 125 or 100,

$AREA_t$ = the area planted to paddy in hectares in year t,

and

$OUTPUT_t$ = rice balance sheet estimate of paddy production in year t.

Table 6A-5 Comparison of Rice Balance Sheet Estimates to Official Paddy Production Figures for Thailand, 1906–1955

Period	Per capita rice consumption (kg.)	Seed rate (kg./ha.)	Population series	Mean of ratio of rice balance sheet to official estimate	Variance of ratio	Range of ratio
1906–40	170	125	Skinner	1.45	0.036	1.11–1.93
	170	125	B-P	1.48	0.034	1.15–1.92
	170	125	Sternstein	1.39	0.035	1.07–1.89
	144	125	Skinner	1.33	0.030	0.99–1.74
	144	125	B-P	1.35	0.028	1.02–1.73
	144	125	Sternstein	1.28	0.029	0.95–1.70
	144	100	Skinner	1.31	0.029	0.97–1.72
	144	100	B-P	1.33	0.028	1.00–1.71
	144	100	Sternstein	1.26	0.028	0.94–1.68
	170	100	Skinner	1.44	0.035	1.10–1.91
	170	100	B-P	1.47	0.033	1.13–1.90
	170	100	Sternstein	1.38	0.034	1.05–1.86
1906–55	170	125	Skinner	1.41	0.042	1.01–1.93
	170	125	B-P	1.42	0.043	1.01–1.92
	170	125	Sternstein	1.37	0.037	1.04–1.89
1930–41	170	125	Skinner	1.55	0.039	1.25–1.93
	170	125	B-P	1.55	0.038	1.25–1.92
	170	125	Sternstein	1.52	0.036	1.25–1.89
1906–29	170	125	Skinner	1.36	0.035	1.01–1.76
and	170	125	B-P	1.38	0.038	1.01–1.82
1942–55	170	125	Sternstein	1.33	0.028	1.04–1.71

Sources: Data on rates of per capita rice consumption, rice exports, and seed rates are taken from Feeny, *The Political Economy of Productivity*; population data are from table 6A-1; paddy area data are from table 6A-3; rice milling rates are discussed in the text of the appendix to this chapter.
Note: B-P means Bourgeois-Pichat.

Estimates were constructed for the 1906 to 1940 and 1906 to 1955 periods. The rice balance sheet estimates were then compared to the official estimates. The mean, variance, and range of the rice balance sheet estimates as a percent of the official estimates are presented in table 6A-5.

Table 6A-6 Regressions of the Ratio of the Rice Balance Sheet Estimates to the Official Production Figures on Time, Thailand, 1906–1955

Period	Per capita rice consumption (kg.)	Seed rate (kg./ha.)	Population series	Regression coefficient		
				Intercept	Time	$\bar{R}^2$
1906–40	170	125	Skinner	− 7.25 (1.21)	0.005 (1.46)[a]	0.03
	170	125	B-P	− 3.38 (0.57)	0.003 (0.82)	0.00
	170	125	Sternstein	− 15.06 (2.82)[c]	0.009 (3.08)[c]	0.20
	144	125	Skinner	− 6.60 (1.21)	0.004 (1.45)[a]	0.03
	144	125	B-P	− 3.32 (0.61)	0.002 (0.86)	0.00
	144	125	Sternstein	− 13.22 (2.68)[c]	0.008 (2.94)[c]	0.18
	144	100	Skinner	− 6.30 (1.16)	0.004 (1.45)[a]	0.03
	144	100	B-P	− 3.01 (0.56)	0.002 (0.81)	0.00
	144	100	Sternstein	− 12.91 (2.64)[c]	0.007 (2.90)[c]	0.18
	170	100	Skinner	− 6.94 (1.17)	0.004 (1.41)[a]	0.03
	170	100	B-P	− 3.07 (0.52)	0.002 (0.77)	0.00
	170	100	Sternstein	− 14.76 (2.79)[c]	0.008 (3.05)[c]	0.20

1906–55	170	125	Skinner	6.70 (1.75)[b]	−0.003 (1.38)[a]	0.02
	170	125	B-P	9.45 (2.50)[c]	−0.004 (2.13)[b]	0.07
	170	125	Sternstein	−0.87 (0.24)	0.001 (0.62)	0.00
1930–41	170	125	Skinner	4.76 (0.14)	−0.002 (0.10)	0.00
	170	125	B-P	8.97 (0.27)	−0.004 (0.23)	0.00
	170	125	Sternstein	− 0.06 (0.00)	0.008 (0.05)	0.00
1906–29	170	125	Skinner	9.15 (2.68)[c]	−0.004 (2.28)[b]	0.10
and	170	125	B-P	11.69 (3.38)[c]	−0.005 (2.98)[c]	0.18
1942–55	170	125	Sternstein	1.30 (0.39)	0.000 (0.01)	0.00

F-Tests of Structural Stability 1906–55 versus 1906–29/1942–55 and 1930–41

Case	Calculated F	Critical F (2,46) at the 1% level
Skinner (170/125)	6.29	5.10
Bourgeois-Pichat (170/125)	5.34	5.10

a. Significant at the 10% level, one-tailed test.
b. Significant at the 5% level, one-tailed test.
c. Significant at the 1% level, one-tailed test.
Note: t-statistics are given in parentheses; B-P means Bourgeois-Pichat.

Clearly, there appears to have been a significant tendency for the official production estimates to understate production. If, as was argued above, the yield per hectare data are roughly accurate, this conclusion implies a consistent understatement of both area and production in the official figures.

The conclusion for the 1906 to 1955 period (and various subperiods) is consistent with the results obtained by the National Accounts Office when they examined the Rice Department's estimates versus rice balance sheet estimates for the 1951 through 1966 period.[54] They concluded that cumulative paddy use over that period exceeded the Rice Department estimates by 21.36 percent. Using an alternate set of assumptions that differed slightly, Ingram found that his rice balance sheet estimates exceeded the Rice Department estimates by 13.4 percent over the 1958 through 1965 period but were in close agreement with the Rice Department for the 1966 through 1968 period.[55] Thus the conclusion that official estimates understate production in the pre-1960s period does not appear to be sensitive to the particular parameter values chosen in constructing a rice balance sheet.

It is natural to expect that the degree of understatement was reduced over time as coverage became more complete, and regression results do give limited support to that hypothesis (see table 6A-6). The ratio of the rice balance sheet estimate to the official estimate was regressed on time. For the 1906 through 1940 period for the Skinner (probably the most reliable) and Bourgeois-Pichat population series, the coefficient on time was seldom significant. For the Sternstein series, the coefficient was positive and significant, but as in the other cases, the coefficient of determination was very low.

When the time period is expanded to 1906 through 1955, the coefficient on time is negative and significant for the Skinner and Bourgeois-Pichat series, but the coefficient of determination is still very low. In examining the time path of the ratio of the two estimates, the ratio appears generally to be quite high over the 1930s, suggesting that the constant rate of per capita consumption assumed in the rice balance sheet estimates may overstate actual consumption in the Depression period. Thus separate regressions were run for the 1906 to 1929 and 1942 to 1955 periods and the 1930 to 1941 period. In the former two periods for the Skinner and Bourgeois-Pichat series, the coefficient on time is negative and significant and the coefficient of determination is improved, but still low. The 1930 to 1941 period results suggest no significant time trend. F-tests performed on the structural stability of the relationship for the two periods (for the Skinner and Bourgeois-Pichat series) suggest structural instability.

Thus one is left with the conclusion that there has been some slight reduction in the degree of underreporting of paddy production over the years, but that the trends reflected in the overall official series are probably roughly reliable.[56] If one takes the estimated coefficient on time seriously (for Skinner, 170/125 case), the mean ratio of the rice balance sheet to the official estimate would be reduced from 1.3589 to 1.2054 over a 38-year period.[57] Revisions in data collection and processing procedures used by the Rice Department in the late 1960s have reduced this apparent underestimation of rice production.[58]

PART 2

LINKAGES:

THE GLOBAL TIMBER TRADE

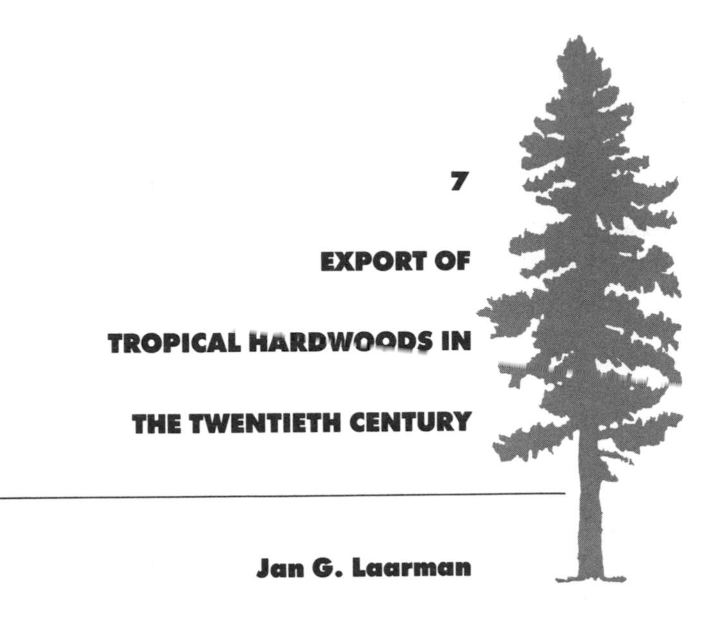

7
EXPORT OF TROPICAL HARDWOODS IN THE TWENTIETH CENTURY

Jan G. Laarman

Since the late 1970s a sea of ink has been spilled on the tropical forests.[1] The issues concern the definition, rate, consequences, and causes of unplanned tropical deforestation. When assessing blame for the degradation and depletion of the tropical forest cover, careful analysts distinguish causes from effects, determinants from agents. Causes and determinants include skewed political power and failures of institutions.[2] On the other hand, effects and agents are those enterprises and persons observed to be the instruments in the conversion process: ranchers and farmers, loggers, road builders, miners, soldiers, shifting cultivators, firewood gatherers and charcoal makers, and so on.[3]

This essay will examine just one of these visible agents: logging for export. Whether directly or indirectly, some of the hands on the axe are Japanese, European, and North American.[4] Through the course of this century, that axe has been wielded relatively more by the Japanese and relatively less by the others. Also, that axe now swings hard in some tropical countries where it was not much in evidence only a few decades earlier. In parts of the tropics, the pattern of timber exports that existed prior to 1920 has been completely turned on its head.

This study is divided into three parts. The first briefly describes the development of tropical hardwood exports from the end of the nineteenth century to the present time. The second assesses our state of knowledge of the deforestation impact stemming from this trade, and the third concludes with observations on supply and demand to the year 2000.

Exports of Tropical Hardwoods, 1900–1985

Timber exports from the tropical countries have been largely logs and wood-based products of the hardwoods (i.e., nonconiferous woods). Exports of tropical softwoods have included modest flows from the pinelands of Central America, Mexico, and some of the Caribbean islands. Paraná pine has been exported from the subtropics of southern Brazil. Indonesia accounts for a small proportion of current softwood log exports. In most cases these softwood exports are (or have been) important at the regional or subregional level, and more so in former years than presently for most of the regions in question. However, this study will focus on trade in tropical hardwoods only, because it is the hardwood trade that has traditionally generated the greatest industrial logging activity in support of external markets.

According to the Food and Agriculture Organization's *Yearbook of Forest Products*, in 1983 the aggregate of developing countries (including the tropical ones) shipped no more than 16 percent (by value) of the total world exports of wood products.[5] The developing countries held more than 10 percent of world exports in only four products: hardwood logs; hardwood sawnwood; veneer sheets; and plywood. Although FAO does not separate softwoods from hardwoods in the two latter categories, the veneer and plywood exported from the developing countries are overwhelmingly tropical hardwoods.

Figure 7.1 shows the enormous growth in the export of tropical hardwoods since World War II when compared with the previous half century. The post-1945 data are estimated from FAO's *Yearbook of Forest Products*, and then plotted as five-year averages. It is necessary to exercise caution in assessing the significance of these figures, however; FAO discusses the difficulties of obtaining comprehensive statistical coverage and accurate conversion factors (between products and roundwood equivalents).[6] Moreover, the definition of the tropics and the tropical countries is a variable one, as is the definition of a tropical wood.

If the statistics since World War II are deficient, the statistics before

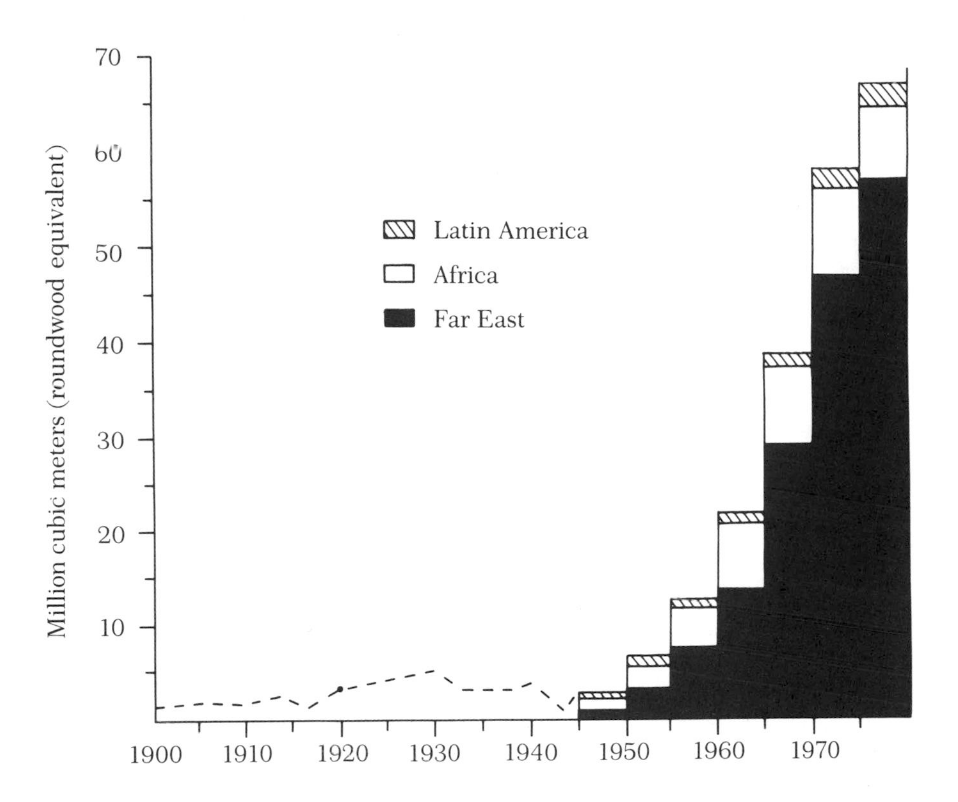

Figure 7.1 World Exports of Tropical Hardwoods, 1900–1980.
Source: FAO, *Yearbook of Forest Products*.

Note: Dashed line represents estimates.

that time are even less reliable. The dashed line for the pre-1946 period signals that these export levels and trends are purely guesses, with no pretense to accuracy. The only datum for pre-1946 is the dot at 3.1 million cubic meters circa 1920, estimated from Zon and Sparhawk.[7] The dashed line before and after that point represents tentative inferences from Zon and Sparhawk (for years prior to 1920) and from statistics of the International Institute of Agriculture (for the mid-1920s through the 1930s).

Exports before 1946

From Zon and Sparhawk we can infer that the export of tropical hardwoods entered a period of transition and realignment as the nineteenth century became the twentieth. Technological change was decreasing the foreign import demand for tropical dyewoods even as the foreign import demand for tropical cabinetwoods apparently was rising.[8] Timber exports began to trickle from such regions as the Philippines—regions that eventually would become major exporters.[9] At the same time, the economically available supply of mahogany, Spanish cedar, and other cabinetwoods was reaching exhaustion in various parts of the Caribbean basin.[10]

In the years prior to World War I, it seems fairly certain that world trade in tropical and subtropical hardwoods had been increasing. Between 1900 and 1914 the United States doubled its imports of tropical cabinetwoods.[11] Argentina's export of quebracho logs (subtropical) for tanning began to decline in 1914, when high ocean freight rates made it more economical to export the extract.[12] Africa's exports of tropical woods to Europe may have attained a prewar peak in 1913.[13] Evidence for tropical Asia is sparse, but timber exports from British North Borneo declined after 1913.[14] Teak exports from Java fell sharply after 1914.[15]

Exports of tropical hardwoods presumably recovered and grew after World War I and through the 1920s, perhaps reaching their next peak about 1929 or 1930. North America, together with the major European importers, registered a precipitous drop in wood imports after 1930.[16] Africa's wood exports were valued at 105 million gold francs in 1930, but subsequently fell to less than half that level for all of 1932 to 1938.[17]

For Europe the onset of World War II interrupted import flows and obligated self-sufficiency.[18] Another consequence was timber raiding in neighboring countries.[19] In such occupied tropical exporting regions as the Philippines, all timber production was quickly commandeered for the war.[20] In the United States the war greatly expanded the demand for wood products, but it is uncertain what effect this may have had on imports from the tropics.[21] The poor data on timber production and trade for the wartime years reflect the obvious strain on the world's statistical services for that period.[22]

Export Growth, 1946–1985

Export expansion after World War II has been uninterrupted, occurring on a scale disproportionate with export growth during all previous periods (figure 7.1). Exports of tropical hardwoods from the principal producing regions grew from an annual average of 2.8 million cubic meters in 1946–1950 to 66.6 million cubic meters in 1976–1980, or by a factor of 24. Exports grow in absolute terms during each successive five year period, although the increase between 1971–1975 and 1976–1980 was less than half the increase that had preceded it. Moreover, export growth since 1980 has been stagnant or even declining, as will be discussed below.

Explanations for the dramatic postwar expansion are many and interrelated.[23] On the demand side are the obvious postwar increases in income and population in Japan, Western Europe, the United States, and other industrialized regions. Another demand shifter has been the dwindling availability of certain species and grades of hardwoods in some parts of the temperate zones, stimulating partial substitution by imports from the tropics.

On the supply side, mechanized logging and timber transport spread rapidly after the war (principally in insular Southeast Asia and West Africa) to permit cutting of species and tracts previously considered inoperable. More recently improvements in pulping technologies enable a large variety of tropical hardwoods to be pulped in a single mixture, leading to "any-tree" or "all-tree" harvesting.[24]

Most critically for hardwood exports from the Philippines, Malaysia, and Indonesia, great strides were made in the late 1940s and early 1950s to improve technologies to produce veneer and plywood from lauan (also traded under such names as Philippine mahogany and meranti). As production costs fell sharply during the 1950s lauan plywood began to compete very successfully in world markets against other utility-grade construction materials and helped create large new markets for flush doors and interior paneling.[25]

Table 7.1 portrays postwar growth in the production and export of tropical hardwoods in the Far East (Southeast Asia and South Asia), Latin America (including the Caribbean islands), and developing Africa (all countries except the Republic of South Africa). Excluded from table 7.1 are the countries of the Near East and Oceania, both of which are relatively small timber producing areas.

Table 7.2 shows the direction of tropical hardwood trade in 1983.

Table 7.1 Postwar Trends in Tropical Hardwood Production and Export in Three Major Producing Regions (annual averages, in millions of cubic meters)

Period	(1) Production of industrial roundwood[a]	Exports, roundwood equivalent[b]				(5)/(1) Ratio of exports to production (in percent)	(2)/(5) Share of logs in exports (in percent)
		(2) Logs	(3) Sawnwood	(4) Veneer & plywood	(5) Total		
Far East							
1946–50	9.4	0.3	0.5	*	0.8	9	37
1951–55	22.0	2.0	1.1	*	3.1	14	64
1956–60	43.4	5.2	1.6	0.4	7.2	16	72
1961–65	54.9	10.4	2.2	1.0	13.6	25	76
1966–70	72.8	21.5	3.7	3.5	28.7	39	75
1971–75	96.1	33.0	6.4	7.3	46.7	49	71
1976–80	113.1	35.6	11.1	9.6	56.3	50	63
1981–83	113.2	23.8	6.1	10.3	40.2	36	59
Africa							
1946–50	5.1	0.8	0.2	*	1.0	20	80
1951–55	10.0	1.8	0.5	0.1	2.4	24	75

1956–60	17.3	3.5	0.9	0.2	4.6	27	76
1961–65	19.9	5.2	1.2	0.4	6.8	34	76
1966–70	24.9	6.5	1.3	0.5	8.3	33	78
1971–75	28.9	7.0	1.3	0.6	8.9	31	79
1976–80	32.3	6.2	1.3	0.5	8.0	25	77
1981–83	21.5	4.8	0.6	0.5	5.9	27	81
Latin America							
1946–50	14.9	0.4	0.4	0.1	0.9	6	44
1951–55	17.7	0.4	0.4	*	0.8	5	50
1956–60	21.0	0.4	0.4	*	0.8	4	50
1961–65	21.5	0.4	0.5	*	0.9	4	44
1966–70	24.4	0.4	0.8	0.2	1.4	6	29
1971–75	30.4	0.3	1.3	0.3	1.9	6	16
1976–80	39.0	0.1	1.5	0.6	2.2	6	4
1981–83	51.8	*	0.9	0.7	1.6	3	*

Source: FAO, *Yearbook of Forest Products 1983*.

a. Hardwoods only.

b. Conversion factors used to obtain roundwood equivalent are from FAO's *Yearbook of Forest Products 1972*, p. lxvi.

*Less than 0.1 million cubic meters.

Table 7.2 Direction of Trade of Major Forest Products Exported from the Tropics, 1983

Exporter	Principal importers
Tropical hardwood logs	
Malaysia	Japan, China/Taiwan, Rep. Korea, Hong Kong, Singapore
Indonesia	Japan, China/Taiwan, Singapore, Rep. Korea, Hong Kong
Ivory Coast	Italy, France, Spain
Gabon	France, Spain, W. Ger.
Philippines	Japan, China/Taiwan, Rep. Korea
Cameroons	W. Ger., France, Spain
Tropical hardwood sawnwood	
Malaysia	Singapore, Netherlands, W. Ger., Thailand, Japan, U.K., Australia
Indonesia	Japan, Italy, Singapore, U.K., China/Taiwan
Singapore	Saudi Arabia, Netherlands, China/Taiwan, Belgium-Lux.
Philippines	Japan, U.K., France, U.S.A., Australia
Brazil	U.S.A., U.K.
Tropical hardwood veneer	
Malaysia	Japan, Singapore
Paraguay	Brazil
Congo	Australia, W. Ger.
Brazil	U.S.A., W. Ger.
Philippines	U.S.A., Netherlands

Table 7.2 Continued

Exporter	Principal importers
Tropical hardwood plywood	
Indonesia	U.S.A., Hong Kong, Singapore, U.K.
China/Taiwan	U.S.A., Hong Kong, Saudi Arabia
Singapore	Very diverse
Malaysia	Singapore, U.K., Hong Kong
Rep. Korea	U.S.A., Saudi Arabia

Source: FAO, *Yearbook of Forest Products 1983*.
Note: FAO estimates for China include "the province of Taiwan." Exports of plywood are most certainly from Taiwan, although imports of logs and sawnwood are likely aggregated over the People's Republic and Taiwan together.

Under each product type the prominent exporting countries are listed in descending order of importance. Likewise the principal importers are arranged from largest to smallest. These rankings shift gradually from year to year, but the pattern observed for 1983 will suffice to illustrate current flows.

The Far East. The Far East has been the world's major source of tropical hardwoods throughout the postwar era (table 7.1). Exports from the dipterocarp forests of insular Southeast Asia have been sold internationally following an intensive marketing effort by the Philippines at the beginning of the century.[26] Teak exports from Burma, Thailand, and Indonesia (Java) comprise a small part of the export total. India and other countries of South Asia account for very little of the exports, since nearly all of their industrial roundwood is for domestic consumption.

Table 7.2 shows that Malaysia (more precisely, Sabah and Sarawak) is currently the world's largest exporter of tropical hardwood logs. This single country accounted for 64 percent of world exports of tropical hardwood logs in 1983. The largest source of exports in the 1950s was the Philippines, joined by Malaysia around 1955–1957, and finally by Indonesia only after 1965.[27] In 1980 Malaysia and Indonesia together accounted for 78 percent of world exports of tropical hardwood logs.

But since 1980 Indonesia's log exports have fallen drastically in response to that government's log export restrictions and promotion of plywood production and exports.[28] In 1983 no less than 58 percent of the log exports from Malaysia, Indonesia, and the Philippines were purchased by Japan.

Additionally, exports of sawnwood and plywood by Taiwan, Singapore, and the Republic of Korea can be traced to the dipterocarp forests of the three log-exporting countries. Beginning first in Japan in the early 1950s, lauan logs were imported for processing into plywood and sawnwood, with subsequent "re-export" of a large share of the processed wood to the rest of the world. With a lag of several years, this model of in-transit processing was adopted by Taiwan, Korea, and Singapore.[29]

Hence even though Singapore has virtually no commercial forest area, it appears in table 7.2 as the world's third largest exporter of both tropical hardwood sawnwood and tropical hardwood plywood. Singapore is both a major importer and exporter of sawnwood, which reflects the remanufacturing of the product "in transit" to ultimate consumers. By 1983 Indonesia already led the developing world in exports of hardwood plywood, rising rapidly from sixth position in 1980.

Two of the more recent developments in this lauan trade are the shift by processors (especially Japan) away from re-export and toward domestic consumption[30] and the successful attempt of the timber-producing countries to process an increasing quantity of logs at home.[31] In 1980 Malaysia, Indonesia, and the Philippines shared 61 percent of world exports of tropical sawnwood, 35 percent of world exports of tropical hardwood veneer, and 28 percent of world exports of tropical hardwood plywood. With the huge expansion of plywood plants in Indonesia after 1980, the percentages for veneer and plywood have been rising rapidly, exceeding 53 percent by 1983.

West and West-Central Africa. The second largest export flow of tropical hardwoods in the postwar era has been from the lowland semi-deciduous forests of West and West-Central Africa. Most of Africa's exports are cabinetwoods intended for use in furniture and interior finishing.[32] In each postwar period roughly three-fourths of these exports have been logs (table 7.1), and the primary destination has been Western Europe (table 7.2). Europe's role as importer can be explained by colonial ties predating the century, as well as by relative geographical proximity.[33]

Africa's exports of tropical hardwoods exceeded those of the Far East and those of Latin America in 1946–1950, but then rapidly lost importance relative to the Far East, especially after 1960 (table 7.1). By 1960 log exports from Ghana and Nigeria had peaked and declined.[34] Production and export shifted so that by 1983 the largest log exporters were Ivory Coast, Gabon, and Cameroon (table 7.2). Together these three countries accounted for 14 percent of world exports of tropical hardwood logs in 1983. Other African log exporters of modest importance were Liberia, Congo, Central African Republic, and Zaire. Africa also had 26 percent of world exports of tropical hardwood veneer, three-fourths of which was purchased by Western Europe.

Latin America. Exports have accounted for no more than 6 percent of Latin America's postwar hardwood production (table 7.1). Even though Latin America has more than 56 percent of the closed hardwood tropical forests, the region has not been a major exporter of tropical hardwoods in the postwar period.[35] This is conventionally ascribed to the extremely mixed species composition and relatively small log size, and to the presently inaccessible location of many of the timber tracts.[36] Another factor is remoteness from both Japan and Western Europe.[37]

Exports and Deforestation

Looking back over the first 85 years of this century, the export of tropical hardwoods from south to north has both created and followed shifting zones of timber cutting. It has created these zones where timber itself has been the primary commodity of interest. It has followed them —or been contemporaneous with them—where penetration to obtain new agricultural land has generated timber as a by-product. In either case the extraction of timber for export is one of many agents—often acting integrally—altering the extent and composition of tropical vegetation locally and sometimes regionally. Analysts correctly view timber cutting for export as one of several links in the depletion or biological degradation of the tropical forest cover in a number of countries.

Adding the export volumes depicted in figure 7.1, nearly 1.2 billion cubic meters of tropical hardwoods (roundwood equivalent) were exported from the tropics during the period 1900 through 1980. Is it important for policy-making purposes to know whether this number is large or small? If it is important, by what scale should we judge the number? If export-led timber cutting has drained some regions of their forest resources too quickly, what slower rate of removals would have been preferable?

These questions, or others like them, have been asked before. But the answers are still elusive, and in recent years the deforestation issue has appeared on the intellectual agenda of thousands of the world's opinion leaders, scientists, and academics.[38] This spread of awareness, primarily since the late 1970s, may in itself be considered an important historical development of the late twentieth century.

As attention has grown and in some quarters matured, so has the definition of deforestation.[39] Secondly, analysts are now careful to distinguish unplanned (or uncontrolled) deforestation from that which is planned (controlled), even if they frequently are less careful to clarify from whose viewpoint these perspectives are defined. Additionally we are becoming more analytical about the different kinds of problems —or classes of effects—arising from deforestation.[40] This in turn ultimately should add precision to our statement of purpose and choice of methods for measuring it.

The State of Knowledge

What then is our state of knowledge on deforestation generated by this century's international commerce in tropical hardwoods? At the risk of introducing much generalization, the following thirteen points briefly highlight a few current perspectives.

1. Selective logging of unmanaged closed forests has transformed undisturbed forests into logged-over forests, but this has stopped far short of "deforestation." Intensive logging of some tracts and species has resulted in forest "degradation," but only in the final stages have the processes of degradation resulted in deforestation.[41]

2. The most extreme forms of degradation leading to deforestation occur when logging works in concert with expansion of agricultural cultivators. Here the contribution of logging to forest degradation is multiplicative, not merely additive, because of the entry to the forest created by road building.[42]

3. Rates of global tropical deforestation estimated only a few years ago now appear too high in view of the better data assembled by FAO and UNEP in their Global Environment Monitoring System.[43] Accord-

ing to the FAO/UNEP data, logging in the tropics penetrates about 4.4 million hectares of previously undisturbed productive closed forests each year. This amounts to 0.65 percent of the total land area so classified. The area opened by logging compares with 7.5 million hectares of closed forests cleared each year for agricultural and other purposes.[44]

4. Despite the enormous amount of effort required to make the global estimates cited above, they are not particularly useful. Rather, the deforestation impacts are best defined and assessed at the levels of countries, subnational regions, or even provinces.[45]

5. Through the long sweep of time, tropical forests recede in some regions and advance in others. We should look back to periods well before the twentieth century to see the ebb and flow of tropical forests in sharper perspective.[46]

6. Even if not renewable, most tropical forests are replaceable.[47]

7. At this point in the twentieth century, deforestation pressures (of all kinds) appear to be least severe on the tropical forests that are biologically the most fragile, that is, the undisturbed forests of the humid tropics.[48]

8. Whatever the ultimate judgment will be on the impact of deforestation attributable to tropical hardwood exports in the twentieth century, much of the logging is still too recent for consequences to have appeared. Assuming that world exports of tropical hardwoods averaged 5 million cubic meters annually for 1900 through 1945 (and this is probably a generous estimate), the cumulative volume of exports for the decade 1971 through 1980 was roughly equal to the cumulative volume for the entire previous 70 years of the century.

9. Technological advance has been one of the primary factors shifting foreign demand and the locus of timber cutting rather dramatically through time. As new technologies (and markets) have arisen and spread, logging enterprise has shifted among different tropical zones to keep abreast of what can be sold. Thus until synthetics gradually displaced them, dyewoods and tannin woods contributed to make parts of Latin America the leading exporters of tropical and subtropical timbers in the early years of the century.[49] On the other hand exports from Southeast Asia were primarily teak because the virtues of lauan were not fully appreciated until the age of plywood.

10. Not only technological progress, but also a number of other determinants have shifted the geographical origins of timber cutting for export. Among them are changes in timber availability, competing demands for timber by local economies, and the development of trans-

portation and other infrastructures. Indonesia, currently the world's largest exporter of tropical hardwoods (valued at about $2 billion annually), did not enter the timber trade in earnest until after 1965. While exports from Indonesia expanded rapidly into the 1970s, those from the Philippines peaked in the late 1960s and then contracted. In British Africa, the leading exporters before World War II were Nigeria and the British Gold Coast (Ghana), two regions whose export levels dwindled considerably after the mid-1960s. In Francophone Africa, the French Congo accounted for more than half of timber exports circa 1920, while the Ivory Coast was a distant second. In the early part of this century such Caribbean islands as Haiti, Jamaica, and the Bahamas were net exporters of wood.

11. Even though most of the export timber trade is concentrated in the humid tropics, it has been argued that the more limited vegetation removals in and at the fringes of the tropical dry forests are qualitatively the more critical.[50]

12. To the extent that there is export-related deforestation, much of it is explained by politics and policies in the producer countries.

13. It is now Japan whose imports overshadow those of either Western Europe or North America in generating a demand for timber cutting in the tropics.

Each of these points requires elaboration and qualification. However, in the remainder of this section my objectives are to focus briefly on the last point only.

Japan Swings a Mighty Axe

Japan currently imports roughly half of the world's volume (roundwood equivalent) of traded tropical hardwoods, a volume exceeding that purchased by Western Europe and North America combined. Yet in the course of this century, Japan's ascendency in the trade is of very recent origin.

In 1950 Japan accounted for less than 4 percent of the world's tropical hardwood consumption.[51] Going back to the years before and after World War I, Japan was still a net exporter of wood.[52] But with an eye toward the future, Japan apparently was studying the possibilities of using tropical timbers as early as the 1920s or before.[53]

Some analysts would hold that Japan's current forest and trade policies are responsible for a large part of export-related deforestation in the tropics. Japan would seem to be holding back its own reserves of timber, shifting timber cutting (and its environmental consequences) to where wood can be supplied more cheaply.[54] Whether or not we assign Japan a moral responsibility in this question, an immediate practical implication is that a boycott of tropical timbers led by the United States or Europe would not be effective.[55]

The moral question is, of course, an intriguing one that may capture increasing political attention in the years ahead. Japan's dominance has been concentrated in the dipterocarp and other forests of the Asia-Pacific region; Western Europe still swings the biggest axe in the tropical moist forests of Africa. Even if smaller volumes of timber are exported from tropical Africa than from tropical Asia, does this leave Europe any less a party to deforestation than Japan (and the United States)? Is the deforestation impact simply a matter of adding roundwood equivalents? Or does it extend to a careful consideration of forest types entered, inventory of resources in potential jeopardy, and character of logging practices?

My argument is that we have not yet developed enough of an analytical framework to lay blame on one trade flow or another as being the more destructive to tropical forests. Sheer appetite for import of tropical wood is not in itself a valid indicator of either the extent or seriousness of deforestation shifted to the tropical timber-producing countries. A useful beginning to establish the necessary evaluative framework is to inventory the tropical forest types and regions of the world most at risk,[56] reckoning moral and political responsibility when imports of tropical woods can be traced to extractions from those particular forests.

Closing the Century

How will we close the century? What is the next chapter? The next chapter is difficult to write—possibly containing an element of mystery—in light of the profound structural change affecting the world trade of tropical wood since the mid-1970s.[57] As world trade has been slowed by one world recession (1974 to 1975) and then another (1980 to 1983), the forecasts of future trade in tropical wood have become ever more difficult to make. In such key consumption regions as Western Europe, the long-term demand trend may have already adjusted downward.[58] Not only for Western Europe but also at a worldwide level, most pre-1974 projections of import demand for tropical wood are proving high.[59] It could be argued that the decline is more complex and far-reaching than that which accompanies an ordinary contraction in the business cycle.

One recent study foresees Japanese imports of "South Seas timber" (i.e., timbers in the lauan group) diminishing from 19.9 million cubic meters in 1986 to 17.3 million cubic meters in 1996. Nevertheless, forecasts of demand for timber products by the Japanese government have been consistently high. Since 1973 economic growth in Japan has generated increasingly less than proportional demand for timber and wood products. Japanese population growth and household formation have been slowing; construction costs as a proportion of personal income have been rising rapidly; and nonwood housing has taken an increasing share of the Japanese housing market.[60]

This downward adjustment in Japanese demand converges in time with Indonesia's post-1980 shift from exporting logs to exporting plywood. The prohibition of log exports from Indonesia has caused severe repercussions in log-importing countries like Japan, Taiwan, the Republic of Korea, and Singapore. Many wood-processing plants in these countries have closed, while others are attempting to substitute softwoods (mainly from temperate zones) for the tropical hardwoods.[61] Indonesia's forest policy, therefore, is likely to generate continuing repercussions for the quantities, forms, and directions of forest products trade in many parts of the world.

In sum, with increasing indications that aggregate demand for tropical timber is passing through a critical transition period the 1980s mark an epoch of fundamental structural and locational change for world trade in tropical hardwoods. Looking ahead to the relatively short period between now and the year 2000, the level and composition of tropical hardwood exports will depend on a complex web of factors in addition to those specific to individual countries like Indonesia and Japan.

A primary factor is surely the strength of competing demand in some of the tropical timber-producing countries themselves.[62] A second is the direction and magnitude of changes in tariff and nontariff trade barriers in the importing countries together with the intent and effectiveness of agreements reached in UNCTAD.[63] A third is the will and capacity of a great number of organizations and governments to work together rather than as isolated entities.[64]

Important also will be the near-term progress in timber utilization technologies, both in the tropical forests as well as in the processing plants. Receiving particular attention—by environmentalists and industrialists alike—will be the rate of any-tree logging and chipping of tropical timbers. While still limited in extent on a world scale, few observers miss the potential significance of this technology.[65] Still another factor will be the level of achievements during the next 15 years in the marketing of logs and processed woods of mixed species and secondary species. Related to this question of economic supply is the changing availability and cost of timber from tropical plantations vis-à-vis the natural forests.

Last but most importantly, timber exports from the tropics will be governed by the content and strength of the institutional controls (or lack of controls) on forest management and protection in the producer countries. This refers in particular to timber pricing systems, contract methods, and the mix of public and private enterprise for timber cut on government lands. It also refers to laws and regulations, enforcement, and investment to protect and replace the commercial timbers.

At present many if not most of the governments of the tropical countries underprice the timber they sell on forest concessions (i.e., sell at a price less than the full cost of timber replacement), administering a patchwork of royalties and fees that frequently work at cross-purposes with respect to progressive timber utilization and forest management.[66] Then too, there is the problem of weak and understaffed forest services, and their relation to the "malpractices" that occur on the timber concessions.[67] The "kleptocracy"[68] within and above the level of the forestry agencies could be expected to find obvious attraction and challenge in the revenues flowing from timber-producing forestry sectors. In sum, much can and should be done within the political spheres of the timber-producing countries themselves that would have major implications for timber exports and their linkages to deforestation.

8

THE NORTH AMERICAN-JAPANESE TIMBER TRADE: A SURVEY OF ITS SOCIAL, ECONOMIC, AND ENVIRONMENTAL IMPACT

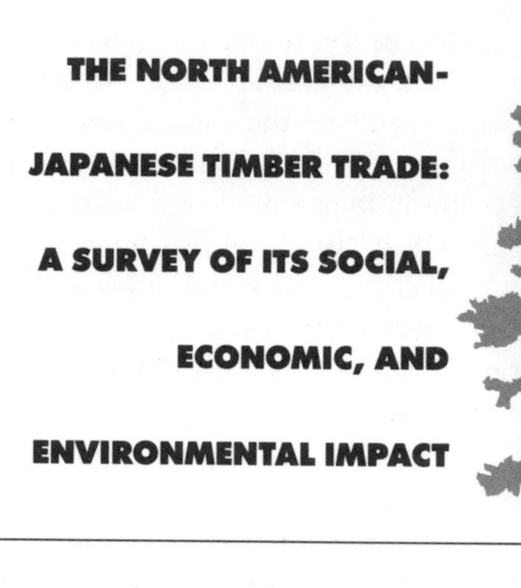

Thomas R. Cox

A major trade in forest products has developed between North America and Japan since about 1950. There was a trickle of such commerce as early as the 1880s, and a large enough flow by the 1920s to become the focus of considerable attention, but not until the period of rapid economic growth that commenced in Japan during the Korean War did the trade truly burgeon.[1] When it did, its social, economic, and environmental effects led to much debate and criticism on both sides of the Pacific.

That these shipments should have generated concern seems, at first glance, not nearly so surprising as that they should have occurred at all. Japan is a well-wooded land; among temperate zone countries, only Finland has a larger percentage of its territory in forests. Alaska, British Columbia, and the Pacific coast states of Washington, Oregon, and California are also richly endowed with commercial timberland. Moreover, the forests of Japan and the west coast of North America are comparable, producing similar species used for similar purposes. A major commerce in timber between Japan and the countries of Southeast Asia developed too. This comes as no surprise, for Southeast Asia has tropical forests that yield woods quite different from those of Japan, but a major transpacific timber trade seems odd indeed. Closer examination explains the apparent anomaly.[2]

Japan and the Transpacific Timber Trade

Accessible Japanese forests were badly overcut during the hectic modernization of the Meiji period (1868 to 1912). Timber supplied ties for railroads, timbers for bridges, and building materials for houses and factories; it also was used to provide exports to Korea and China so as to earn badly needed foreign exchange with which to buy machinery and other fruits of Western technology needed to bring the schemes of Meiji planners to fruition.

The situation was not much better in the years that followed. Imports from North America, Siberia, and elsewhere grew during the 1920s and 1930s but were too small to prevent further overcutting (see table 8.1). Professional forestry advanced rapidly in Japan during the period, bringing with it increased efforts at improving forest utilization. This too proved inadequate for the needs at hand, while the opening of new areas to logging through improved methods of transportation, supplies from colonial possessions, and the development of long-isolated Hokkaido served temporarily to mask the dangers of what was transpiring.[3]

World War II created fresh problems. There was extensive cutting to supply building materials, containers, and other items needed in the war effort, and to furnish the charcoal used to cook, heat homes, and fuel numerous small manufacturing enterprises. Wartime diversion of manpower and funds from forestry to military demands contributed to poor harvesting practices, to a sharp decline in replanting after harvest, and in other ways to a failure to husband stands as well as they had been previously.[4] Postwar efforts to correct the problems spawned by overcutting and mismanagement were hampered by a number of factors. Unable to meet booming demand for timber from domestic stocks, the Japanese turned increasingly after 1950 to foreign sources of supply (see figure 8.1).[5]

By 1972 over 50 percent of the softwood timber being used in Japan was coming from abroad—most from the west coast of North America, but a considerable quantity also from Soviet Siberia. The effect of this on Japanese timberland and timber owners was uneven—depending on local forest types, their nearness to ports and major markets, and the nature of the operations involved. Soviet timber was nearly all imported through ports on the Sea of Japan coast; it competed primarily with karamatsu (*Larix leptolepis*), a species of marginal commercial

Table 8.1 Lumber Production in Japan, 1880–1960 (thousands of cubic meters, roundwood equivalents)

	Exports	Imports	Domestic Production
1880	17	4	7,474
1885	19	3	8,135
1890	32	3	7,354
1895	34	13	9,782
1900	248	40	13,494
1905	411	55	14,046
1910	892	61	15,599
1915	953	83	17,023
1920	788	533	21,021
1925	689	6,195	20,662
1930	696	5,500	18,721
1935	1,269	3,827	26,339
1940	4,130	1,776	40,881
1945	300	13	25,235
1950	104	105	43,722

Source: K. Ohkawa, M. Shinohara, and M. Umemura, *Choki keizai tokei: Suikei to bunseki* [Long-term Economic Statistics: Estimates and Analysis] (Tokyo: Toyo Keizai Shinposha, 1966), 9:238–39.

value that was common in Hokkaido and high mountain areas further south. On the other hand, Douglas fir (*Pseudotsuga menziesii*) and western hemlock (*Tsuga heterophylla*) primarily entered through Japan's east coast ports, which were nearer the main timber markets, and competed directly with sugi (*Cryptomeria japonica*) and hinoki (*Chaemaecyparis obtusa*), the primary commercial species of Japanese softwood.[6]

The loss of so much of the domestic market to foreign suppliers was on the whole traumatic. Douglas fir became the price leader in the Japanese timber market, and its impact was felt throughout the country. Even in the interior of isolated Shimane prefecture, Douglas fir was, as one timber producer there put it, "a terrible headache."[7]

Prices for domestic logs lagged consistently behind those for imported ones (see figure 8.2). This was but one manifestation of the fact that forest industries were less competitive with their American counterparts than were most Japanese industrial enterprises. Reflect-

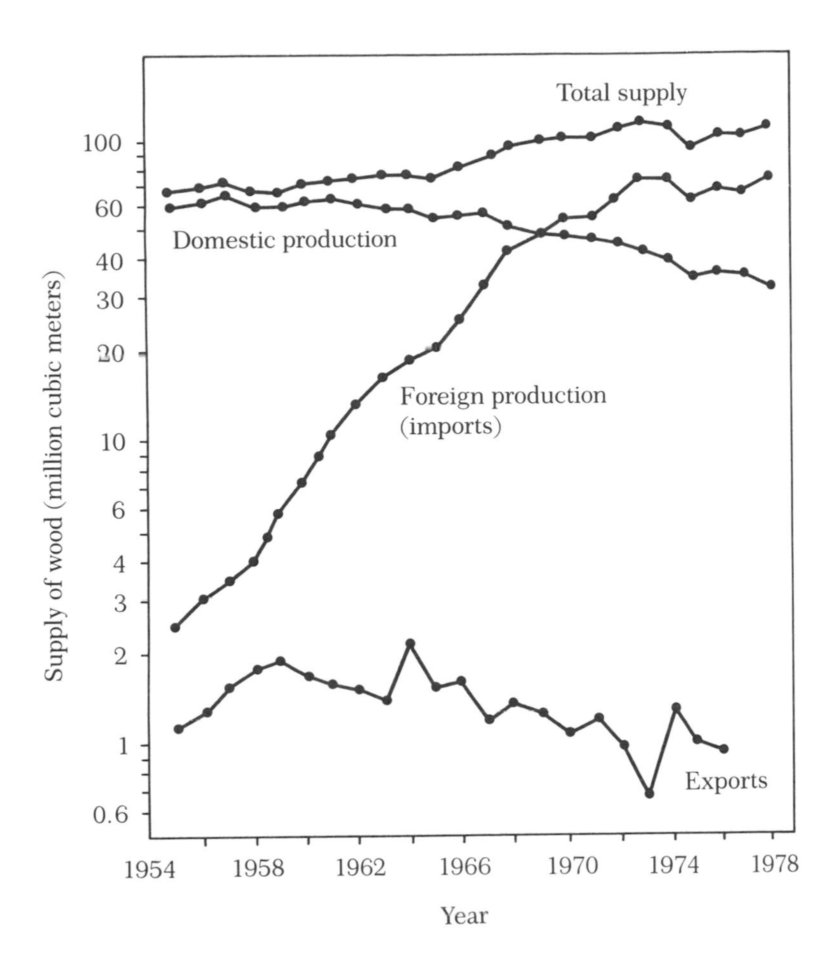

Figure 8.1 Supply of Wood and Wood Products in Japan, 1954–1978.
Source: Japanese Forestry Agency.

ing this, wages in forest areas dropped further and further behind those in other areas and types of employment. Forest-related jobs became more scarce, and mountain villages faced an accelerating out-migration to the cities even more extreme than that for rural Japan as a whole. Initially Japanese officials spoke of these areas as having surplus population and applauded the movement to urban areas, but by 1964 they were referring instead to the depopulation problem of mountain areas. Much of this emigration was among younger

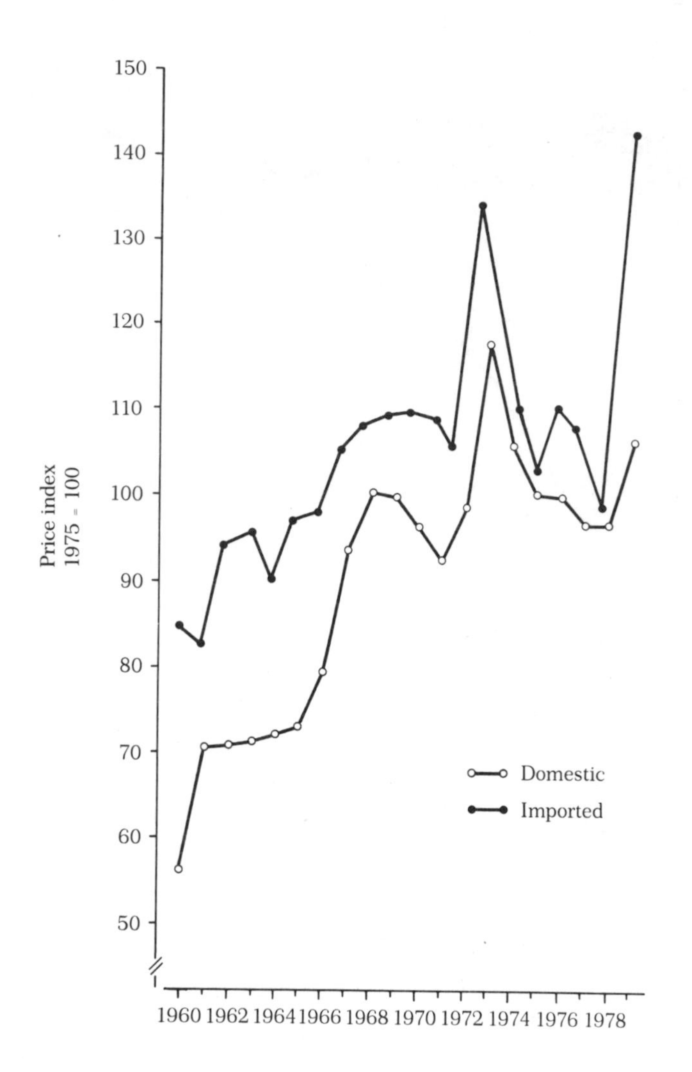

Figure 8.2 Price of Domestic and Imported Timber in Japan, 1960–1978. *Source:* Adapted from Mori Yoshiaki, "An Econometric Model for the Japanese Timber Market," in *The Current State of Japanese Forestry*, ed. Handa and Nomura, 60.

Table 8.2 Percentage of Forest-owning Japanese Households with Holdings of over 1 Hectare by Class Size, 1960–1980

	Size of forest holdings (ha.)				Total number	
Year	1–5 ha.	5–20 ha.	20–100 ha.	100 ha.	(%)	(in thousands)
1960	77.3	19.3	3.1	0.3	100.0	1,132.9
1970	74.5	21.3	3.9	0.3	100.0	1,144.5
1980	74.0	21.5	4.2	0.3	100.0	1,112.6

Source: Kumazaki, "Trends in Small Private Forests of Japan," 79.

workers, leaving behind in the forests a work force growing steadily older. Forest management and utilization patterns changed as these shifts occurred.[8]

Concerned over the social costs of the depopulation of forest villages, desirous of decreasing dependence on foreign sources of supply (especially after the trauma of the Arab oil boycott of 1973), and wishing to placate Americans who were bitterly protesting the effects of the growth of log exports on the United States, the Japanese government stepped up efforts to increase domestic timber production and make Japanese timber competitive with imports. New laws were drafted, studies of the nation's forests and forest industries were launched, improved roads were built into formerly isolated forest areas, and appropriations for the Forestry Agency were increased.[9] On the other hand officials remained lukewarm toward American and rural Japanese pleas for tariffs or quotas to stem the influx of logs. Sawmills cutting imported logs were an important source of employment, the owners of large mills on the ports lobbied effectively to keep their access to raw materials unrestricted, and the cheapness of the resulting product encouraged domestic building that the government desired. Under the circumstances, efforts to encourage domestic timber production enjoyed only limited success. Still, as stands planted in recent years mature, dependence on imports is expected to stabilize or even decline.[10]

Other factors were also at work, making the problems faced by Japan's forest policy makers, forest owners, and domestic timber industry intractable. Forest ownership is and has long been highly fragmented (see table 8.2). In 1960 the average family forest holding was some 2.4 hectares, while company holdings averaged approximately 225 hectares per unit. Only 15 percent of company holdings were over

50 hectares. The holdings of villages and other local public entities also tended to be small. To complicate matters further, the tiny public and private holdings were frequently divided into separate tracts. National forests, although much larger and less fragmented, were usually located in relatively inaccessible, mountainous areas. There has been some consolidation in the years since 1960, but the basic pattern remains.[11]

Attitudes as well as ownership patterns stood in the way of the adoption of forest management practices that would maximize production. Most forest owners and their families were not primarily dependent upon forests for their support. Forests represented a secondary, or supplemental, source of income. As one authority has noted, many looked on their timberlands as a treasure box to be tapped at a time of special need—to pay a son's university expenses or to provide a daughter's dowry, perhaps—rather than as a resource to be managed for maximum long-run return. Forests harvested for such reasons were often cut at a point when professional foresters would consider harvesting undesirable.[12]

Nor were the forest extension agents, provided in some areas to aid smallholders in managing their forestlands, much help. Trained in forestry, most had little knowledge of farming, which was the basic support of most owners of small tracts. Narrowly educated, the majority were unprepared to provide the kind of advice that would have helped to bring about the integration of forest and agricultural holdings into a viable whole while at the same time increasing the production of forest products.[13]

The cost of bringing timber to commercial size in Japan is higher than in the Douglas fir region of North America. Rugged terrain makes replanting after harvest difficult and expensive. The climate and Japan's major forest soils are such that in many areas rank growth of brush and weeds springs up and must be periodically cut back by hand to free young trees. One recent study suggests that it costs several times more to regenerate stands of sugi in Japan than it does Douglas fir in western Washington. The advantage that the American species holds over sugi, its main Japanese competitor, is increased by other factors, especially the high internal transportation costs in Japan. Even when stumpage costs are comparable, the expense of transportation frequently adds so much to the price of domestic logs that they lose out in major urban markets to imported logs, brought there by less expensive maritime means. Revaluation of the Japanese yen in

1973 exacerbated the problem by making American logs an even better bargain than before.[14]

An additional problem is that much of Japan's forestland is not stocked with the species of greatest commercial value. Vast acreages are given over to karamatsu, which yields lumber that tends to split and warp, and akamatsu (Japanese red pine, *Pinus densiflora*), which is useful for erosion control, grows rapidly, and can flourish in poor, dry soil but which seldom has a straight trunk and is thus costly to transport and to manufacture into lumber and seldom yields large-dimension stock. Moreover, pine stands have been swept in recent years by infestations of a nematode (*Aphelenchiodes xylophilus*) carried by a species of longhorn beetle (*Monochammus alternatus*) apparently introduced into Japan on pine logs from the United States. An estimated eight million trees per year are killed by pinewood nematodes, devastating the warmer, drier forests of such places as Hiroshima and Okayama prefectures.[15]

Karamatsu and akamatsu are not the only problem species. Extensive stands of mixed hardwood also exist, especially in the southern half of the country. These forests long had great worth. Entire villages depended upon them, producing charcoal to meet the nation's fuel needs. But beginning during the 1950s, as Japan shifted from charcoal to petrochemical fuels for nearly all its domestic cooking and heating needs, the value of hardwood stands plummeted. Most Japanese broad-leafed species have limited value to lumber producers. Only in areas such as Hokkaido, where rising demand for pulpwood took up the slack, was the impact of the fuel revolution muted.[16] Elsewhere conversion to sugi, hinoki, and other valuable lumber species was soon under way, but it will be years before recent plantings of these varieties will be ready for harvest. The costs of replanting after harvest or of conversion to more desirable species were often difficult for smallholders to meet, although various government programs designed to help with these expenses have alleviated this problem somewhat.[17]

Private forests were the most afflicted by problems hindering proper management, but publicly owned forests faced them too. Public holdings were larger (see figure 8.3), not so apt to be managed as a "treasure box," and not as likely to be overlooked while their owners focused on managing their farmland. But like private holdings, public forestlands have suffered from the effects of torturous terrain; from high internal transportation costs; from the need for expensive replanting, brush and weed clearing, and converting to more commercially valu-

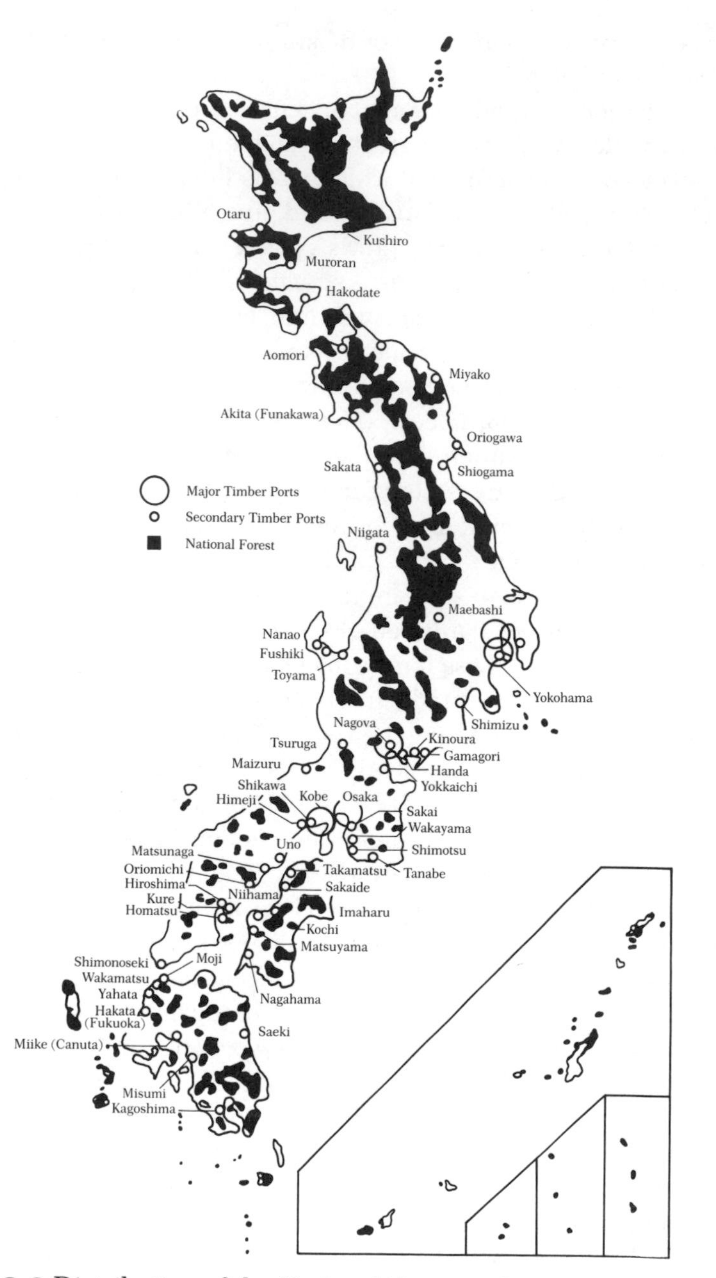

Figure 8.3 Distribution of the National Forest in Japan. *Source:* Japanese Forestry Agency.

*Not including ports importing only tropical woods.

Table 8.3 Number of Sawmills in Japan, Classified According to Power Use, 1960–1977 (kilowatts)

Year	Under 7.5 kw.	7.5–22.5 kw.	22.5–37.5 kw.	37.5–75 kw.	Over 75 kw.
1960	4,178	14,807	5,594	3,046	782
1965	2,625	11,495	6,646	6,874	1,788
1970	1,395	8,125	6,704	6,349	3,368
1975	699	6,201	5,579	7,035	4,815
1976	599	5,878	5,463	7,159	4,982
1977	511	5,442	5,396	7,172	5,126
1978	444	5,077	5,302	7,215	5,200

Source: Takahashi, Shiota, Tanaka, and Schneiwind, "Recent Trends in the Wood Industry of Japan," 22.

able species; and from the results of earlier overcutting. Moreover, the managers of publicly held forests have been subjected to pressures seldom felt by smallholders. They are regularly urged to manage the stands so as to protect watersheds, to preserve opportunities for recreation, and to maintain greenbelts and in other ways take aesthetic considerations into account.[18]

Forest management, both public and private, has suffered from the small size of Japanese sawmill and logging operations. Whether carried out by companies or individual entrepreneurs, operations have nearly always been too restricted to allow significant economies of scale (see table 8.3). Occupation authorities sought to overcome this problem through forest owners' cooperatives. These have continued since the Occupation, but cooperatives and other attempts to increase the competitive position of domestic lumber products by encouraging larger, more rationalized operations have had limited success. The nation's jumbled terrain results in fragmented stands that are difficult to manage as large units. Moreover, individual, family, and local traditions die hard in Japan and, hanging on, often hamper efforts of planners from the Forestry Agency and cooperatives to institute management and utilization compatible with present-day realities and needs.[19]

Imported logs have complicated matters further, disrupting established price and market mechanisms in a variety of ways and introducing elements of the unfamiliar into the picture. Among the changes encouraged by imports was a system of marketing logs through wholesale middlemen at central markets. The new timber markets, better suited for supplying large mills and for handling the large consignments of logs that importers could make available, posed major problems for smallholders with timber to sell. Recognizing that imports were exacerbating their problems, the smallholders repeatedly petitioned the government for protective tariffs or quotas on imported logs and lumber, but with little success.[20]

The social and economic changes taking place in forest areas and among forest-dependent people in Japan are by no means all caused by imported logs and the associated decline of the market share of domestic timber. In many ways these changes simply reflect those in Japanese society and business as a whole: centralization, economic integration and rationalization, and internationalization and urbanization of the economy. Change would surely have come to the forests of Japan without any log imports at all. Nonetheless, timber imports have added to the problems of forest owners, and of forest-dependent people and firms, and have accelerated the pace of change. The rapid loss of a major share of the market has generated problems demanding immediate attention. Because these problems came on so suddenly, adjustments to them have been especially jarring.

The environmental effects of log imports and associated changes have been less noted than the socioeconomic effects, in part because environmental concern is less developed in Japan than are its social and economic equivalents.[21] Furthermore, it may be years before some of the environmental results of these changes are manifest.

Still, some such effects are already evident and have been commented upon by both Japanese and foreign observers. Plantation forestry in Japan goes back to the Tokugawa period, but monocultural plantings and even-aged stands are becoming more common than ever. The shift to sugi, hinoki, and other high-value coniferous species that can reasonably be expected to compete successfully with imported logs is reducing the environmental diversity in Japan's forests. Pinewood nematodes and the almost complete cessation of commercial charcoal production have worked, and are continuing to work, in the same direction.[22]

Meanwhile, the decline in the real price of domestic timber and the increasing cost and scarcity of labor in mountain areas have resulted in many smallholders giving less attention than formerly to their forestlands. Many have sold off their holdings, seeing little chance of future profits from them. Just who has been buying these lands is not entirely clear, but it would appear that many buyers are speculators anticipating good future returns from sale of the timber—or perhaps from land sales to real estate developers. As absentee owners, such buyers could not give the care to their forestlands that village smallholders long provided, nor would they be as apt to tap grass, mushrooms, and other minor forest products. The shift to less labor-intensive management of stands that has resulted from out-migration from mountain villages has had the same kind of effect even where changes in title to the land have not occurred. Thus, while the overstory is becoming less diversified as a result of monocultural plantings, the understory of Japan's forests may be becoming, through neglect, more dense, tangled, and complex. It also seems that long-accessible areas of marginal commercial value for the production of timber will, for the same reason, revert to a more nearly natural state than they have been in for years—perhaps even for generations.[23]

Continuing demand for domestic logs of large enough size to compete successfully with imported ones should also have its environmental fallout. On the one hand, Japan's remaining old-growth forests are apt to dwindle steadily in the face of cutting pressure, for in spite of the difficulties presented by the country's rugged terrain, modern road building and logging methods leave few places inaccessible. From now on, pressure for the protection of recreational resources, watersheds, and amenity values, rather than isolation, seems destined to provide the bulk of what protection old-growth forests get. On the other hand, the availability of imported woodstuffs will mean that there will be alternative sources of supply that may help those who would protect remaining old-growth forests to gain a more open hearing than they would otherwise get. Thanks to imports, policy makers will not have to make so clear-cut a choice between economic and noneconomic needs. All in all, it is not yet evident whether in the long run imports will do more to protect Japan's old-growth forests or to speed their demise. Because Japan's forests are themselves quite diverse, it may well be that the record on this, as on so many things, will be mixed. Old-growth stands in Hokkaido may well have a far different future than those in interior Kyushu, far to the south.[24]

Timber imports have exacerbated problems for forest owners and for workers and lumbermen dependent upon domestic stands; at the same time they have unleashed forces that may bring extensive environmental change. Nevertheless, they have on the whole been welcomed by both industry and government. Planners have recognized that the nation's growing economy will continue to require more wood than domestic forests can supply, while businessmen have found importing timber to be profitable. Both the government and private sectors have invested heavily in facilities to serve the trade.

Recognizing that the importation of huge quantities of logs through the few major ports that long handled the bulk of the nation's international commerce would result in excessive crowding of harbors, governmental agencies spent large sums to improve a number of secondary ports. Wakayama was receiving shipments by 1958; a host of other secondary ports followed (figure 8.3). For their part, in order to handle imported timber, private investors erected large, modern sawmills and processing plants near both major ports and growing secondary ones. The result has been an increase in the average size of sawmills and other wood processing plants, greater concentration of facilities in port areas, and more economic development near a number of secondary ports.[25] While domestic forest owners and woods workers have been suffering from log imports, many a businessman and urban resident has benefited from them. By encouraging such changes in order to aid economic development, government policies have in fact encouraged the depopulation of forest areas in spite of all the official statements deploring the movement of people from mountain areas and the efforts of the Forestry Agency and others to stem it.

North America and the Transpacific Timber Trade

The impact of the North American-Japanese timber trade has also been great on the eastern side of the Pacific. In the Northwest, especially, the social, economic, and environmental effects of log shipments to Japan have been profound. In North America, as in Japan, social and economic results have received more attention than environmental ones.[26]

Initially, shipments to Japan were universally welcomed in the forested regions of western North America. Chronically overbuilt, the West Coast's lumber industry long depended on maritime exports to consume a portion of its cut, to provide a cushion of demand when consumption at home dropped, and to supply outlets for grades and sizes in little demand in the United States. The so-called cargo mills, plants that dispatched their cut to market by sea, laid the foundations of the area's lumber industry during the half-century that followed the California Gold Rush.[27] Lumber shipments to Japan began in the 1880s but long remained only a tiny portion of this ocean commerce. During World War I, high maritime freight rates brought shipments by sea to a standstill, but soon thereafter they began anew. The Pacific coast's lumber industry was badly depressed during most of the 1920s. Consumption in Japan offered at least some relief, and when the Tokyo-Yokohama area was devastated in the great Kanto earthquake/fire of 1923, cargo mill operators hoped that the demands of reconstruction would prove a turning point. In fact, they provided only limited, temporary relief.[28]

Gradually shipments to Japan became more important. During the twenties a number of operations emerged that specialized in the shipment of so-called "Jap squares." Although there was a modicum of manufacturing involved in the production of squares, they were in fact little different from the logs whose shipment was to become so important during and after the 1950s.[29]

Japanese demand for American timber increased rapidly during the 1930s. In 1929 a total of only 57.5 million board feet of Douglas fir logs were exported. By 1936 the figure had risen to 132 million board feet, 74.6 million of which went to Japan. In the latter year 16 million board feet of Port Orford cedar (*Chamaecyparis lawsoniana*) logs were also exported, 11.7 million of them to Japan.

These shipments were welcomed at first, as sales anywhere were wont to be in the depression-plagued 1930s. Protests soon began to arise, however. One cause was undoubtedly the growing estrangement between the two countries, but other factors were at work too. Most of the Douglas fir being sent to Japan was in the form of peeler logs used to make plywood. Softwood plywood manufactured in Japan from American logs was capturing an increasing percentage of the international market, thanks to then-low Japanese wage scales, while many plywood plants in the Pacific Northwest stood idle. Exporting Douglas fir peelers was the equivalent of exporting jobs, critics argued. Commerce Department figures, they pointed out, showed the labor content of a thousand square feet of Douglas fir plywood was $19.12, only $2.11 of which came through logging operations. Therefore, critics claimed, the exportation of peeler logs meant a loss in wages of $17.01 for every thousand square feet of plywood they yielded. Moreover Japan had no major domestic sources of peeler logs; without imported Douglas fir,

America's overseas markets for plywood would be safe from Japanese competition.[30]

The situation with regard to Port Orford cedar was somewhat different. Demand for this species was growing because its wood was one of the few materials then available suited for use as battery separators. In Japan it was also used in residential construction. However, stands were quite limited and were being cut faster than they could grow. There was increasing fear that Port Orford cedar would soon cease to be available.[31]

Responding to the concerns of their constituents, Senators Rufus Holman (R., Ore.) and Lewis Schwellenbach (D., Wash.) introduced legislation into Congress in 1939 that would have virtually halted the exportation of Douglas fir peelers and all Port Orford cedar logs. Senator Charles McNary (R., Ore.) offered an amendment also to embargo Sitka spruce (*Picea sitchensis*) logs, valued in the construction of airplane frames. Economic arguments combined with those of conservation and national defense as champions of the bill sought its passage. Support came from a variety of sources. President Franklin D. Roosevelt expressed sympathy for the bill; Governor Clarence D. Martin (D., Wash.) came out in support; both the American Federation of Labor and the Congress of Industrial Organizations went on record in favor of passage; and the lumber industry's trade press lauded the measure's virtues. In spite of the objections of Secretary of State Cordell Hull, an advocate of reducing restrictions on trade, the Senate soon passed the bill. However, it languished in the House, pushed into the background by the more pressing concerns of approaching war.[32]

Prewar arguments quickly reemerged when shipments to Japan began expanding during the 1950s. With this expansion, the protests took on even greater force than before. On the one hand, available stands were now more genuinely inadequate for all the demands placed on them. On the other hand, with new ships and technology for handling materials, the exportation of squares was largely replaced by that of whole logs. The argument that shipments represented an exporting of jobs from the region thus took on greater force than ever. By the mid-1960s, log exports had become perhaps the single most hotly debated and deeply felt issue in the Pacific Northwest.[33]

Log exports were to continue to be a central issue on the West Coast for years—indeed, until high interest rates brought a precipitous drop in the number of housing starts with the result that demand for lumber plummeted, bringing mill closures and near-depression to many parts of the region during the late 1970s. In the face of these new economic realities, log exports began to appear more acceptable, offering a level of employment in the woods and ports of the region that domestic demand could not. For the owners of extensive timber stands—including The Weyerhaeuser Company and the State of Washington—shipments to Japan promised a way of ensuring an acceptable level of sales and income in spite of the state of the domestic economy. Some who had once bemoaned log shipments to Japan now found themselves regretting that economic conditions had led to slumping demand for lumber in Japan too, a slump that reduced log shipments and prices below what they would otherwise have been (see figure 8.4). Reflecting this shift, in February 1983 Congressman Donald Bonker (D., Wash.), long a foe of log exports, declared flatly that the issue of limiting log exports was dead. The fact that the Japanese were no longer significant exporters of softwood plywood in competition with Americans helped to defuse the issue. Meanwhile Canadians, faced with slumping demand for lumber in Japan and elsewhere, began to talk of exporting logs as well as sawnwood to buttress the economy of British Columbia.[34]

Most log shipments to Japan during and after the 1950s came from a handful of ports—most notably from Coos Bay in Oregon, and Grays Harbor, Tacoma, and Longview in Washington. But their impact was felt far into the interior. As competitive bidding between American and Japanese buyers drove up the price of logs in the vicinities of these ports, domestic operators turned increasingly to interior stands for the sawlogs to keep their plants running. As they did, stumpage prices rose there too. Government action banned the exportation of logs cut on federal land, but while this insured a supply of sawlogs to domestic mills, it failed to hold down log prices. Certain tracts were set aside for bidding only by small operators in an attempt to insure that federal timber would be available to them at reasonable prices, but the practice made little real difference. By reducing the supply of nonfederal timber available to domestic buyers, log exports insured that overall competition at Forest Service and other federal timber sales would increase. Stumpage prices were soon climbing sharply as far inland as Idaho and Montana.[35]

The impact on stumpage prices of Japanese buying and of the dispatch by American firms of logs that would later be sold in Japan, either as logs or manufactured wood, was mitigated somewhat because much of the demand, especially initially, was for species little valued in the United States and because for a time after 1962 much of it came from

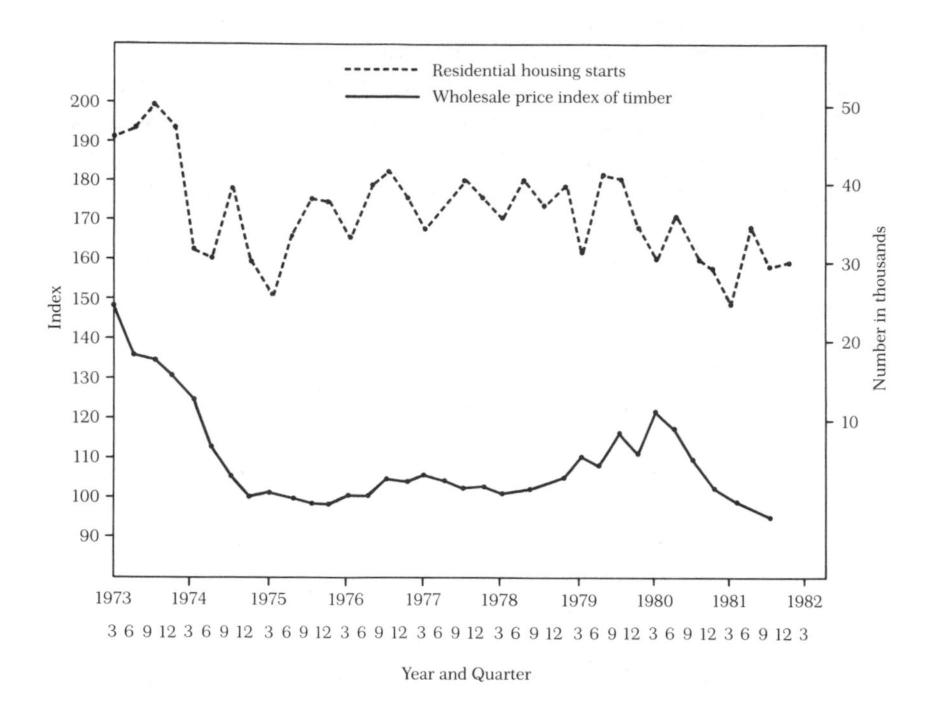

Figure 8.4 Housing Starts and Wholesale Price Index of Timber (deflated) for Japan, 1973–1982. *Source:* Adapted from Nomura Isamu and Yukutake Kiyoshi, *Timber Price Movement in Japan since World War II and Its Theoretical Causes* (Tsukuba: Ringyo Shikenjo, 1981), 12.

blowdown from the disastrous Columbus Day storm that had glutted the market with logs. Port Orford cedar had been replaced as the main source of battery separators; domestic demand had plummeted. Nevertheless, Port Orford cedar, western hemlock (*Tsuga heterophylla*), and spruce (*Picea* spp.) were valued in Japan because of their color and grain; frequently they were substituted in interior finish for traditional, but more expensive hinoki and sugi. American buyers, by contrast, were most interested in Douglas fir, redwood (*Sequoia sempervirens*), and pine (*Pinus* spp.).

Gradually the windthrown Columbus Day timber was salvaged and shipments of Douglas fir grew—partly because available quantities of the other species were inadequate to meet growing demand, partly because sharply rising prices for hemlock and Port Orford cedar made them less attractive vis-à-vis Douglas fir than they had been, and partly because Japanese consumption simply diversified (see table 8.4). A group of American representatives to the IUFRO Forest Congress, held in Kyoto in 1981, was appalled to see no. 1 peeler-grade Douglas fir logs being sawn into lumber at a tiny sawmill not over a mile from the congress headquarters. Such logs would have been used for plywood in the United States, being too valuable to manufacture into lumber with its smaller value added by manufacture; moreover, such logs were at the time in short supply on the West Coast, and many plywood mills were again shut down because of a combination of low demand for plywood and high prices for peelers.[36]

Behind the debates in the United States were internal divisions within the forest products industry, especially between small sawmills and large, integrated operations on the one hand and between old, marginal plants and efficient modern ones on the other. Small, old, or marginal operations were hurt more by rising stumpage prices than were their opposites, which were better equipped to maximize returns by utilizing all portions and sizes of the raw logs that came to them. Occasionally, timber owners or other champions of exports who welcomed the rising price of stumpage stated baldly that the country and the industry would be better off when marginal sawmills had been squeezed out. In any case, they argued, the days of such inefficient operations were numbered; by raising the price of sawlogs, exports were simply bringing closer what was inevitable anyway.[37]

But while such arguments may have made economic and even environmental sense, they overlooked social and political realities. Many of the small mills were located in single-industry towns, sometimes in towns dependent upon a single company. In such cases, the closure of sawmills had a devastating effect. Plant closures brought death or near death to a long list of communities, including Valsetz, Westfir, and Kinzua, Oregon. The concentration of forest industries in a smaller number of plants in a few favored locations (see table 8.5) could be achieved only at the cost of social and economic displacement and human suffering. Moreover, the individuals adversely affected by such shifts had political influence both as voters and in some cases as articulate spokesmen for the virtues of small-town life. Oregon's Senator

Table 8.4 Log Exports from Oregon and Washington to Japan by Species and Average Prices, 1961–1974 (MBF, log scale, and dollars per MBF)

Year	Douglas fir		Port Orford cedar		Other softwoods[a]		Total	
	MBF	Dollars	MBF	Dollars	MBF	Dollars	MBF	Dollars
1961	60,949	75.76	50,135	123.88	217,032	64.12	328,116	75.41
1962	42,630	79.09	35,458	144.34	208,567	56.08	286,655	70.42
1963	65,996	68.88	52,042	116.07	520,684	57.31	638,722	63.30
1964	64,891	69.38	40,212	132.51	634,589	64.93	739,692	69.90
1965	72,880	76.00	29,407	174.93	672,621	78.09	774,908	81.57
1966	122,723	74.04	34,141	176.41	650,831	82.39	1,023,242	84.52
1967	262,067	81.55	32,688	262.07	1,199,478	87.81	1,494,233	90.52
1968	351,307	98.31	31,112	319.84	1,477,626	100.24	1,860,045	103.55
1969	328,586	113.71	34,472	359.29	1,426,257	113.93	1,789,315	118.61
1970	430,634	120.87	41,254	329.86	1,672,444	124.57	2,144,332	127.77
1971	390,978	120.12	34,390	317.70	1,284,931	125.81	1,710,299	128.36
1972	692,308	136.08	36,907	327.56	1,661,948	138.03	2,391,163	140.39
1973	822,160	245.63	20,966	736.97	1,612,359	277.21	2,455,485	270.56
1974	638,225	278.84	17,342	1,009.12	1,320,008	283.36	1,975,575	288.27

Source: U.S. Forest Service.

Note: MBF = million board feet.

a. Largely western hemlock, but includes some true fir, spruce, etc.

Table 8.5 Number of Sawmills in Oregon by Log Consumption Capacity, 1968–1976 (millions of board feet per 8-hour shift)

Size	1968	1972	1976
Over 50	21	27	24
25–50	72	68	61
5–25	106	95	96
Under 5	90	65	53
Average size[a]	76	97	104

Source: U.S. Forest Service.
a. based on lumber production capacity, not log consumption.

Wayne Morse and Governor (later Senator) Mark Hatfield were among those especially responsive to expressions of concern from and for the residents of these endangered communities.[38]

Log exports also sparked protests by home buyers, builders, real estate developers, and planners concerned over the shortage and high cost of housing. By driving up stumpage prices, they argued, shipments to Japan increased the cost of new housing, in some cases to the point where potential buyers were forced out of the market. While such claims were of doubtful validity, because lumber and other wood products represented a relatively small portion of the cost of a new home, they had widespread appeal at a time when housing costs were skyrocketing. In response to such protests, legislators in California convened hearings aimed at bringing log exports from the state—or at least the state's own lands—to a halt.[39]

As in Japan, opposition to log shipments spawned many apocryphal tales. One rumor circulated in the Pacific Northwest held that the Japanese had large factory ships anchored off the west coast where they manufactured American logs into plywood, which was then shipped back to the United States to undersell the domestic product. Not only did no such ships exist, Japanese plywood had held a brief and relatively minor place in the American market in the 1930s and conditions had changed since; Japanese plywood made from North American logs could no longer compete in the United States market. By the mid-1960s the sheets of laminated wood being sent to the United States from Japan were veneers of tropical hardwoods (albeit for a time with a core of low grade Douglas fir). Such products provided only limited competition to American plywood, which was normally used for different purposes. If Americans were going to find a logical cause for closed plywood plants, they would have to find it somewhere other than in competition from Japan or from Japanese factory ships.[40]

No thorough comparison has as yet been done between developments in the Pacific Northwest, British Columbia, and Alaska, but preliminary investigation suggests a pattern. In both British Columbia and Alaska policies have been implemented to restrict the exportation of raw logs. The bulk of what has been shipped to Japan has been lumber and other manufactured wood products from British Columbia and woodpulp from Alaska. The production of these items generates more jobs and more added value than the log trade or its forerunner, the trade in squared timber. As a rule only the exportation of surplus logs, of pulp, or of sawn material has been allowed. Under the circumstances it is hardly surprising that in these places there has been little opposition to timber exports on the grounds that they represent an exporting of jobs. However, because the control of timberland in Oregon, Washington, and California is divided among federal, state, and private owners and because the Constitution of the United States prohibits the levying of export duties, such controls as those in force in British Columbia and Alaska have been impossible to put into operation in the "lower forty-eight."[41]

In North America, as in Japan, the environmental impact of log, lumber, and woodpulp exports is less apparent and has been less commented upon than social and economic ones. Here too it seems certain that not all the environmental effects are as yet apparent.

The exports have led to a number of developments that have environmental implications. They have resulted in increased demand for certain species, thereby encouraging logging in stands that might otherwise have been considered economically submarginal. They have caused higher stumpage prices than would otherwise have existed and thus probably led some timber owners to sell stands they would have left untapped. And they have lessened the fluctuations in timber harvest by generating demand for logs beyond that furnished by domestic consumption.

Still it would appear that the environmental impact of the transpacific wood products trade may in the long run prove to be less extensive in North America than in Japan. Exports have caused no basic changes in the patterns of timber harvest, but have simply speeded and accentuated developments already taking place. The conversion of West Coast forests from old-growth stands to managed second growth, a basic change long under way, has continued since the late 1970s despite the collapse of demand for building materials that occurred as interest rates rose sharply. Indeed, it is clear that with or without exports, the West Coast's commercial stands are destined for such conversion.[42] Similarly, the trend toward modern, integrated plants that are capable of handling small logs and utilizing all portions of the trees harvested may have been speeded by the high prices for timber that exports helped bring, but the trend was under way before exports burgeoned and would have occurred without them. Also, the inadequacies of traditional sources to meet an ever-more-literate world's rising demand for pulpwood would sooner or later have attracted paper manufacturers or those who supply them to the stands of coastal Alaska, even if Japanese investors had steadfastly ignored their presence.[43]

Log exports brought extensive timber harvesting to stands of hemlock and other species little valued in the United States, but even here they would not appear to have made much difference environmentally. The exports have probably most affected timing. With available old-growth stands of other species dwindling and with sawlogs expected to be in short supply until much more second growth timber reaches a harvestable size, it seems certain that lumbermen would have soon been turning to such stands in any case.

Of course, by increasing the overall commercial demand on the forests of the West Coast, exports have increased the difficulties faced by environmentalists and others seeking to save old-growth stands, wilderness areas, and the like. But even this seems only a matter of degree, rather than the introduction of something new. The contest between those who value the utilitarian contributions of the forest and those who emphasize the aesthetic and noneconomic values was well developed long before log exports to Japan became significant.[44]

Conclusion

In the final analysis it would appear, in spite of differences, that the trade in forest products between the United States and Japan has had remarkably similar effects on both sides of the Pacific. In both countries the economic and social impact has been more evident than the environmental impact. In both countries internal divisions over the trade have appeared. No systematic analysis of the opposing forces has as yet been done, but this preliminary survey suggests the presence of a sharply defined pattern: in both places large operations appear to have tended to profit from and support the trade, small operations to have been hurt by and oppose it; industrialists appear to have tended to favor it, woods workers and small-time entrepreneurs to oppose it; and those in port areas seem to have tended to profit from and support it, those in more isolated interior locations to oppose it.[45]

Whether one looks at Japan, British Columbia, Alaska, or Oregon and Washington, one detects the growing internationalization of economic activities that is transforming the present-day world. The economies of the nations involved in the transpacific timber trade are being steadily interpenetrated and increasingly influenced by those of each of the others. In the process, nations, and the individual citizens and firms therein, lose a bit of their independence. Weyerhaeuser Timber Company has become a major multinational corporation. Its large investments aimed at supplying the Japanese market with logs from North America insure that it must now consider Japanese as well as American factors and pressures in its corporate decision making. Japanese investors with pulping plants in Alaska and purchasing agencies, shipping facilities, and other interests in the Pacific Northwest find themselves in essentially the same position.

As a result of these changes, the residents of small communities, workers in old, now-marginal sawmills, environmentalists, and those concerned with purely local problems find themselves more isolated than ever from the seats of power, less able to influence decision making. The result may be a more economically integrated, productive, and careful pattern of natural resource utilization, but it is also one that has less room for the participatory democracy and citizen control that many Americans cherish—and, in their own ways and forms, many residents of Japan's forest-dependent mountain villages value too.[46] In the long run the role log exports play in encouraging this shift to a more bureaucratized, centralized, and economically rationalized (but less egalitarian) order, may well prove to be the most important effect of the North American-Japanese timber trade. Still, it must be admitted that these shifts, given the economic and political philosophies prevailing on either side of the Pacific, would also surely continue to take place even if not a single additional log, stick of lumber, or boatload of woodpulp were to cross the Pacific. The transpacific timber trade has not caused these changes, it has merely exacerbated them.

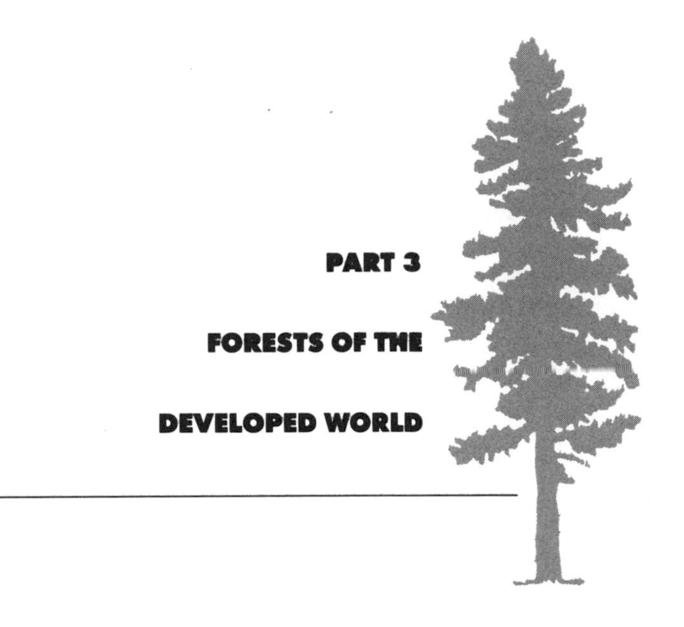

PART 3

FORESTS OF THE DEVELOPED WORLD

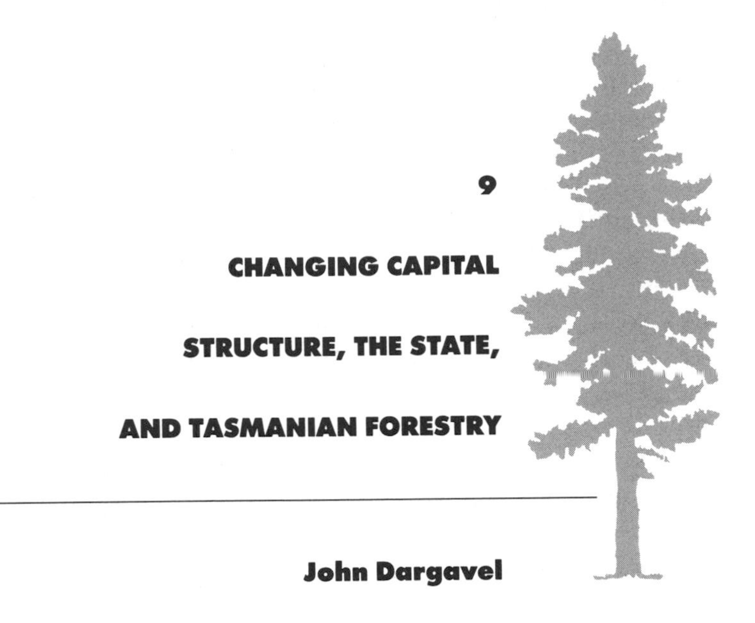

9

CHANGING CAPITAL STRUCTURE, THE STATE, AND TASMANIAN FORESTRY

John Dargavel

The capitalist world system changed turbulently during the twentieth century. Wars bloodied the struggles for resources, markets, and hegemonic power, and empires collapsed into new nations. Not only the size but the structures of economies changed too, either commanding or accommodating the realities of power.[1] A few of the competitive capitalists who had organized industrial production during the nineteenth century grew into national monopolists, and fewer still generated the giant corporations that now net the world.[2] Labor was reorganized or redivided and new means were found to transmit the resulting wealth across national boundaries. Although in places development did yield shared rewards, overall poverty grew no less than the rarer abundance that it fed. Many resources were depleted recklessly, yet others were marshaled most ingeniously. Some forests, for example, were obliterated, yet others of unrivaled productivity were planted. The promising ideas of modern forest management—to protect productivity, to regenerate felled stands, and to sustain or increase long-term yields—were disseminated widely. Although these ideas were often opposed, ignored, or but partially implemented, in some places forest destruction was halted or reversed. Such cases warrant close examination.

What then are the processes at work when forest depletion is countered by deliberate measures? What forces must be overcome? How can the construction of modern forestry in a particular society be related to economic changes? Specifically what changes in the structure of capitalism permit or require the perpetual reproduction of the forests—or the integration of multiple claims within them? This chapter attempts to answer such questions through an analysis of forestry in Tasmania, Australia's island state. The case is not a trivial one, for Tasmania has about the same area as Sri Lanka or Ireland, commercial forests still cover one-third of the land, the wood industries are fully developed, and the struggle to implement modern forest practices has had partial success since the beginning of this century. In this case, particular attention must be given to the operation of state institutions, for the majority of Tasmania's forests are public ones and the struggle for forestry has been fought in that arena. After first detailing the setting, this essay will describe how a series of changes in the structure of the privately owned industries are related to changes in the public forests and their management.[3]

The Setting

Only the narrow coastal fringe of mainland Australia and most of Tasmania receives enough rain annually (over 750 millimeters) to grow good forests. Parts of this zone, which is only about 15 percent of Australia's land area, are mountainous and parts barren. Into it are now crowded the commercial forests, much of the arable agricultural land, almost all the population, the major cities, and the manufacturing industries. Drier zones support woodlands—some used commercially—wheat growing, sheep, and cattle, though the interior one-third of the continent is virtually desert. The continent was first occupied by aborigines who managed extensive areas of the hardy eucalyptus forests and woodlands by extensive burning in order to create grassy, parklike conditions that favored the native grazing animals they hunted.

Australia was at first passed over by the Dutch, who dominated the core of the world system in the seventeenth century, for they sought only preciosities in distant lands.[4] A century and a half later, however, the world system had changed. Abroad, Britain was struggling with France—not Holland—for hegemonic power, which it was soon to win. As "ruler of the waves" Britain wanted a trading post from which to exploit her China trade as well as Australian timber for masts, and flax for sails. At home, British industrialists and estate owners kept the urban and rural masses at bare subsistence, while the British state repressed struggles and riots by sentencing thousands to transportation. With the United States inconveniently securing her independence, Britain had to find a remote destination for those convicted. Australia was a solution, and in 1788 a settlement was made in New South Wales.[5]

The stable scene of aboriginal management was smashed as Britain announced that all the Australian land was hers. Nowhere was this process bloodier than in Tasmania; within 73 years of the first penal settlement there virtually all of the 5,000 aborigines had been exterminated and their lands appropriated for grazing 115 million sheep. It was a startling example of how changes in the core of the world system can have devastating effects at the periphery.

The effects on the forests varied across the island. For the first few years the tiny settlements of convicts, guards, and a handful of free settlers did little more than scratch for their subsistence, but from the 1820s, when wool suddenly became a profitable export to the booming British mills, the economy expanded rapidly. Convicts and paupers were sent from Britain, flocks were bred, and broad areas of land demanded. The plains and grassy forests that had been kept open by the aborigines were ideal, and settlement proceeded rapidly along the central area between Hobart and Launceston (see figure 9.1). Some of the forest was cleared by felling the trees and "grubbing" out the stumps, some by fire, and much from the 1830s on by girdling or "ring-barking" the standing trees.

At first the economy rested on the labor power coerced from the convicts with lash or promised pardon. The brutal punishment regime served not only to profit the colonial graziers but also to deter the British poor from migration. In time, continual transportation created a pool of wifeless, homeless, rootless ex-convicts who competed with immigrant and colonial born labor and kept wages low. Legal self-government, obtained in 1856, placed power in the hands of the landed gentry, who firmed their grip on the grazing lands by leasing huge areas and expelling such petty producers as the ex-convict timber workers. Notably, much of the land remained wholly or partly forested. Even though transportation ended, the gentry also managed to strengthen their grip on the labor force, which remained the most dispirited proletariat in the Australian colonies.[6]

Settlement was not confined to the broad acres of the "pastoral prince"; smaller-scale production in family farms, particularly of fruits

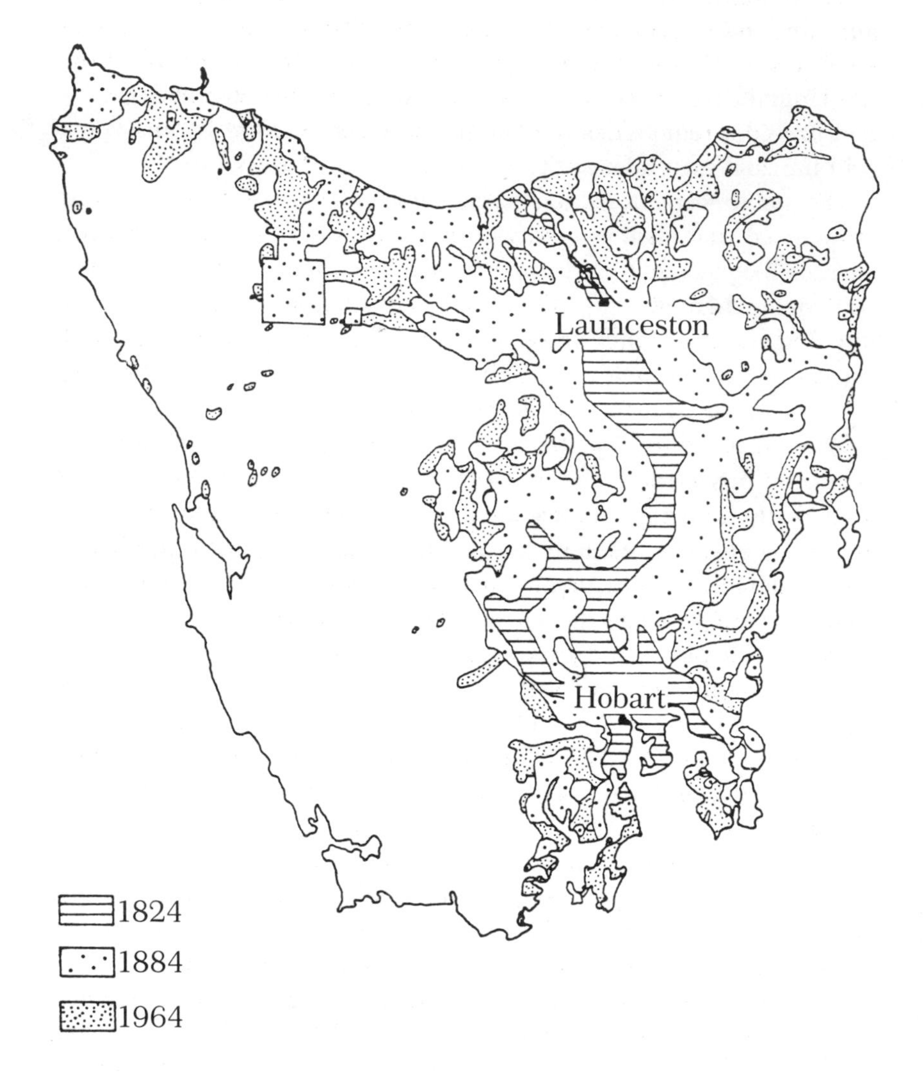

Figure 9.1 Land Transferred to Private Ownership in Tasmania by Various Dates (Bass Strait islands omitted). *Source:* Adapted from J. L. Davies, *Atlas of Tasmania* (Hobart: Lands & Surveys Department, 1965).

and vegetables, developed along the north and northwest coast and south of Hobart in the Huon district. Both regions grew magnificent eucalyptus forests right down to the edge of the many rivers and sheltered inlets. Although timber was plentiful for local use throughout the island, it was from these two areas that an export trade developed. This trade received a huge boost in the 1850s when the rush to Victorian gold created a sudden demand for timber to build a booming Melbourne. The boost was, however, short-lived, and although four sawmills had been built before the gold rush and 19 more were built during it, the low wages and the particularly depressed situation of the ex-convict proletariat in the forests retarded further mechanization until the mid-1870s.

The settlements left most of the western third of the island untouched, apart from some mining on the west coast and forays into the remote rivers of the southwest for the valuable Huon pine. Most of the region is inaccessible and wet with an average rainfall ranging from 1,300 to 3,600 millimeters (50 to 140 inches). Temperate rainforest and sedgelands predominate. Even today, with the cancellation in 1983 of a scheme to dam the Franklin River—a famous cause for conservationists—the bulk of the area remains a wilderness.

The Prelude

The structure of the capitalist world's economy took a new turn in the last quarter of the nineteenth century, with important consequences for Australian development. Although the first significant steps toward modern forestry were not implemented until the start of the twentieth century, the period from 1875 to 1896 was an important prelude to what followed.

By the 1870s metal working had become so advanced in the industrial nations of the core that the use of machines to make machines —notably steam engines—had become standard. Important railways had been constructed in both Europe and North America and the cost of transporting materials had fallen. The flood of cheap manufactures from Britain, "workshop of the world," and other core states had captured world markets and secured raw materials, such as Australian wool, in exchange. Until that time, investment had been directed mainly to industrial expansion within the core while colonial development had relied very largely on colonial accumulation. A sudden fall in prices and profits in Britain in 1873 caused a redirection of investment much to Australia's advantage, for quite suddenly funds became available for new mines, railways, ports, and a host of other ventures.[7]

British rifles marched across new lands—sometimes leading, sometimes protecting, sometimes trailing British pounds—and competed, most successfully, with France and Germany in the grab for Africa. The mines on The Rand (Witwatersrand) were opened and new railways started. The demand for timber was intense and Australia—particularly Western Australia with her durable red jarrah—was well placed to supply it.

Tasmania was less well placed for several reasons. First, few of her timbers could match the durability of jarrah for use as railway sleepers, or ties, and they were unknown on the African markets. Second, many of the stands readily accessible to water had been picked over by the manual timber cutters whose rough tracks left no permanent access from which larger-scale logging could be organized. Finally, the sawmills were mostly small and had no security of tenure in the forests.

The last point is the most important here. The relief from economic stagnation in the mid-1870s brought increases in timber exports to the mainland and rises in prices and wages, all of which led to a second wave of mechanization and the end of the manual production of sawn timber. The number of sawmills doubled, from 26 to 52, between 1874 and 1879 and then gradually increased to 73 by the 1890s. The expansion was based on local, not British, capital and the new mills were still small. A typical mill was powered by a 10 to 20 kilowatt steam engine and equipped with a vertical frame saw to break down the large logs, and circular saws for resawing and crosscutting. Each new mill cut about 40 to 90 cubic meters of sawn timber each week and employed about 25 to 30 men, of whom at least half worked in the forest. As the stands immediately surrounding the mills were cut, wooden tramways were built along which horse teams could bring in the logs. Although these were still short—two or three kilometers would have been typical—they represented major but immovable investments to the local sawmillers. Hence the millers needed security of tenure in the forests while they cut the stands out and realized their investments. In a state governed by the interests of pastoral and agricultural capital, where every influence was exerted to take up more land and every endeavor made to rid it of trees, the interests of the few sawmillers counted little. Indeed it was the sawmillers' actions in felling trees and providing access, or "opening up the country," that often attracted settlers or speculators to the very patches in which tramways had been built. In 1875 sawmillers proposed that settlement should be kept two years behind cutting, and in 1881 legislation was passed to create reserves that would not be cleared until the timber was cut. Little was done, however.[8] From 1885 sawmillers were allowed to lease areas in the reserves of up to 200 hectares for five-year periods, but by 1886 only 21,000 hectares had been reserved, and by 1897 only 13 leases over 1,787 hectares had been applied for.[9] Although G. S. Perrin was appointed as the first Conservator of Forests in 1886, he could do little more than write a few reports. He left in 1892. Thus the needs of small-scale local capital gained scant support.

Imperial Capital and Forest Concessions

As the economy of Southern Africa and Britain's investment there expanded, such enterprising Tasmanian merchants as H. E. Day canvased the use of the better Tasmanian timbers with the railway authorities and gained a toehold in the trade.[10] The opportunities did not slip the notice of British capitalists, however, and in 1897 a group of them proposed to build a cement works and a sawmill on Maria Island, off the east coast of Tasmania. The response of the state was immediate and in 1898 legislation was passed to give the new venture much greater support than the local millers had obtained. The maximum area that could be leased was increased from 200 to 2,000 hectares and the period from 5 to 21 years.[11] Regulations were introduced that provided for the leases to be divided into 10 blocks to be worked sequentially and released for selection when cut out.[12] The Boer War (1899 to 1902) brought the venture to a halt, but thereafter Britain's unrivaled confidence, commercial prosperity, and need to reconstruct the South African economy and put the goldmining industry back on its feet promised expanded trade. Optimistic British capitalists invested in Tasmania, and the state took more decisive action to protect the capital they invested in the forests.

Two companies were formed in London to exploit the forests of southern Tasmania: the Tasmanian Timber Corporation and the Huon Timber Company. The former was a subsidiary of the British Transvaal and General Finance Company and the latter, promoted by R. A. Robertson, was funded by a group of Glasgow capitalists and timber merchants. The Tasmanian Timber Corporation built the "specially powerful" Hopetoun sawmill at Dover in 1901–1902 at an initial cost of about $112,000 (hereinafter figures are in Australian dollars). It was far larger than any sawmill previously built in Tasmania. However, it was the Huon Timber Company's mill at Geeveston that was most remarkable. It was claimed to be the largest sawmill in Australasia and was certainly the most sophisticated, being equipped with the latest Canadian machines. Instead of a frame saw, it had two large band saws for breaking the logs into flitches, and much of the timber was handled by conveyors rather than manually. Notably, much of the equipment was powered by electricity generated on site. A substantial railway with iron rails for steam locomotives was engineered for 10 kilometers of lighter construction. Steam haulers replaced bullocks to winch logs to the lines. The mill opened in 1903 and by 1911, $210,000 had been invested.[13]

None of the existing legislation met the need to secure such large-scale capital and, although local sawmillers complained of unfair advantage, both companies successfully negotiated private acts for individual forest concessions.[14] The major provisions of both acts were similar; very much larger areas were made available—10,000 and 8,000 hectares—than had been envisaged before and for far longer periods—21 years with the right of extension to 42 years. By comparison the local sawmillers had only been able to obtain rights to 200 hectares for five years at best. Despite the generous concessions, neither mill was a financial success. The Huon Timber Company's mill was over-capitalized and after a long strike in 1921–1922 and changing hands, it closed in 1924.

Other forests in southern Tasmania were exploited for the international trade on a smaller scale. A specialist business developed there in providing very long, heavy, and durable blue gum piles for such imperious works as the Admiralty docks at Dover, England, and Simonstown, South Africa.

Imperial influence was not limited to capital or works but affected deeply, though sometimes in mutually contradictory ways, the whole of Australian life. For example, while contingents of loyal young men sailed cheerfully to the Boer War, they did so with an independent air. Constitutionally too, the federation of the six states into the Commonwealth of Australia in 1901 was both a step to independence and a change in the path of British influence.

Public lands remained the state's responsibility, hence imperial views on the management of public forests could be expressed at either the state or federal level. Most importantly they could be expressed across the empire through personal channels. Forestry was advocated in Tasmania from the start of the twentieth century, not by individual capitalists concerned with production, but by Tasmanian and British officials in government reports, papers to learned societies, conferences, correspondence, and personal contacts.[15] The potential of Tasmania's forest wealth was clearly recognized and policies were advocated of planting exotic species, such as English oak and radiata pine, and protecting regrowth stands. The state's response was leisurely. A corner of the Botanic Gardens was declared an arboretum. Two small nurseries were started in 1908, but trial plantings were not made until 1916. More usefully, in 1909, 47,000 hectares of regrowth stands on isolated land not required for selection were placed in reserves.

In 1916 the strongest case was put by D. E. Hutchins, who had served in the Indian Forest Service and had recently retired as conservator of forests for South Africa.[16] He castigated Tasmanian forestry as the worst in Australia; grants of forest land had been made recklessly, deliberate or uncontrolled fires were destroying the forests, and slipshod felling was wasting valuable timber. Tasmanian politicians and officials became involved increasingly with functionaries from other states and the Australian Commonwealth by attending interstate conferences on forestry. At first only the government botanist went, but after Hutchins's report the premier and the surveyor-general also attended.

In spite of all this, the Tasmanian state took no significant action prior to 1920 to conserve the productive capacity of the forests or control the rate of resource depletion, other than in the relatively short term and on the localized scale necessary to protect the investments of individual capitalists while they cut the forests.

Dominion Capitalism and Tasmanian Forests

The great upheavals of the world system in the two world wars, the Russian revolution, and the Great Depression of the 1930s changed the structures of power and the economy. It was the United States that emerged from World War I as economic leader and from World War II as capitalism's superpower. Between the wars Britain sought to restructure her economy within the defensive walls of empire. Her markets were protected by the tariffs of "Empire Preference," her capital was channeled to her own dominions, and emigration from her populous cities to less crowded lands was supported.

The Australian economy was critical to the British strategy: it was stable; capital for electrification, irrigation, and industrialization was keenly sought; there was space for immigrants; the expanding and comparatively affluent population provided a growing urban market; and some of the major concentrations of capital were already British associated with Australian interests.[17] As a result Australia became the major destination of British overseas investment in the 1920s.

The economic structure that resulted in Australia, like that in New Zealand and Canada (as well as Uruguay and Argentina, although those were only informally integrated with Britain), has been described as "dominion capitalism" to distinguish it from that in the core and from colonialism.[18] The structure is particular in that although it has many aspects in common with advanced industrial nations (urbanization, a white proletariat, wage labor, technical sophistication in production, high wage rates, and good social conditions), it has other aspects similar to those of most peripheral nations (exports mostly in primary products, small internal markets, industrialization depending on first-stage processing or import substitution that has to be protected to survive international competition, and foreign capital owning the commanding heights of the economy).

While the new structures just described characterized the salient and emerging features of the world and Australian economies, the old structures—such as the many small sawmills in Tasmania—continued to operate. How then did the rising new economies affect the forests during the transitionary period before the old economies fell?[19]

Although systematic forestry had been ardently advocated for Britain herself, as for Tasmania before, it had only been established exceptionally, as in India or South Africa. Suddenly forestry was recognized as an urgent need throughout the Empire. Britain could no longer rely on exploiting distant resources; the productivity of a self-sufficient Empire had to be secured. A number of measures were taken in quick succession. In 1919 a commission was formed to rehabilitate Britain's own woodlands and plant her wasted uplands. In 1920 dominion representatives and colonial officials were called to the first Empire

Forestry Conference in London. In 1921 the Empire (now Commonwealth) Forestry Association was formed to communicate ideas across the empire and informally, but importantly, monitor progress. One of its major objectives was "to secure the early extension in all countries of the British Empire of a constructive forest policy whereby the natural sylvan resources of the Empire may be scientifically conserved and prudently exploited for the mutual benefit of the British Commonwealth of Nations."[20]

In 1923 the Imperial Forestry Institute was established in Oxford to provide higher training for forestry officers (undergraduate training had been offered at Oxford since 1905) and to conduct or coordinate research. While the formal steps to establish forestry could be taken through governments, as 75 percent of the empire's forests were still in state control, the informal links of shared training, professional literature, Empire Forestry conferences, intercolonial postings, and the like created a common set of values within the small "brotherhood" of foresters.

Australian forestry was well connected to the imperial measures. One key figure was Sir Ronald Munro-Fergurson, who served as governor-general from 1914 to 1920. As a Scottish estate owner with knowledge, experience, and enthusiasm for afforestation, he encouraged the development of forestry in Australia.[21] On his return to Britain he became, as Lord Novar, the first chairman of the Empire Forestry Association. Another important figure was C. E. Lane-Poole, who attended the first Empire Forestry Conference as conservator of forests from Western Australia and urged the formation of the association. He later became Australia's first director-general of forestry and head of its Forestry School.

Lane-Poole's artistocratic mein doubtless smoothed many of the connections between forestry and the new institutions set up to implement Australia's development strategy.[22] Herbert Brookes, a director of Australian Paper Mills, became a member of the Federal Bureau of Commerce and Industry. Herbert Gepp, an able technocrat who managed the Electrolytic Zinc Company for Collins House, became chairman of the influential Development and Migration Commission. Russell Grimwade, whose companies strove to monopolize the chemical and glass markets, was an ardent advocate for forestry. He led the Australian Forest League from 1912 and endowed a scholarship to send an Australian forester for higher training in the Imperial Forestry Institute. There were many other connections.

Australian officials calculated that Australia should reserve 9.9 million hectares of indigenous forest and manage them to ensure a permanent supply of timber; Tasmania's share was 0.6 million hectares.[23] Such advocacy as this, with imperial and industrial connections, could no longer be ignored, especially in 1920 when it was Tasmania's turn to host an Interstate Forestry Conference. Tasmania appeared to respond dutifully. Certainly the Forestry Act of 1920 was passed "to establish a Forestry Department and to provide for the better management and protection of the forests." The act specified that the required area was to be selected from forested crown lands and "dedicated" as "State Forest" within seven years. Moreover, once such forests had been selected and proclaimed, their status could be revoked only with the agreement of both houses of the Tasmanian parliament.

The legislation met the needs of the new economic structures, but its implementation encountered strong opposition from the old structures that comprised the bulk of the Tasmanian economy. The dedication of state forests proceeded slowly because of conflicts between different departments over the use of crown land for pastoral, agricultural, mining, or forestry development. By 1927 only 0.13 million hectares of the targeted 0.6 million hectares had been reserved. In the next year, under the influence of an incipient pulp and paper industry, and a visit of the Empire Forestry Conference to Australia, a further 0.37 million hectares were proclaimed, but the full area was only reached in 1939.

The new Forestry Department was too starved of funds to control cutting in the forests. This was in spite of a report in 1929 by Lane-Poole with recommendations introduced by Gepp, hence carrying the imprimatur of both the Australian government and central capital, that "the fundamental policy recommended is the acceptance of the principle of continuous yield which foresters term *sustained yield* in place of the present system of exhaustive exploitation."[24]

Even though a new conservator of forests, S. W. Steane, prepared an appropriate forest policy the next year that was formally adopted by the state, little was done. By 1937 Steane was to look back and reflect that "the regime of *laissez faire* had become so firmly established that any mention of control or restriction was regarded as heresy of the most dangerous order and one might almost say that it was only as such opposition was overcome that the policy could be formally recognized by Government."[25]

The sawmillers' opposition to the Forestry Department was partly to state control over the logging areas and partly to economic effects. They did not want to pay the costs of forestry, and in the severe economic conditions of the 1930s most could not afford to. And many who habitually cheated on their royalty, or stumpage, payments did not want an efficient department that could collect them.

The Forestry Department struggled to map and select the best forests, make cutting more orderly and less wasteful, control fires, collect royalties, and plant test areas. Being continually opposed, underfinanced, and understaffed they achieved little. It was not until World War II ended and the structures of capitalism changed within Tasmania that any significant forestry was undertaken. The change within Tasmania was mainly due to the emerging paper industry.

The Start of the Paper Industry

The paper industry suited the new Australian development strategy of the 1920s. Local mills made only 20 percent of the paper used and concentrated on wrapping paper and cardboard, both made from local waste and imported wood pulp. Yet to be exploited were the large and expanding market for newsprint, which was being met by imports, and the smaller market for the fine paper used for printing and writing. To industrialize, forest resources had to be secured, capital mobilized, markets organized, and a way found to make wood pulp from the refractory eucalypti.[26]

Collins House was very interested, and the Australian government's researchers in the new Institute for Science and Industry were directed to the task. They worked intimately with Collins House men and by 1924 could make eucalypti into fine paper, at least experimentally, and by 1927 into newsprint. They found *Eucalyptus regnans*, the mountain ash of Victoria or swamp gum of Tasmania, to be the best species, for it is light in color and its fibers are longer than in other hardwoods.

Tasmania too was very interested. The state had adopted a policy of "hydro-industrialization" by which it hoped to attract industry with cheap electricity and whatever subsidies or concessions could be managed. Tasmanian forests contained extensive stands of the favored swamp gum and these grew magnificently. In the inaccessible Florentine Valley there were trees of giant dimensions—the tallest hardwoods in the world. Other forests in the northwest of the state had been barely explored, let alone mapped or logged, but they were known to contain a wealth of timber.

Throughout the 1920s and 1930s, three groups of capitalists planned, schemed, and wrangled with each other and governments until, at the end of the Depression, each could build a mill to monopolize a sector of the market. The several existing paper makers concentrated their capital to form Australian Paper Manufacturers (APM), which built a pulp mill in Victoria in 1939. This fed its paper machines and enabled it to command the packaging market. The two major newspaper proprietors managed to rake together enough capital to build a newsprint mill at Boyer in Tasmania in 1941. This was only possible with the agreement of Canadian newsprint mills that allowed the Tasmanian mill to supply 20 percent of the Australian market (presumably in exchange for the purchase of the Canadian softwood pulp that had to be blended with the eucalyptus pulp).[27] The major capitalist company, Collins House, was left with the worst market but built a mill at Burnie in northwest Tasmania to make fine paper from 1938.

Each company secured a forest concession from the state under an individual act (see figure 9.2). Far larger concessions for far longer periods were obtained than had been obtained by the largest sawmills. The Burnie concession granted to Collins House's Associated Pulp and Paper Mills (APPM) provided exclusive rights to all the pulpwood in the area, apparently in perpetuity, at a nominal royalty rate and without any lease payment.[28] A quarter of the concession was declared state forest for which the Forestry Department was made responsible, but the rest had to be managed by the company. The concession granted to Australian Newsprint Mills (ANM) for 88 years over 124,000 hectares of the fabulous but remote forests of the Florentine Valley provided exclusive rights to all the timber, not just the pulpwood, again for trivial royalties and no lease payments. The company did, however, have to protect and regenerate all the forests.[29]

In 1924 the conservator wrote to his minister to the effect that a pulpwood company must practice sound forestry principles if it is to have prolonged tenure of existence whereas sawmilling, which looks only to the rapid cutting of the mature timber and then moves elsewhere, has no such economic restrictions.[30]

The beleaguered Forestry Department seems to have passed these huge areas over almost with a sense of relief from not having to struggle with recalcitrant sawmillers. In time both APPM and ANM were to achieve the conservator's expectations.

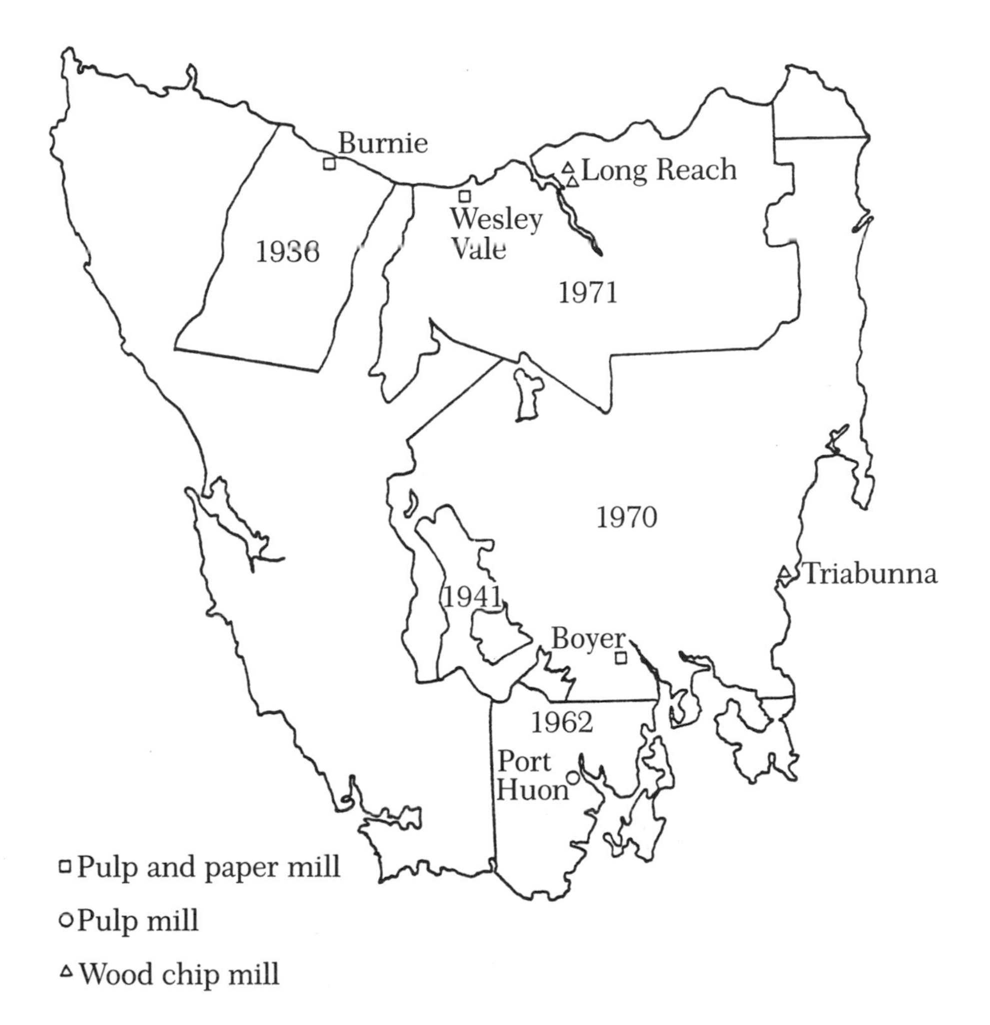

Figure 9.2 Pulpwood and Wood Chip Concession Areas in Tasmania by Date Mills Started. *Source: Tasmania Yearbook* (Australian Bureau of Statistics, 1979).

The new mills profited quickly from enforced protection from international competition during World War II and were well placed for expansion at its end.

The Expansion of the Paper Industry

Australia, the "lucky country," thrived in the 1950s and 1960s. The economy grew quite rapidly but inflated little. Some 2.5 million refugees and immigrants arrived, and the population increased two to five times faster than in the industrialized core. Although total population reached 12.8 million by 1971, employment was very full. Wool prices climbed to record heights, primary production rose, new finds of minerals were exploited, and most significantly manufacturing doubled its share of the economy; Australia became an industrial nation.

The new factories were built or bought by foreign capital to an extent exceeded only in Canada or developing countries.[31] Although Britain continued to invest, it was the United States, capitalism's new imperialist, that became the major investor in mining and finance.[32] The older Australian and Anglo-Australian competitive and monopoly firms with a grip on their markets endured and expanded, while foreign investment—shuffled round the globe by multinational corporations—concentrated on expanding sectors or new products. For example, while the existing papermakers expanded, it was only in making tissue paper that United States capital was able to enter the Australian paper industry.

Not only did the population increase so much that the stock of houses had to be doubled, but good times brought increased consumption. People used twice as much paper as before the war and they could afford larger homes. The wood industries boomed. In Tasmania, both the paper mills expanded and APM set up a smaller mill to supply pulp to its mainland machines. The number of sawmills increased phenomenally as people of all sorts rushed to buy a motor, a sawbench, a simple carriage, and a truck to cart rough timber to the hungry markets. By 1951 about 150 new mills were recorded. Of these, 20 larger mills (with 21 or more employees) captured most of the trade from the 1960s onward.

The industrial boom placed huge demands on the forests and their administration. New concessions had to be granted, new logging areas opened up, and the demands of the two sets of capitalists—paper makers and sawmillers—integrated. The ineffectual Forestry Department was foreseen not to be equal to the task and Australian officials were again asked for their advice. They recommended strengthening and professionalizing the Forest Service, giving it greater freedom from ministerial direction, implementing policies to protect and regenerate the forests, and planting at a greatly increased rate. Although the sawmillers opposed planting pines, fearing they might have to pay for them, they were too politically weak to object strenuously; a royal commission had just found that one firm had bribed the minister, and others some forestry officials.[33] A new Forestry Commission was established in 1946 to implement the new order.[34]

As the paper industry expanded, new forests were conceded. The ANM concession in the Florentine Valley was extended in 1966 when it realized its international connections from Canada to New Zealand, secured more of the domestic newsprint market, and added another paper machine. The extension was governed by the existing conditions; the company had to protect and regenerate the forests but could continue to grind fine sawlogs to pulp if it chose.[35] Elsewhere the demand for sawlogs forced saner solutions.

The Forestry Commission worked hard around the world to "sell" a timber concession in southern Tasmania. This proved surprisingly hard, but eventually APM took it up and in 1962 built their Port Huon pulp mill on the site of the Huon Timber Company's old band mill.[36]

At Burnie, APPM expanded its production of fine paper and diversified into hardboard manufacture and sawmilling. The company drew logs from its own extensive forests as well as from its concession. Gradually APPM took over the independent sawmills and a plymill that also drew logs from the concession until by 1980 production was fully integrated within the firm. Excited by the rapid growth, Collins House planned to expand APPM still further. It absorbed the British partners in two small joint ventures and also an independent mill that had opened on the mainland specializing in making cigarette papers. In doing so it granted a 25 percent direct shareholding to the British American Tobacco Company. More importantly, it planned a large new venture at Wesley Vale and obtained another concession in 1961.[37] APPM proceeded cautiously, not building a paper mill until 1970 and not building a promised large pulp mill even yet (1984).

The new concessions had new provisions. First, the expanded and energetic Forestry Commission could take responsibility for constructing the major roads, protecting the forests from fire, and regenerating the cut over stands. Second, having mapped the forests and assessed the quantities of the different products, it was better placed to integrate the operations of the different capitalists. As the forests were spattered with jealously preserved patches held by sawmillers under Exclusive Forest Permits for up to 15 years, the expansion of the pulp and paper industry was hardly possible unless it did integrate operations. Yet there were potentially complementary benefits if the costs of roads, logging, and so forth could be shared. Detailed state planning was introduced for the new concessions, but this was arranged between the commission and the paper companies; the sawmillers, although invited, took little part in an enterprise beyond their scale.

The sawmillers' opposition to forestry was overcome not only by the continual pressure of "professional" advocacy and the generation of some real benefits, but also because they did not have to bear its costs. Royalties for sawlogs were kept at the lowest level in Australia and those for pulpwood remained at a concessionary level. In consequence the commission's expenditure leapt ahead of income. In the early 1950s the commission had returned $70,000 annually, but by the early 1970s it was drawing over $500,000, and by 1983 the drain had escalated to $9.1 million annually. The drain was funded by loans and justified because roads were held to have a long life and plantations were expected to more than offset declining sawlog yields from the overcut hardwood forests.

Thus the establishment of effective forestry in Tasmania was closely and casually related to the expansion of the monopoly structure of the pulp and paper industry with its connections to the powerful concentrations of British and Australian capital. It was established only with the eventual and in part reluctant agreement of the competitive structure of Tasmanian sawmillers. And it was only established at substantial cost to the taxpayers.

Wood Chip Exports

The balance of the capitalist world's economy gradually shifted from the mid-1960s. Japan and the European Economic Community improved their relative positions and some formerly peripheral nations were able to industrialize in ways often arranged by multinational corporations at the expense of older regions where workers fared better. The United States' economy notably weakened and her hegemony became less assured, but as ruthlessly sought, after the North Vietnamese victory and OPEC's success in raising oil prices.

The balance of the Australian economy changed too. Industrialization halted and in some industries reversed, while the export of raw materials increased. Japan became Australia's major market for such things as coal, iron ore, and wood, and sold cars, appliances, and a host of other consumer durables to Australia in return. Foreign capital continued to flow in, and foreign companies, often absorbing Australian capital, strengthened their grip on the most dynamic sectors. The less exciting wood industries stayed mostly in Australian hands, though the effect of the economic depression after 1974 was to force small mills to close, to centralize production and to concentrate ownership—a trend evident in many Australian manufacturing industries. All these structural changes were expressed in the Tasmanian wood industries and affected the forests there.

The search for raw materials for its rapidly expanding paper industry led Japan to seek eucalypti from Australia in addition to softwoods imported from North America. The structure in which the Australian trade developed differed from that in which a typical United States or European multinational firm operated but still had features that enabled the supplying country to be exploited through mechanisms of unequal exchange. The key place in the wood chip trade was occupied by giant Japanese trading companies that owned or arranged the shipping and negotiated the contracts, but did not necessarily own either the pulp mills in Japan or the wood chip mills in Australia. Critically, their price negotiations and their investments in shipping were coordinated by an Overseas Pulp Material committee, which functioned as a cartel under the guidance of the Japanese Ministry for International Trade and Industry.

In 1971 and 1972 three wood chip mills were built in Tasmania. They were large—one was the largest in the world—but comparatively simple and added little value to the logs carted from the forests. The first was built at Triabunna on the east coast by a consortium of Tasmanian and mainland sawmillers and timber merchants, Tasmanian Pulp and Forest Holdings (TPFH), who arranged to sell 610,000 tonnes of wood chips a year to the Jujo Paper Company. The state quickly provided a concession over forests on the east coast with initial pulpwood rights for 18 years, but with further rights for 80 years if a pulp mill was built or the matter renegotiated.[38] Logging for pulpwood was integrated with that for sawlogs through state planning, and the Forestry Commission was responsible for protecting and regenerating the forests. The second mill was built by APPM at Long Reach on the Tamar estuary in the north. It could use the wood from the concession granted for its Wesley Vale site. APPM made sales of 910,000 tonnes a year to two Japanese paper mills through the trading companies Mitsubishi and Sumitomo. The third mill, first known as Northern Woodchips, was built alongside the APPM mill and was designed to draw its logs from private forests. Its ownership changed during construction so that it became a subsidiary of the oil company H. C. Sleigh, which at the time was partly owned by the multinational corporation Caltex. It negotiated a sales contract for 710,000 tonnes a year with the Japanese trader Jamamoto Sangyo and later arranged a sale of 300,000 tonnes to South Korea.

The total quantity of wood being cut in Tasmania doubled within three years with dramatic effects on Tasmania's forests. Suddenly whole hillsides were bared by clear felling, well-loved views were despoiled, rural roads were broken by the pounding of heavy trucks, and the peace of quiet townships was shattered. The effects were particularly conspicuous in the settled districts. The sudden market for timber was very attractive to many landowners.[39] However, the $0.50 per ton they received was only a quarter of what would have been needed to induce them to regenerate their land to forest again; instead they burnt and grazed the cut-over blocks, hoping for grass but generally unable to afford sufficient cultivation to guarantee it. Their depressed condition was itself a reflection of the new world economy, in which primary commodity prices had fallen and the European Economic Community had shut the traditional British market to Australian producers. Thus the century-long gradual depletion of Tasmania's private forests suddenly became a rapid degradation; it was found that only a quarter of the private forests being cut would become fully productive again.[40]

By 1975 it was clear that unless more of the private forests were regenerated, their long-term yield would fall to an extent that might depress Tasmania's total production of pulpwood by one-quarter and affect industrial production. The state then acted to introduce forestry to the private lands by setting up a Private Forestry Division within the Forestry Commission and providing loans and subsidies to private growers. By 1981 these were costing Tasmanian taxpayers nearly $0.5

million a year. The state scheme was paralleled by industry schemes that paid better rates for the wood if landowners would agree to regenerate their forests and sell them a second crop of pulpwood. Thus, as with public forestry, private forestry was introduced in response to the needs of mainly non-Tasmanian capital to the cost of Tasmanian taxpayers. By contrast a number of measures recommended in official inquiries to control the environmental thump of the forest operations have yet to be covered by legislation.

The impact on the public forests was dramatic too, as the two chip mills that drew on the Wesley Vale and Triabunna concessions (those belonging to TPFH and APPM) cut several times more wood than did all the 87 sawmills there together. Clearly the forests could not be managed by continuing selectively to pick over numerous small patches as the sawmillers did; only extensive clear felling would feed the ravenous chip mills. The sawmillers relinquished their exclusive permits but were supplied from the large operations planned by the Forestry Commission with the companies. The sawmillers, although losing their independence, gained somewhat as the extra sawlogs gleaned partially offset inevitably declining yields.[41]

The Forestry Commission generally regenerated the cut-over areas with little difficulty. However, the new forests being created will differ markedly from the old. In place of the mosaic of small stands or patches of different ages and species, much of the new forest will be in large, even-aged blocks, less diverse but more productive. Although the Forestry Commission hopes to grow sawlogs in the new forest, the future balance between different structures is uncertain. Nevertheless it is clear that the forests being created in Tasmania serve an industrial future.

The Question

At the start of this essay, questions were asked about instances in which modern forestry had been constructed to halt or reverse deforestation. Then in the case study of Tasmania, steps in the construction of forestry were related to specific conjunctures in the changing structure of capitalism and to actions of the state. Now in conclusion it is useful to reflect on the complexity of the terms used and questions raised earlier in discussing economic structures, the state, and forestry.

First, capitalism was not seen as a monolithic economic system that moved through a series of universal stages but as a number of analytically distinct structures that operated on different scales and that were articulated together in various ways. Hence the construction of forestry was related to the needs of distinct structures rather than to capitalism as a whole. Second, the state in this case was seen to respond fairly simply to the needs of the dominant structure, though acting ambiguously—as it did in the 1920s and 1930s—while dominance was changing. The state can act in more complex ways, however, and should be thought of more as an arena for struggle than as a domain of the dominant structure alone.[42] The situation is further complicated by the interactions between different states and levels. Finally, forestry too has to be considered as a complex and specific construct. In Tasmania, forestry served a mix of interests—local companies and Japanese multinational trading corporations. The nature of the mix changed in ways that ended in favoring, but never exclusively so, the dominant structure. Hence, we may observe that the type of forestry and the interests it served were specific to the structure it supplied and the state that constructed it.

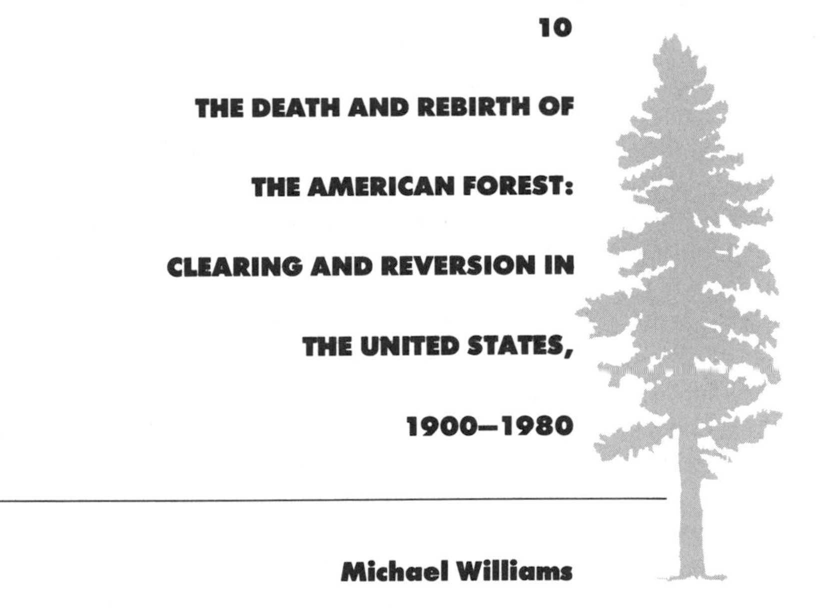

10

THE DEATH AND REBIRTH OF THE AMERICAN FOREST: CLEARING AND REVERSION IN THE UNITED STATES, 1900–1980

Michael Williams

The history of European impact on the American forests until the early twentieth century can be summed up in one word, "devastation." It can be calculated conservatively that the amount of forest cleared for agriculture and industrial felling and lumbering amounted to about 113.7 million acres before 1850; it then climbed to 223 million acres by 1879, and to 296.3 million acres by 1909. Clearing decreased in intensity after 1880 as prairie lands became relatively more attractive for agricultural settlement compared with forest lands because they took far less energy and labor to bring into cultivation. In addition the pace of agricultural settlement and expansion lessened everywhere throughout the continent.[1]

Concern about forest destruction was being voiced as far back as 1860 when Starr, Lapham, Marsh, Hough, and others pointed to the deleterious economic, environmental, and aesthetic effects of excessive clearing.[2] Most Americans knew firsthand, or at least could appreciate, the experience of John Thomas of Union Springs, New York, who wrote in 1864, "during a ride through a portion of the country I had well known twenty years ago, but had not seen in this interval of time, I was struck with the havoc that had been made in the woods."[3]

Everywhere the forest was being reduced in size by the combined impacts of agricultural clearing and industrial felling and lumbering. Many thinking people were beginning to question the conventional nineteenth-century assumption of the inexhaustibility of the resource that had always been accorded such an important place in the life, livelihood, and landscape of the country. After the Civil War some began to take stock of the situation. "The nation has slept because the gnawing of want has not awakened her" wrote Frederick Starr in 1865. "She has had plenty and to spare, but within thirty years she will be conscious that not only individual want is present, but that it comes to each from permanent national famine of wood."[4] Starr's use of the word famine was wholly appropriate because wood was next to food in being a basic commodity in American life. A dearth of wood would not merely be a shortage but the equivalent of the withdrawal of a major life support commodity. The abundance of forests, however, pushed the image of the wastage, emaciation, and death of the forest to the back of the minds of those people who strove to achieve such immediate and concrete goals as, for example, the establishment of the Division of Forestry, the creation of the National Forests on the public lands of the West, and the development of management policies. This held true for the broadly protectionist policies of the nature lovers, personified by John Muir, as well as the broadly utilitarian, "wisest use for the greatest number," policies of Pinchot and his adherents.[5]

But by the closing years of the nineteenth century, as agricultural clearing continued steadily and lumbering activity climbed dizzily, the realization was dawning that Starr's threatened famine was imminent. Between 1900 and 1910 Gifford Pinchot in conjunction with President Theodore Roosevelt skillfully used the coming timber famine theme for all it was worth in order to further the general policy of revitalization of the nation through conservation, and for the regulation of monopolies and big business.[6] At the grand American Forestry Congress staged by Pinchot in Washington, D.C. in January 1905, in order to influence Congress to transfer the forest reserves from the Department of the Interior to the Forestry Bureau in the Department of Agriculture, Roosevelt's opening speech set the tone for thinking about the forest for the next couple of decades:

> Our country . . . is only at the beginning of its growth. Unless the forests of the United States can be made ready to meet the vast demands which the growth will inevitably bring, commercial disaster, that means disaster to the whole country, is inevitable. The railroads must have ties . . . the miner must have timber . . . the farmer must have timber . . . the stockman must have fence posts. If the present rate of forest destruction is allowed to continue, with nothing to offset it, a timber famine in the future is inevitable.[7]

By 1910 Pinchot was convinced that the country had already "crossed the verge of a timber famine so severe that its blighting effect will be felt in every household in the land." The famine was the direct result of a "suicidal policy of forest destruction which the people of the United States have allowed themselves to pursue."[8]

How Much Forest Is Left?

With all this talk of famine, devastation, dearth, and death, it was natural enough that some people tried to calculate how much forest was left. In 1906 Defebaugh pointed out that up until the beginning of the twentieth century the country had been "drawing on the surplus" but that now it was starting "to draw upon the capital fund" of the timber.[9] The analogy of a financial statement was a good one because politicians, professional foresters, and the public alike wanted to be able to see the debits and credits at any given time. The debits or drain were reasonably well known; the credits or growth were a mystery.

In order to produce the balance sheet, two major investigations had to be carried out. First, the location and extent of the forest land had to be determined, and second, the stock of timber in the forest areas had to be calculated, as did its rate of growth and depletion.

Location and Extent

The history of locating the extent of the forest is not dealt with here. Suffice it to say the western forests were not mapped accurately until the beginning of the twentieth century. The importance of mapping lies in the fact that only when that was done could acreage estimates be made for the stand and also estimates of the forest volume and growth, which had to be multiplied by the acreage of forest of different types and age structure. Locational knowledge was the basic prerequisite for estimating the forest reserves.

Marion Clawson has listed the estimates of forest acreage (see table 10.1).[10] Broadly, the estimates fall between 820 and 950 million acres of original forest, the most commonly accepted figures being 850 million acres of commercial forest and 100 million acres noncommercial forest. By 1907 Greeley had estimated that a mere 515 million acres of commercial forest were left, together with 65 million acres of inferior

Table 10.1 Estimated Area of Commercial and Noncommercial U.S. Forest and Total Standing Saw-Timber Volume

Date	Commercial forest (million acres)	Noncommercial forest (million acres)	Standing saw-timber volume (billion board ft.)
1630	850	100	7,625
"Original"	850	150	5,200
1895	562	—	2,300
1902	c. 494	—	2,000
1907	515	65	—
1908	550	—	2,500
1920	464	150	2,215
1930	495	120	1,668
1944	461	163	1,601
1952	495	248	2,412
1962	508	249	2,430
1970	500	254	2,421
1977	488	252	2,569

Source: Clawson, "Forests in American History."

forest. Although later estimates varied considerably, there was little doubt that the forest had diminished by between 300 and 350 million acres by the opening decades of the century. Projections carried forward by Greeley to 1950 suggested that another 100 million acres would disappear in new farms, and that the total area of the unproductive cut-over would reach 182 million acres, leaving a mere 298 million acres in forest.[11] It was an alarming prospect, and one that struck at the very heart of Americans' image of the boundless resources of their country; the nation might be reduced to the state of some impoverished Mediterranean country.

The situation was not helped by Chief Forester Greeley's professional but almost obsessional concern about the virgin forest, as he called it repeatedly. For him the virgin forest seemed to represent the major and perhaps the only true resource remaining in the forest.[12] While it is true that the old, mature trees were the source of the large merchantable timbers, they did not represent a source of growth for the future. In his writing Greeley did not seem to recognize that distinction, and he stoked the fires of concern about general depletion by compiling and publishing three maps of the declining acreage of the virgin forest in 1620, 1850, and 1920, which were alarming in their stark and seemingly simple message of devastation and denudation. These maps were, and often are still, interpreted as showing the diminution of the total forest cover rather than the diminution of a small but important part of it.[13] Second- and third-growth forests more than compensated for the decrease in "first" growth timberlands.

The Timber Stand and Its Rate of Growth

Acres of forest really told one little about the other attributes of the forest—its stock of timber and its rate of growth. Such calculations were fraught with difficulties. Subjective assessments, differing measurements, deliberate underestimations by large landholders in order to avoid taxation and competition, and the changing value and use of different timbers all generally led to pessimistic interpretations and underestimations of the forest resources during the early part of the century.

Our best estimates show an original volume of 5,200 billion board feet of standing timber in the original forests of the country reduced to between 2,000 and 2,800 billion board feet by the end of the nineteenth century (table 10.1). In these varied estimates the work of Zeigler was important because he pointed out something that had been known intuitively before by most people but was rarely appreciated:[14] simply, that timber harvest could not exceed net timber growth indefinitely; neither could net growth exceed harvest for very long because the standing timber accumulated to a level where no further net growth could take place.[15] It was all very well to extol and guard the virgin forests as Greeley tried to do, but other than being the source of large constructional timbers they registered zero production per acre. "For all practical purposes" wrote Zeigler, the virgin timber could be "regarded as nonproducing capital." Growth could only take place after storm, damage, or fire, or if old decayed trees were replaced by young ones. He went so far as to suggest that if all the mature timber was removed and replaced by young second growth then yields would be raised from nothing to between 30 and 110 cubic feet per acre per annum, depending on locality and species.[16] In other words, the careful production management of the forests could increase yields and be the answer to the impending famine.

Although Zeigler had estimated that annual growth was 6.75 billion cubic feet, which was barely more than a quarter of the estimated annual drain, it remained for Greeley in 1920 to refine the ideas and calculations further (table 10.2). The result was, if anything, more alarming than the estimates of the number of acres cleared. Growth was down to 6 billion cubic feet and the drain was up to 26 billion cubic feet, 4.3 times the rate of growth. It was, said Greeley, "the steady wiping out of the original forests of the country."[17] It seemed as though the doomsday men from Thomas Starr to Theodore Roosevelt and Gifford Pinchot were correct; the country was heading for a timber famine.

In this gloomy scenario sketched by Greeley there was one bright spot. Experience in the United States and elsewhere had proved that protection from fire and from extensive grazing after cutting, together with the careful restocking of the cut-over land (rather than trying to sell it off as poor-quality farms) could push up yields dramatically —even enough to exceed current consumption. But that sort of fire control and planting program needed political and legislative provisions that were beyond the powers of any government or department of the day, and their implementation needed some degree of either coercion or cooperation. Indeed, forestry in the United States throughout the 1920s was dominated by these opposing and deeply held views about the management of the forest. On the one hand Pinchot thought that coercion of private forestry was necessary, on the other Greeley thought that cooperation and example were preferable. Ultimately, Greeley's view prevailed.[18]

By the time of the publication of the massive Copeland Report in 1920, annual growth had risen significantly to 8.9 billion cubic feet, while consumption had dropped to 16.3 billion cubic feet, still more than annual growth but significantly less than it had been before.[19] The greater growth was never explained at the time, but with hindsight we can say that better measurement had something to do with it, as had natural regeneration. Few people realized the capacity of the forest land to regenerate timber even without fire suppression or replanting, nor did they appreciate that massive areas of farmland were being abandoned to regrowth forest, particularly in the eastern half of the country, and that this was leading to a great increase in the amount of timber. In just over a decade, between about 1935 and 1945, the whole idea of a timber famine was reversed as timber growth slowly began to catch up with timber drain. The Copeland Report had not

Table 10.2 Estimated Volume and Annual Growth and Drain in the United States, 1909–1977 (billion cubic feet)

	1909 Zeigler	1920 Greeley	1933 Copeland	1950 Copeland Forecast (1)	1950 Copeland Forecast (2)	1952 F.S.	1962 F.S.	1970 F.S.	1977 F.S.
Stock of all standing timber	545	746	487	634	634	603	648	680	711
Growth per annum									
Actual	6.75	6.0	8.9	—	—	13.9	16.7	19.8	21.7
Potential									
cutovers restocked	—	19.5							
virgin timber cut	—	8.2		10.6	21.4	27.5	—	—	c. 36.0
all forest well stocked and managed	—	27.7							
Drain									
Firewood and lumber	23.0	24.3	14.5	15.3	15.3	11.8	12.0	14.0	14.2
Fire and insects	c. 2.0	1.7	1.8	1.2	1.2	3.9	4.3	4.0	3.9
Total	25.0	26.0	16.3	16.5	16.5	15.7	16.3	18.0	18.1
Ratio									
Growth to drain	1:3.5	1:4.3	1:1.8	1:1.5	1:0.77	1:1.13	1:0.97	1:0.90	1:0.83

Note: F.S. = U.S. Forest Service; c. = circa.

been unmindful of the potential of the forest to increase its stock. In two projections of growth to 1950, based on models of varying mixes of extensive and intensive management of the forest stand (forecasts 1 and 2, table 10.2), the report had estimated that growth could be pushed upward to either 10.6 or 21.4 billion cubic feet depending upon the protective measures and degree of replanting undertaken.[20] The upper estimate was optimistic, but by the early 1950s the lower prediction of 20 years before had been surpassed. In 1952 the actual annual growth was 13.9 billion cubic feet, by 1962 it was 16.7 billion cubic feet, by 1970 it was 19.8 billion cubic feet, and by 1977 it was 21.7 billion cubic feet (see figure 10.1). The forest famine and the specter of death were over, and the rebirth of the forest had begun. Its potential growth is thought to be 36 billion cubic feet.[21]

The Rebirth of the Forest

Whatever measure is taken, the conclusion that the forest is building up today rather than declining is unavoidable. The amount of land in commercial forests has risen from a low point of about 461 million acres in 1943 to nearly 500 million acres in 1977. The amount of land in noncommercial forest has increased even more from a low point of 160 million acres in 1930 to 252 million acres in 1977. Standing saw timber has risen from 1,601 billion board feet in 1944 to 2,569 billion board feet in 1977. The annual drain on timber is down from 23 billion cubic feet to 15.7 billion cubic feet (tables 10.1 and 10.2). The visible evidence is even better than statistics. Regrowth can be seen everywhere, and one is struck by the robustness of the forest.[22]

The causes of the change from decline to growth, from death to rebirth, are many. Not all can be dealt with here, but a few stand out for special mention. They are the past cutting that allowed a new harvest to grow; the control and suppression of forest fires; the abandonment of farmland and its subsequent reversion to forest; the planting of new trees; and the falling demand for lumber and lumber-derived products.

The New Harvest

The enormous growth of timber had been possible only because the old forest had been removed. Without the harvesting of the timber

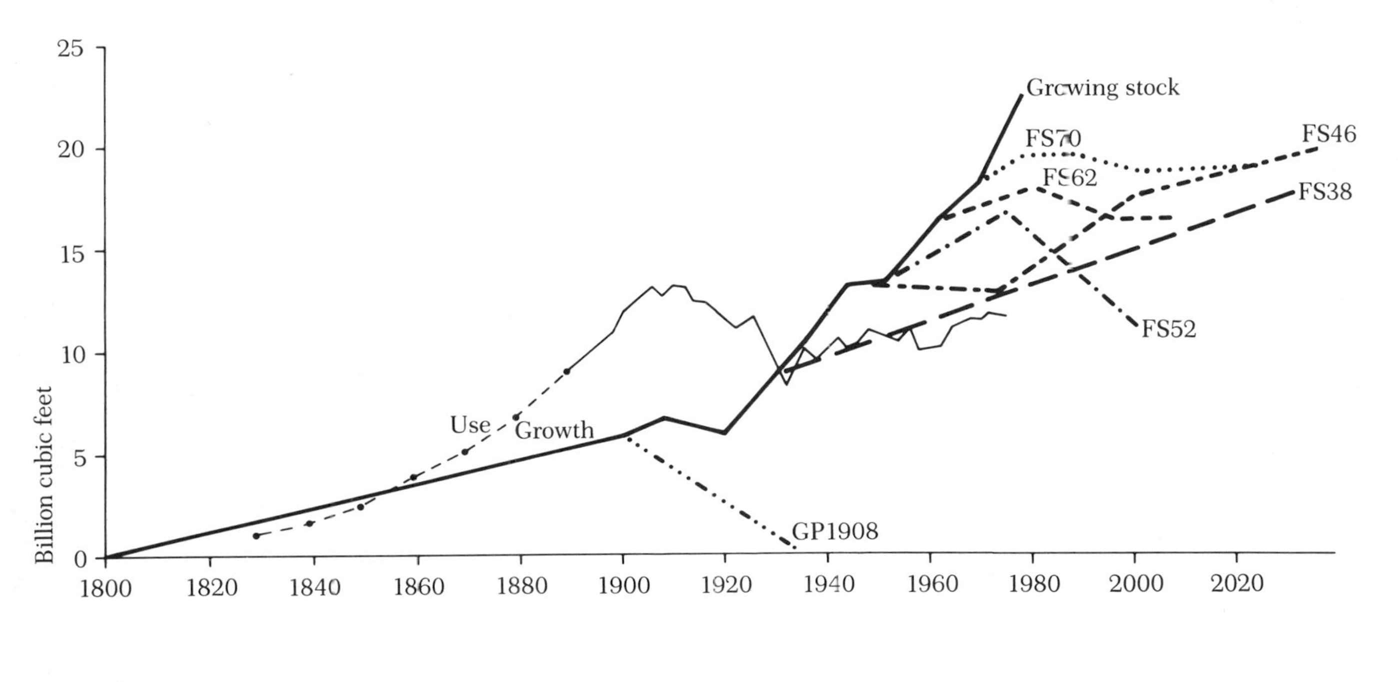

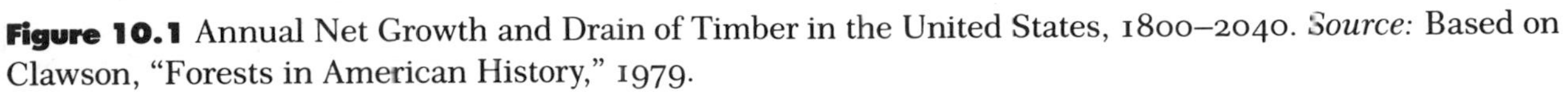

Figure 10.1 Annual Net Growth and Drain of Timber in the United States, 1800–2040. *Source:* Based on Clawson, "Forests in American History," 1979.

crop in the past, whether done carefully or carelessly, there would be no growing wood today. The only major proviso to that relationship is that forest fires must be suppressed and controlled to allow young saplings and trees to colonize the cut-over land. This is essential if the forest is to regenerate successfully.

Suppression of Forest Fires

It was Chief Forester William Greeley, building on the ideas of his predecessor Henry S. Graves, who pursued the idea of cooperative fire control legislation during the 1920s in order to stop the devastation that swept away millions of acres of good timber annually and prevented new growth from taking hold. Greeley had been district officer in Montana at the time of the Great Idaho Fire of 1910, which wiped out over 2 million acres of prime timber forest. The experience of that fire was seared into his soul, so that achievement of fire control became a passion with him. He believed that the suppression of fire, which everyone could see as an obvious destroyer of timber stock, property, and even lives, would be a practical beginning of the implementation of better forestry practices and regulation and would encourage the lumbermen to be more careful and not to leave after-lumbering, a thick debris of highly combustible slash on the forest floor and the cut-overs.[23]

Statistics of fires were collected spasmodically from about 1910, and afterward more regularly. They showed that there was a dramatic increase in the amount of forest burned, from 8.5 million acres per annum in the decade to 1920, to 23.7 million acres per annum in the next decade, to an astronomically high 39.1 million acres per annum during the 1930s—53 million acres being burned in 1931 alone (figure 10.2).[24] The increase was both apparent and real—apparent in that the fires that had gone undetected in the past were now recorded, but real in that depressed prices for lumber during the years leading up to the Depression had caused lumbermen to cut costs and corners, so that they left a legacy of litter that was to burn fiercely.

Greeley campaigned vigorously for the measures that ultimately became incorporated in the Clarke-McNary Act of 1924, which among other things provided funds for a cooperative federal/state fire control system.[25] The initially modest expenditures (later increased) and particularly the new awareness of the need for fire control that they engendered began to bite into the fire problem after 1920, as did the pro-

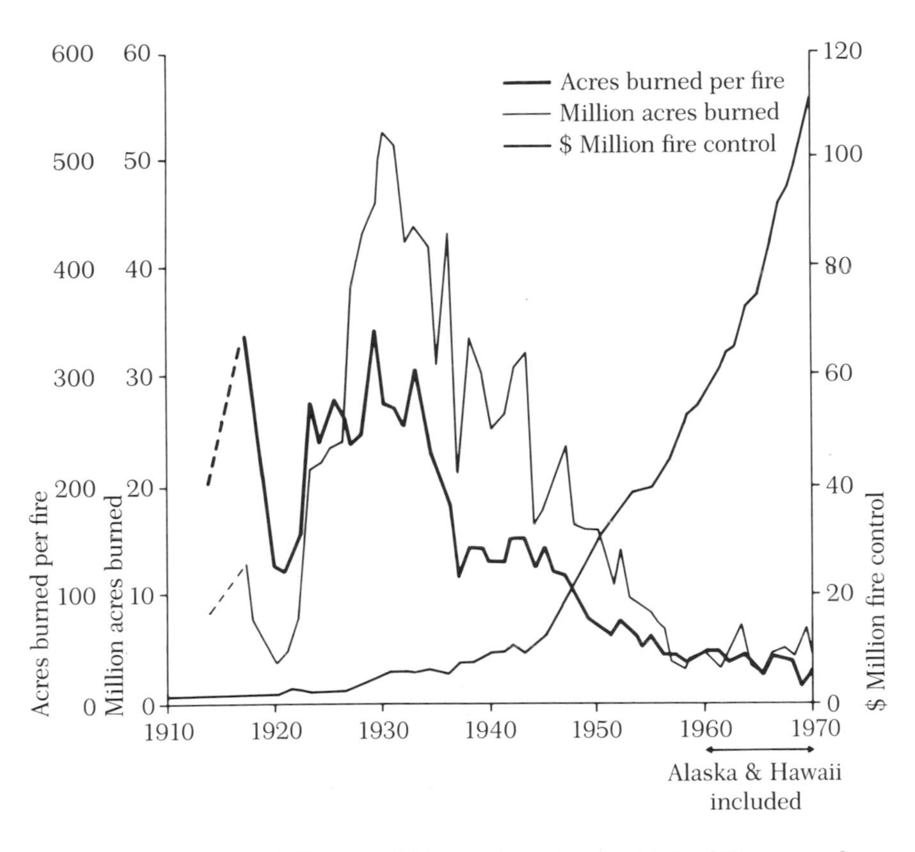

Figure 10.2 Annual Extent of Forest Fires in the United States and Average Size of Area Burned per Fire, 1910–1970. *Source:* U.S. Bureau of the Census, *Historical Statistics*.

vision of new techniques and equipment such as the system of watch towers, telephone lines, spotter aircraft, and forest trackways. After 1920 the number of fires and the area burned began to fall steadily so that the average annual burn during the decade of the 1960s was a mere 4.6 million acres, below which it is not likely to fall markedly.

However, the really significant statistic that can be teased out of the data is the average size of these fires (figure 10.2). This has dropped from 253 acres per annum during the 1920s to 38 acres per annum during the 1960s. Simply, detection and fire fighting organization are quicker and more effective, and fires do not get a chance to spread. The cumulative effect of this reduction in fires has been to allow the young stock to grow vigorously and without interference, and this, above all, has contributed to the massive increase in the inventory of standing stock that is now evident.

The Abandonment of Farmland

Even as Pinchot and his clique were proclaiming the coming timber famine during the early years of the century, the forest was making its comeback in the eastern half of the country.

From as early as 1840 farmland was being abandoned in New England and gradually reverting to forest as farmers left lands that were difficult to farm and either moved to better farmland farther west or migrated to urban industrial centers. After 1880 the process began in the mid-Atlantic states of New York, Pennsylvania, and New Jersey, and after 1910 in the east central states of Ohio, Indiana, and West Virginia. In later years the incidence of abandonment spread down into the South as old cotton and tobacco fields reverted to pine forest.[26]

Considering the importance of the question of reversion, it is surprising how difficult it is to find out how much forest has regrown and why reversion has happened in a particular location. Regrowth statistics do not appear in the agricultural census; they can be deduced only by comparing the amount of total farmland less farm woodland—and then only meaningfully in the predominantly wooded counties of the 31 eastern states of the country.[27] Even then complications (either the land cleared in one period reverted to forest and then was cleared again at a later time, or land abandoned on one farm that reverted to forest may be compensated for subsequently by clearing elsewhere in the original forest) make it difficult to be exact. One thing is certain, however: aggregate statistics for the nation as a whole can be very misleading, as can statistics of the total acreage of farmland, because woodland already covers about one-quarter of the total area. Again, the causes of abandonment and the ultimate use of the land vary, some abandoned land being taken over for urban, industrial, and even strip mining purposes and not allowed to revert to forest.

One important reason for our sparse knowledge about the gain in forest land that farm abandonment represented is that abandonment was either ignored or even disapproved. In a society imbued with the frontier ideals of development, progress, and the virtues of forest clear-

Table 10.3 Cleared Farmland in the United States, 1910–1979 (millions of acres)

	Coterminous U.S.	Thirty-one eastern states		
	Total farmland	Total farmland	Farm woodland	Cleared farmland
1979	—	351.1	72.5	278.6
1975	1,013.7	343.4	77.7	265.7
1969	1,059.6	369.4	78.2	291.2
1965	1,106.9	396.1	95.6	300.5
1959	1,120.2	415.4	113.2	302.2
1954	1,158.2	455.0	133.5	321.5
1950	1,161.4	470.5	135.9	334.6
1945	1,141.6	467.6	118.5	349.1
1940	1,065.1	459.9	106.1	353.8
1935	1,054.5	474.5	132.2	342.3
1930	990.1	443.4	110.1	333.3
1925	924.3	445.7	108.7	337.0
1920	958.7	478.0	130.4	347.6
1910	881.4	490.3	144.3	346.0

Source: Hart, "Loss and Abandonment of Cleared Farm Land."

ing, abandonment was retrogressive, difficult to comprehend, and even sinful to contemplate. It was something to be ignored tactfully rather than praised blatantly. Although there were hints in the major reports on forestry in the pre-1940s era that reversion was happening, the implications for forest growth were never spelled out.

However, an analysis of the census data shows that in the 31 eastern states the net loss of cleared land between 1910 and 1959 was 43.8 million acres (table 10.3), with 65.5 million acres having been abandoned, but 21.7 million acres gained, largely from forest, mainly in Florida, Minnesota, Iowa, Arkansas, and Louisiana. Since 1959 and up to 1979, another 16.9 million acres net have been lost to agriculture or gained by the forest. In other words 865,000 acres have been lost to agriculture and added to the forest in every year between 1910 and 1959, and a slightly lesser amount of 845,000 acres has been added since then. These later data have not been mapped in detail but preliminary results suggest a pattern similar to that of the previous 60 years.

Figure 10.3 Abandonment of Agricultural Land to Forest and Conversion of Forest to Agricultural Land, Carroll County, Georgia, 1937–1974. *Source:* Hart, "Land Use Change," 1980.

Most of the figures mentioned have been mind-boggling in their millions and billions, and therefore difficult to comprehend and appreciate. An example of a single county tells the story graphically. Carroll County is situated about 30 miles southwest of Atlanta, Georgia.[28] A comparison of air photos between 1937 and 1974 reveals the extent to which the land has been abandoned and the forests have taken over (figure 10.3). The change has been gradual rather than dramatic, and that tells us a great deal about the incremental, scattered nature of abandonment by individuals that is so difficult to detect in the aggregate statistics. Like much of the South, there is a long tradition in Carroll County of turning livestock into the woods, so that many pastures and fields do not have sharp boundaries or even fences but merge gradually into the woodland grazing zones. In the past farmers have had to be constantly vigilant and expend a great deal of labor in keeping the brush and saplings from creeping out of the ravines, first onto the steeper slopes and then onto the gentler slopes. In some places land is being changed from forest to agricultural land, but that is about one-ninth of the total change (table 10.4). Much of the pattern reflects the energies and inclinations of the individual farmers. Older men

Table 10.4 Major Land-Use Changes in Carroll County, Georgia, 1937–1974

	Acres
Agriculture to forest	90,807
Agriculture to built-up	6,171
Forest to built-up	515
Forest to agriculture	8,496

Source: Hart, "Land Use Change."

may be content to let the woods encroach slowly on land that was cleared when they were younger and more vigorous; younger men may want to extend their holdings. Abandonment and regrowth are just as individual and difficult to detect as was initial clearing in the forest during the last three centuries.

Planting New Trees

The planting of trees—afforestation—in once-abandoned land and cut-over land has gone on at a greater rate than has perhaps been appreciated, and it has been a positive contribution to the rebirth of the forest. The massive planting programs of 50 million acres suggested in the Copeland Report in 1933 seemed a wild dream and extravagances that would never be fulfilled. And yet a close look at the topic suggests that the replanting (and seeding) has been greater than acknowledged, particularly if private afforestation is taken into account.

Under the Clarke-McNary Act of 1924 small appropriations were made for the growing, distributing, and planting of stock on farms to provide for the extension and more efficient management of wood-lots and shelter belts. The Knutson-Vandenberg Act of 1920 provided funds for planting trees in national forests and in cut-over areas. With these acts the groundwork was laid for federal/state cooperation in the distribution of trees and seedlings for forestry and windbreak purposes, culminating in the Cooperative Forestry Act of 1950.[29] In the 31 years between 1939 and 1970 at least 12.6 billion trees were distributed, 14.8 million acres planted, and 1.7 million woodland owners affected.[30] But this is the tip of the iceberg. A variety of other programs, such as the Agricultural Conservation Program (1930), the Conservation Reserve Soil Bank Program (1956), and its successor,

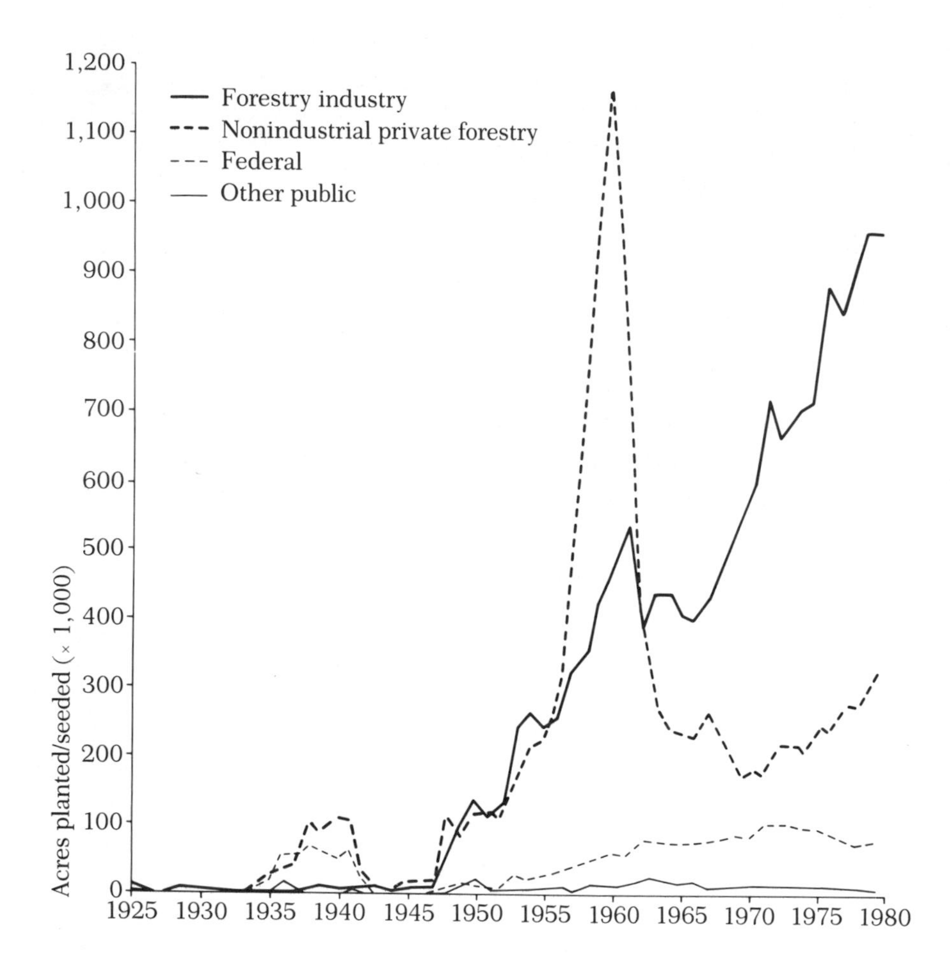

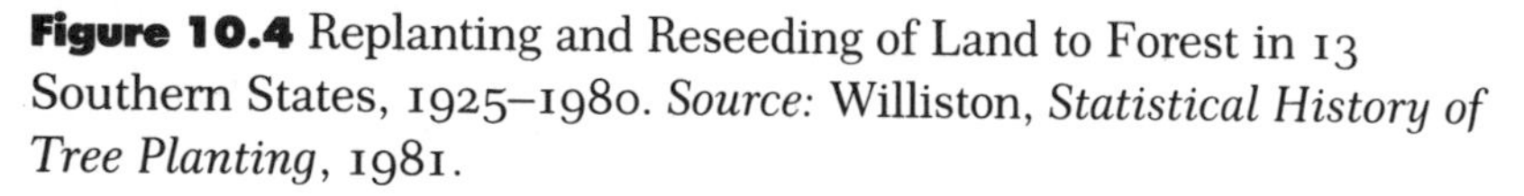
Figure 10.4 Replanting and Reseeding of Land to Forest in 13 Southern States, 1925–1980. *Source:* Williston, *Statistical History of Tree Planting*, 1981.

the Conservation Soil Bank Program (1960), have added many millions more acres, some of which have not been detected yet for the country as a whole. They are mainly minor plantings that would not affect materially the nation's timber supply, but they would certainly affect the forest cover and help to push up the existing estimates of growing stock.

A better and more complete picture of the extent and impact of replanting comes, however, from considering the impact of the forestry industry on the forests. The vast majority of lumber companies long ago shed their image as ruthless woodland destroyers, and although they are clearly aiming to be profit-making concerns in the long-term and difficult business of growing trees, they have done much to put the forest back. The situation has reversed from that which prevailed in the nineteenth and early twentieth centuries. It is no longer cheaper to abandon a plant and move on to new stands (even if there are any); it is now cheaper to replant the forest and maintain the expensive capital equipment and its social and economic infrastructure. In the 13 southern states south of and including Virginia, Kentucky, Arkansas, and Oklahoma, the graph of replanting and reseeding is impressive (see figure 10.4). In gross terms nearly 28.9 million acres of forest were replanted or reseeded to 1980, and the forest industry contribution to that is over 53 percent, now running at just under one million acres a year.[31] A word of caution is perhaps necessary in interpreting this graph. The acreages planted are gross acreages and there is no way of knowing exactly how many acres were replanted in order to come up with net acreages. For example, in the Yazoo-Little Tallahatchie Flood Prevention Program in Mississippi, where nearly 800,000 acres have been planted on blocks owned by small-scale owners, it has been necessary to replant 18 percent of the acreage during the last 30 years. Checks made when the plantations should have been 15 years old revealed that 13 percent of them no longer existed.

Per Capita Consumption of Timber Products

The early alarmists had based their predictions about the ultimate death of the forest largely on the per capita consumption of timber products. Their assumptions of a coming famine were reasonable at the time because per capita consumption was rising and the population itself was increasing at a rapid rate. Per capita consumption was at its highest in 1906 with a figure of 82 cubic feet (see figure 10.5), but with the introduction of such substitute materials as steel, aluminum, and (later) plastics it dropped steadily to little more than one-third of

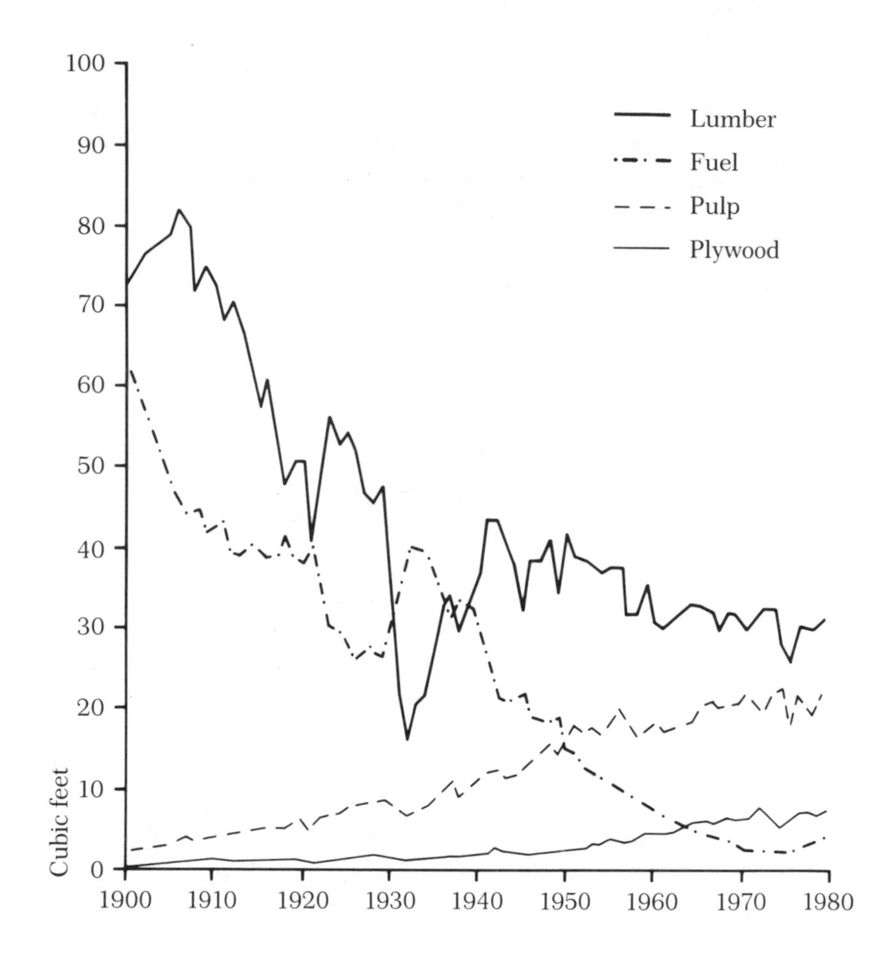

Figure 10.5 Per Capita Consumption of Major Timber Products in the United States, 1900-1980. *Source:* Clawson, "Forests in American History," 1979.

this level by 1955, and it has stayed fairly stable ever since. Fuel wood consumption has dropped even more dramatically with conversions to natural gas and electricity. Some of the decline has been offset by the increase in plywood (a substitute for timber in many cases) and, of course, by pulp wood, which is converted into paper.[32] The remarkable thing is that when all these types of consumption are added together, the total for 1980 is still less than that for 1900, although the population has nearly doubled from 75.9 million to 116.5 million in the meantime.

Summary and Conclusion

The turn of the century predictions of a coming timber famine based on the current experience were reasonable, though a little overdrawn. Nevertheless, the concern was probably instrumental in bringing about a new awareness of the importance of the correct forest management to achieve something approaching a sustained yield. Most commentators and investigators, however, seriously underestimated the regenerative powers of the forest, and they did not appreciate that net growth was a factor of harvest. They could not have foreseen the changes taking place in the economic life of the nation that had led to a dramatic increase in the abandonment of farmland in the East and its reversion to forest, or the substitution of raw materials that led to a lowering of per capita consumption. They had ignored the earlier evidence of abandonment in New England and the Middle Atlantic states.

Taking the long view, it may well be that Greeley's advocacy of fire control was the important trigger for regeneration of timber stands and, as he hoped, a means of raising awareness of "good" practices among forest users. The role of the forest industry in replanting and reseeding has been underestimated.

In large areas of the East and to a lesser extent in the western United States the Jeffersonian ideal of a "middle" landscape—that landscape that struck a balance between the civilizing and domesticating hand of man and the glory of wild nature—is disappearing. The intensely humanized landscape is giving way to wilderness, albeit mature second or third growth, "unnatural," and in relatively small parcels, but wilderness of a sort all the same, a change that would be in accord with current sentiments.

11

PERSPECTIVES ON DEFORESTATION IN THE U.S.S.R.

Brenton M. Barr

Deforestation in the U.S.S.R. in the twentieth century is the product of several processes: state mismanagement of forest land (particularly since the commencement of accelerated logging with the introduction of central planning in 1928); alienation of forest land for industrial development and related infrastructure; and the depredation and exigencies caused by World War I, the civil war, foreign intervention, and World War II. The most obvious manifestation of deforestation is the existence in many Soviet European regions of extensive areas of relatively unproductive and uneconomical forest land primarily comprising poor-quality deciduous species and mixed-age stands of conifers.

Periodic inventories of the forest resource suggest that the total forested area and the volume of growing stock are steadily increasing but that greater accuracy in survey techniques and inclusion of hitherto excluded area in successive inventories may inflate the true dynamics of the timber volume. Regional imbalance, however, in timber quality, age composition of stands, relative importance of commercial species, and volume of growing stock indicates that the spatial characteristics of the resource are more important than total national figures because of the escalating costs of supplying established centers of consumption in European Russia and the Urals with roundwood from increasingly distant northern and Siberian logging enterprises.

Deforestation in the nominally planned economy of the U.S.S.R. appears to arise from poor organization and management of principal and intermediate cuts, from inadequate silviculture, and from conflicting objectives and jurisdictions among various organizations responsible for the forest. As many processing facilities in European Russia and Siberia discover that their areas of planned timber supply have been prematurely harvested by competing organizations, the pressure mounts on central planners to offset the high incremental production and transportation costs of roundwood from peripheral forests by more intensive reforestation, management, and utilization of growing stock in accessible, primarily central-European Russian forests. Annual growth rates in these forests are at least three times those in peripheral regions, and extensive transportation and settlement systems already have been developed.

The main deforestation issue in the U.S.S.R. at the present time, therefore, appears to be the speed with which better management and enhanced utilization will be introduced into the European-Uralian forests. Traditional expectations by many Soviet and Western analysts that further growth of forest utilization could be achieved mainly through greater harvesting of peripheral forests, such as those in the Baykal-Amur Mainline service area of eastern Siberia and the far east, probably should be reduced in relation to domestic consumption of timber. In the decades ahead, however, these forests are likely to play a modest but nevertheless significant role in Soviet exports to markets in China, Japan, and the broad Pacific Rim.

Perspective

Deforestation resulting from the expansion of agriculture or the resettlement of impoverished peoples is not associated with changes in forest areas of the U.S.S.R. Deforestation as a general resource-related problem also is not perceived as an important phenomenon in the U.S.S.R.; although deterioration of the growing stock through depletion of mature stands and replacement of conifers by usable deciduous species is frequently listed as one of the adverse characteristics of forestry that must be overcome in the planned future growth of the Soviet economy. State planning, Marxist-Leninist ideology, and a belief that man should exert a positive influence on the natural environment as natural resources are used to his advantage combine in the U.S.S.R. to produce a climate of thought that sees environmental problems as temporary aberrations in the march toward a superior economic and social system.[1]

If deforestation in the Soviet Union is conceptualized as stemming from man's utilization of the forest for timber, transportation rights of way, reservoirs associated with hydroelectric power stations, opencast mines and spoil heaps from underground mines, and overutilization of forests by recreationists and those pursuing illegal economic activities, then serious problems do exist in the U.S.S.R. The preservation and upgrading of forest areas suffer adversely from numerous factors.

The U.S.S.R. has slightly more than a fifth of the world's forest land and forested land, nearly a quarter of its growing stock and just over half of its volume of coniferous forest.[2] Over half the U.S.S.R. is classified as forest land, while nearly one-third is forested land. The total state land includes agricultural land, land for settlements, industry, resorts, and preserves, state forest lands, state water areas, and general state reserve lands.

Soviet forests suffer depredation due to their sheer size. The abundance of timber, the ubiquity of forests, especially in Russia, and the traditional human struggle to survive in the forests are but a few of the factors that appear to hinder the conservation of timber and the development of practices that might be expected to prevail under conditions of scarcity. Nevertheless, some regions of the U.S.S.R.—themselves larger than many nations of the world—do have shortages of timber due to chronically poor management throughout periods of rapid national growth or recovery in this century and because of the role of expediency in present economic management. Furthermore, problems associated with forest resources have much in common with those related to all other resources in which the valuation offers no impediment to use; consumers of such resources are not penalized by their cost or directly concerned with their optimal utilization. The recent introduction of charges for utilization of water, for example, is unlikely to have any greater effect on that resource's use than those of timber stumpage fees or of mineral and fuel royalties. These charges are simply incorporated into final administrative prices by the state and are deemed to be independent of the costs of production in industrial enterprises.

Deleterious changes in Soviet forests, therefore, do not seem to be a function of urbanization, low return on investment, adverse economic conditions, or competition from other sectors as in much of the Third World, for example. Rather they stem from the practices followed by managers and producers in a strongly centralized economy and society in which financial rewards and production bonuses often occur from behavior that is contrary to the long-term interests of the society at large or the nation as an economic system.

Most of our knowledge about the Soviet forest has been gained from the Russian literature of the past 30 years as Soviet forestry has developed widespread comprehensive analytical facilities and attendant expertise. Deforestation, however, is not a topic addressed in the Soviet literature, although changes in volume and composition of growing stock are discussed, almost parenthetically, in the sources cited in the present analysis. Effective discussions have appeared in the western literature, by Pryde and Sutton.[3]

Pryde demonstrates that major problems in Soviet forests exist through defective administration and poor managerial practices that are not dissimilar from those affecting other industrial uses of the environment in the U.S.S.R.[4] As suggested in the present analysis, Pryde's observations published at the beginning of the environmentally conscious 1970s are shown in his retrospective assessment (1983) still to be acutely relevant.[5] Numerous cases and examples cited in the Soviet literature since the publication of Pryde's comprehensive work suggest that the deteriorating quality of the growing stock and the cumulative legacy of managerial practices employed in its conservation and exploitation do not augur well even for the limited degree of preservation and improved utilization of forests envisioned in Soviet industrial policies and plans.[6] The devastating criticism by Boris Komarov of Soviet use of the physical environment suggests that deforestation is but one element in widespread despoliation of flora, fauna, and landscape in the U.S.S.R.[7]

Deforestation not only has long-term significance for the environment, but also has immediate consequences for the wood-processing industries and for export earnings. Sutton argues convincingly that future timber supply "is likely to be most critical for coniferous sawn timber because of past exploitation and because of the failure of past regenerations"; timber surplus "will only be in the region of Eastern Siberia and the Far East and in low quality species of larch and birch."[8] Although more pessimistic about the ability of the U.S.S.R. forest resource to contribute significantly to future world timber supplies than the present author, Sutton, like North and Solecki, demonstrates that the Soviet economy's priorities and practices place a burden on the forest resource that will retard its future utility at home and abroad.[9]

The issue of deforestation discussed here, therefore, has relevance for numerous sectors of the Soviet economy and for the inherent quality of the resource itself.

This analysis presents the basic area, volume, and administrative characteristics of national and regional forests in the U.S.S.R., discusses the importance of war and planned industrialization to deforestation in the twentieth century, identifies the major changes recorded in recent Soviet forest inventories, assesses differences between Soviet western and eastern forests, and cites examples of factors and processes acting against the long-term stability of Soviet forests. In this way the dimensions of deforestation in the U.S.S.R. can be related to those associated with other studies of this global phenomenon, and opportunities can be identified for more intensive investigation of specific topics in the considerable Soviet forestry, environmental, and geographical literature.

The Forest

State forests comprise forested, nonforested, and nonforestry forest lands (see table 11.1). The total area of state forest land (excluding forests in the general state land reserve) at 1 January 1973 was 1,229.7 million hectares (at 1 January 1978 the total was 1,257.3 million hectares, but no detailed Soviet analyses of this total have been published to date); of this total state forest land reported in 1973, collective farms exercised jurisdiction over 1.8 percent, various ministries and administration controlled 3.7 percent, and the remaining 94.5 percent was under the jurisdiction of central state forestry agencies.[10]

The true area of forest administered by central state forestry agencies, however, which is likely to be "managed" according to some principles of sustained yield, must be reduced from 94.5 percent to 85 percent of total forestry land. The difference of 9.5 percent represents forest land administered by nonforestry ministries and agencies, collective farms, and other agricultural enterprises as long-term leases associated with reindeer and sheep herding. Of the "managed" forest lands administered by central state forestry agencies, only 64 percent represents forested land while 12 percent is comprised of nonforested forest land (table 11.1). This nonforested forest land, nearly 125 million hectares, represents an area of regeneration backlog that may be described as forest land currently in a deforested condition. The ratio of forested to nonforested (i.e., deforested) forest land is thus 5.5 to 1; the current regeneration backlog is approximately twice the size of the total area that has undergone some kind of regeneration since the revolution in 1917 (see table 11.5). The relative importance of nonforested lands varies among regions and is particularly notable in peripheral regions of Siberia and Central Asia (table 11.1). When related to major resource areas, however, the regeneration backlog is significantly large throughout the forests of the Russian S.F.S.R. and extends from the northwest eastward to the Soviet far east.

The regional distribution and characteristics of state forested land administered by central forest agencies are evident in table 11.2. The size of forested area and the total volume of growing stock for the country and its regions is net of forests held by nonforestry ministries and agencies, collective farms, and in long-term leases by reindeer and sheep herding enterprises; major recent Soviet studies of the forest resource analyze in detail only those forests held by central forest agencies.[11] We can conclude, therefore, that a significant portion of forest is utilized by nonforestry agencies without any regard for its long-term preservation and that the total area subject to depredation in the U.S.S.R. is larger than that estimated in table 11.1 (which pertains only to land "managed" under central state forest management).

Land under long-term lease, for example, comprises 10.2 percent of the total land held by central state forestry agencies, although due to its location it only represents 5.4 percent of the central forest agencies' forested land. In the European U.S.S.R., virtually all the leased forest land lies in the northwest, where it is used for reindeer range. In the Asian U.S.S.R., 80 percent of the total leased land lies in the northern part of the far east, where it is also used for reindeer grazing; the rest is mainly in Central Asia where it supports sheep.

The proportion of land held by collective farms and nonforestry organizations varies among the major regions of the U.S.S.R. In the European U.S.S.R., collective farms administer 6.2 percent of the state forest and 7.4 percent of the forested land; various ministries and administrations account for 9.3 percent of the state forest area and 10.3 percent of the forested land area. Collective farms and other ministries and administrations comprise 31 percent of the state forest land in the center, 19 percent in the Volgo-Vyatka, 29 percent in the Black

Table 11.1 Land-use Composition of State Industrial Forests Administered by Central State Forestry Agencies, U.S.S.R. (percent of area total)

		Forestry forest lands					
			Forested lands			Nonforested lands	
				Including plantations			
	Total forest lands	Total forestry lands	Total[a]	Closed	Un-closed	Total[b]	Un-speci-fied
U.S.S.R. Total[d]	100	76	64	1	<1	12	1
European	100	83	78	5	2	5	2
Asiatic	100	74	62	>—	<1	12	<1
R.S.F.S.R. Total	100	76	64	<1	<1	12	1
European	100	82	77	3	<2	5	<2
Asiatic	100	74	62	—	>—	12	—
Northwest	100	76	71	>—	<1	5	—
Center	100	94	87	5	3	7	4
Volgo-Vyatka	100	96	86	4	3	10	7
Black Earth	100	92	85	19	6	7	2
Volga Littoral	100	92	85	5	4	7	2
North Caucasus	100	89	81	4	3	8	2
Urals	100	88	82	1	2	6	—
W. Siberia	100	61	55	—	—	>6	—
E. Siberia	100	83	76	—	—	7	
Far East	100	73	55	—	—	18	
Ukraine	100	91	84	35	4	7	6
Moldavia	100	95	81	23	8	14	10
Baltics							
Estonia	100	71	65	13	3	6	4
Latvia	100	82	77	12	3	5	3
Lithuania	100	87	84	13	2	3	2
Belorussia	100	93	88	15	3	5	3
Transcaucasia							
Georgia	100	92	86	1	1	6	3
Azerbaijan	100	90	80	2	2	10	7
Armenia	100	85	66	3	10	19	14

(Forestry forest lands)			Nonforestry forest lands						
(Nonforested lands)			Miscellaneous usable lands						
Under-stocked	B & DS[c]	Cut-over	Total	Arable lands	Natural hay lands	Grazing lands	Water bodies	Rights of way	Unutilized lands
6	4	1	24	—	<1	<1	<2	<1	21
<1	<1	>2	17	>—	<1	<1	<2	1	13
7	4	<1	26	—	<1	<1	<2	<1	23
6	4	1	24	—	<1	<1	<2	<1	22
<1	1	<3	18	>—	<1	<1	<2	<1	14
7	<5	<1	26	—	<1	<1	<2	<1	23
<1	1	4	24	—	<1	—	3	<1	21
>—	>—	3	6	2	1	—	—	1	2
—	—	3	4	—	1	—	<1	1	2
1	—	4	8	<1	2	—	1	>1	4
1	—	4	8	>—	2	—	1	<1	4
1	—	5	11	<1	1	1	<1	<1	8
—	<1	<6	12	<1	2	<1	<1	<1	9
4	2	<1	39	<1	<1	—	<5	—	33
>3	<4	>1	17	—	—	<1	1	>—	16
9	8	>1	27	—	<1	>—	1	—	26
>—	>—	1	9	<1	1	<1	<1	2	5
—	—	4	5	—	—	—	—	—	4
1	—	>1	29	1	>1	1	1	3	22
—	—	2	18	1	2	1	<1	3	11
—	—	1	13	>1	>1	>1	—	>3	5
—	—	<2	7	<1	1	<1	<1	2	4
3	—	—	>8	<1	<1	>3	—	<1	4
3	—	—	10	1	1	3	—	1	4
5	—	—	15	—	—	>5	—	—	8

Table 11.1 Continued

	Total forest lands	Total forestry lands	Forestry forest lands: Forested lands: Total[a]	Forested lands: Including plantations: Closed	Forested lands: Including plantations: Unclosed	Nonforested lands: Total[b]	Nonforested lands: Unspecified
Central Asia							
Kazakhstan	100	76	48	4	3	28	11
Turkmenia	100	71	20	5	3	51	37
Tadhikistan	100	46	24	2	2	22	13
Kirghizia	100	45	31	2	3	14	7
Uzbekistan	100	72	27	6	5	45	18

Sources: Compiled and adapted from Vorobyev et al., *Ekonomika Lesnogo Khozyaystva SSSR* (Moscow, 1980), 31, 32, 35; and Sinitsyn (1976), 8–23.

a. Percentages in this column have been rounded: they represent values of forested area shown in table 11.2, column 1. Thus total U.S.S.R. forested lands administered by central state forestry agencies, excluding long-lease grazing, comprise 64.7% of total forestry forest land or $675{,}206 \times 10^3$ hectares.

b. Figures in this column represent an area of regeneration backlog. Thus 12% of total

Earth, 25 percent in the North Caucasus, 46 percent in the Ukraine, 24 percent in Belorussia, 24 percent in Georgia, 18 percent in Azerbaydzhan, 29 percent in Lithuania, 22 percent in Moldavia, 34 percent in Latvia, and 42 percent in Estonia. In Soviet Asia, however, these organizations exercise jurisdiction over only 3 percent of the state forest land. The share of state forest land amenable to coherent national planning by central state forest management administrations thus is greatest in northern European forests, in the Urals, in western and eastern Siberia, and in the far east.

The most valuable part of the state forest is the forested land area, which averages 62 percent of the forest land and 59 percent of the Asian forest land. The lowest portion of forested land per unit of forest land is found in Central Asia: in the Russian S.F.S.R., for example, the far east's portion of forested land in forest land is 51 percent; in the northwest the portion is 71 percent.

(Forestry forest lands)			Nonforestry forest lands						
(Nonforested lands)			Miscellaneous usable lands						
Under-stocked	B & DS[c]	Cut-over	Total	Arable lands	Natural hay lands	Grazing lands	Water bodies	Rights of way	Unutilized lands
14	<1	<2	24	<1	2	9	1	<1	11
14	—	—	29	—	—	>17	—	—	11
9	—	—	54	—	—	11	>2	—	40
6	—	<1	55	—	<1	26	<1	<1	27
25	>—	2	28	<1	>—	13	<1	<1	14

forestry forest land, or 124,653 × 10^3 hectares, administered by central state forestry agencies represents an area of regeneration backlog or an area currently existing in a deforested condition.

c. B & DS = Burned and Destroyed Stands.

d. 100% = U.S.S.R. state forests administered by central state forestry agencies, excluding long-lease grazing lands, comprise 1,043.49 million hectares (including long-lease grazing lands, they comprise 1,162 million hectares).

The Early Decades of Soviet Forests

The major commercial coniferous forests of the Soviet Union did not begin to undergo large-scale industrial utilization until the introduction of central planning in 1928 (see table 11.3) although forests were a major source of domestic fuel for rural and urban residents until the discovery and exploitation of large gas and oil fields in the 1950s. The first decade of Soviet rule entailed utilization of the forest in much the same manner and location as in czarist times to satisfy the peasants' need for fuel (over four-fifths of the population lived in rural areas) and to sustain the lumber, pulp, and paper industry that was found chiefly in the west and northwest from Leningrad Oblast to the northwestern Ukraine. Most commercial timber was harvested in areas tributary to Leningrad for conversion in adjacent mills or for export to western

Table 11.2 Distribution of Forested Area and Characteristics of Industrial Stands, U.S.S.R.

	Central state forest management		Percent of forested area by age group of stands			
	Forested area (ha. × 10³)	Total growing stock (M³ × 10⁶)	Juvenile	Medium age	Approaching maturity	Mature & overmature
U.S.S.R. Total	675,206	74,872	15	18	10	57
European	146,826	17,009	28	23	8	41
Asiatic	528,380	57,863	11	16	10	63
R.S.F.S.R. Total	646,102	72,216	14	17	10	59
European	128,360	14,670	26	21	8	45
Asiatic	517,742	57,546	11	16	10	63
Northwest	68,806	7,366	20	17	5	58
Center	12,446	1,542	38	36	13	13
Volgo-Vyatka	9,911	1,209	42	25	9	24
Black Earth	1,113	125	48	35	10	7
Volga Littoral	8,925	1,023	27	31	14	28
North Caucasus	2,853	391	23	33	15	29
Urals	24,306	3,014	31	20	9	40
W. Siberia	75,716	9,655	7	15	15	63
E. Siberia	216,026	27,191	9	17	10	64
Far east Total	226,000	20,700	14	16	9	61
South	70,216	8,948	—	—	—	—
North	155,784	11,752	—	—	—	—
Ukraine	5,810	805	53	30	10	7
Moldavia	210	18	36	48	11	5
Baltics[b]	3,986	506	38	43	11	8
Estonia	997	121	39	38	12	11
Latvia	1,714	228	35	47	10	8
Lithuania	1,275	157	41	43	10	6
Belorussia	5,378	535	59	28	10	3
Transcaucasia	3,082	475	11	48	17	24
Central Asia	1,538[b]	25[b]	19	31	20	29
Kazakhstan	9,100	292	n/a			

Sources: Compiled and adapted from Vorobyev et al., *Ekonomika Lesnogo Khozyaystva SSSR* (Moscow, 1980); Timofeyev, *Lesnaya Industriya SSSR* (Moscow, 1980), 17, 18; *Lesnoye Khozyaystvo SSSR* (Moscow, 1977), 57.

Note: C. = conifers; S.I.H.S. = Shade Intolerant Hardwood Species; S.T.H.S. = Shade Tolerant Hardwood Species; n/a = data not available.

Percent of mature & overmature growing stock			Mature & overmature timber as percent of growing stock volume	Utilization (harvesting of calculated allowable cut for each species, percent of total)			
C.	S.I.H.S.	S.T.H.S		C.	S.I.H.S.	S.T.H.S	Total stock
85	13	2	69				
77	19	4	53				
87	11	2	74				
86	14		71				
n/a	n/a		58				
n/a	n/a		74				
89	11	—	72	104	41	—	87
45	53	2	22	104	79	62	87
57	42	1	44	114	80	62	94
19	47	34	13	102	92	92	95
20	62	18	38	89	58	32	61
19	11	70	44	114	82	80	99
77	23	—	61	110	67	57	93
68	32	—	72	44	16	—	30
92	8	—	74	48	6	—	36
—	—	—	—	—	—	—	—
77	15	8	70	53	13	15	45
99	1	—	74	12	—	—	13
36	9	55	13	102	98	—	100
			9	n/a	n/a	n/a	102
80	20		12	97	98	—	97
n/a	n/a		18	n/a	n/a	n/a	n/a
n/a	n/a		11				
n/a	n/a		9	97	—	—	97
43	43	14	6	56	—	—	64
35	65		36				
100	—	—	46	57	58	66	58
71	29	—	48				

a. Values are net of forests administered by collective farms or nonforestry ministries and agencies, or held under long-lease terms by sheep and reindeer herding enterprises.

b. Excluding Kazakhstan.

Table 11.3 Timber Harvesting in the U.S.S.R., 1913–1983 (cu. m. × 10^6)

Year	Industrial removals			Nonindustrial removals	
	Total roundwood	Commercial roundwood	Fuelwood	Total roundwood	Fuel-wood
1913[a]	60.6 [67.0]	27.2 [30.5]	33.4 [36.5]	232.0	204.0
1928	61.7	36.0	25.7	242.0	215.0
1929	95.5	60.0	35.5	n/a	n/a
1930	147.2	96.7	50.5	n/a	n/a
1935	210.1	117.0	93.1	223.0[b]	201.0[b]
1940	246.1	117.9	128.2	136.0	130.0
1945	168.4	61.6	106.8	140.0	130.0
1950	266.0	161.0	105.0	100.0	92.0
1955	334.1	212.1	122.0	70.0	63.0
1960	369.5	261.5	108.0	40.0	30.0
1965	378.1	273.6	104.5	38.0	32.0
1970	385.0	298.5	86.5	37.0	30.0
1975	395.0	312.9	82.1	35.0	29.0
1980	356.6	277.7	78.9	n/a	n/a
1982	357.0	275.0	82.0	n/a	n/a
1983 (est.)	357.0	274.0	83.0	n/a	n/a

Sources: Promyshlennost SSSR. Statisticheskiy Sbornik, 249; *Narodnoye Khozyaystvo SSSR v 1967 g.* (Moscow, 1968), 272; *Narodnoye Khozyaystvo SSSR. 1922–1982* (Moscow, 1982), 200–201; *SSSR v Tsifrakh v 1982 godu* (Moscow, 1983), 68; Blandon, *Soviet Forest Industries*, 248 (for nonindustrial removals of roundwood).

Note: n/a = data not available.

a. 1913 data pertain to the territory of the U.S.S.R. included within the boundaries that existed until 17 September 1939 (data enclosed by brackets pertain to territory included in the Russian Empire in 1913).

b. Values for 1932.

Europe in exchange for manufactured timber products, equipment, chemicals, technical expertise, and British coal. Major administrative and industrial consumers of paper were concentrated in Leningrad (formerly St. Petersburg); despite the shift of national administration to Moscow in 1918, the czarist imprint on consumption patterns of commercial timber was not immediately lessened (prerevolutionary levels of industrial production generally were not reached again until the end of 1926).[12]

With the advent of national central planning and administration in 1928, forest in the traditionally peripheral northeastern and eastern regions of European Russia, which had played an essentially "colonial" role in czarist times, became the focus of new investment in the timber and wood-processing industries.[13] By the outbreak of World War II two-fifths of national timber harvesting occurred in the European north, the upper Volga, and the Urals.[14] The remainder chiefly took place in the northwest (17 percent), the Central Industrial Region (13 percent) and in districts of eastern Siberia and the far east (10 percent).[15]

Major regional rivers and their tributaries facilitated movement of timber for industrial consumption and export, but the railway system that had been put in place in the last decades of czarist rule and expanded in the first two decades of Soviet administration permitted large-scale movement of timber from numerous rail-water intersections to emerging centers of national industrial consumption in the Central Industrial District, the eastern Ukraine, and the Urals.[16] The expanding rail system also ensured that traditional areas of consumption in the Leningrad region and Belorussia and markets for Soviet timber in western Europe could be satisfied with additional amounts of timber from previously unexploited forests.

Forest industries, despite their significance to such industrialized countries as Finland, Sweden, Canada, and the United States, did not receive (and have never subsequently received) high priority in the prewar central allocation of Soviet investment funds. Myriad competing demands by sectors of the economy deemed crucial for forced industrialization, autarchy, and national survival have traditionally left the forest industry in an inferior technological position with many unrealized opportunities for product and regional growth. Pre–World War II timber production was intended mainly to satisfy domestic demand for fuelwood, for roundwood in unprocessed form during construction, and for roundwood that had received minimal processing (into such items as rough planks, beams, and squared building logs). Export commitments chiefly took the form of roundwood, but the volume of export on the eve of World War II comprised less than 1 percent of commercial timber harvested in the country (export of all forms of roundwood in 1975 at the peak of post–World War II annual timber felling still comprised less than 6 percent of the commercial roundwood produced in the U.S.S.R.), although in some early years of forced industrialization timber was the most important export commodity and contributor to the national budget. (The significance of timber exports declined however, during the worldwide depression of the 1930s.)[17]

The timber industry prior to World War II, or in the years immediately following that conflict, did not engage significantly in comprehensive forms of processing or expand into major forests peripheral to the European-Uralian industrial axes. Product growth and spatial expansion of the wood-processing industry were greatly hampered by shortages of skilled workers in the major regional forest, inadequate surveys of timber reserves for quality, volume and species composition, deficient facilities for product movement and handling by river and rail, and crippling shortages of fuel for industrial heat and processes of material conversion. Furthermore, the major accessible forests of central and eastern European Russia acted as significant intervening opportunities between the established regions of industrial timber consumption and the vast forest that lay in northern European Russia and east of the Urals.

Following postwar recovery, the significant growth in the country's industrial prowess, the demand by administrative and industrial sectors for forest products, and the necessity to finance greater technological transfer from the West by continued sales of such raw materials as timber (second in importance only to petroleum) have led to a weakening of many economic, institutional, and geographical barriers that previously trammeled the forest industry. By the mid-1970s, half of the U.S.S.R.'s timber originated from the European north, the Volga-Western Urals, and the Urals, but one-fifth was harvested in central Siberia.[18] The national share originating in the northwest, center, and far east together was less than that of central Siberia.

The extent and legacy of overcutting and devastation during the first four-fifths of the twentieth century is extremely difficult to evaluate. Reasonably comprehensive arrays of regional forest data were published by the U.S.S.R. between 1955 and 1975, a period of relatively abundant release of information that has subsequently abated. Although often extended back to earlier periods of Soviet rule, particularly to 1940 to provide a baseline for evaluating postwar Soviet growth, these figures were primarily expressed in physical units of production of roundwood and types of forest products by major economic region or administrative area. Some early statistical handbooks in this period, however, reported levels of overcutting by region as a percent of annual allowable cut; the data, unfortunately, could not be analyzed for such information as the depletion of regional or local forests by age and species or the extent to which specific logging areas were subsequently reforested.

These figures notwithstanding, there is no reliable evidence to suggest that timber harvesting prior to 1928, when central planning began, or during the five planning periods prior to 1956 (the first complete analysis of all Soviet forests) was conducted with any reference to allowable cut and sustained yield management. Prerevolutionary logging by up to 2 million peasant loggers in European Russia and by forced laborers in forests of the European north and the Urals has been described as rapacious, predatory, and completely ignorant of management principles.[19] Tseplyaev has reported that despite the 1888 law by the czarist government to conserve the forest, between 1888 and 1908 in European Russia alone 3 million hectares of forested land were cleared and the general forest area declined by over 9 million hectares, "in prerevolutionary logging anarchy prevailed: the exchequer, the aristocracy, mining enterprises, military authorities, and private landowners treated the forests with impunity and with total disregard for their conservation and replacement."[20] Vorobyev and colleagues, citing Tsvetkov's analysis of changes in the extent of European Russian forests from the end of the seventeenth century to 1914, note that about 70 million hectares of forest were cleared in European Russia during that period.[21]

Soviet practices until well after World War II probably were not fundamentally different from those of czarist times or from contemporary practices in most of the world's forests. Soviet concern in the 1920s was with recovery to levels existing before 1914; goals in the 1930s and 1940s were related to expediency, rationalized if necessary by the general opinion that forest reserves were "incredibly immense" and that economic growth and national survival justified short-term exploitation of the forests. If the figures published in the first coherent postwar statistical handbooks are true indicators of planners' estimates of the forest resource, then we can conclude that even the 1950s were marked by large-scale ignorance of sustained-yield management because the annual allowable cut (discussed below) against which annual harvesting was measured exceeded the annual increment of the forests by a factor of 2.5 (by 3 in the early 1950s and was drastically reduced following the comprehensive inventories of the 1960s).[22] If levels reported for 1955, for example, are strengthened by a factor of 2½, then forests were regularly being overcut in the northwest (58 percent), the center (130 percent), the Volga-Littoral (128 percent), north Caucasus (40 percent), and the Urals (42 percent). In the non-Russian western regions, however, the factor of overcutting was extreme: Ukraine (4 times), Belorussia (3 times), Lithuania (5 times), Latvia (2.5 times), and Estonia (2 times).[23]

Although these estimates are extremely crude, they suggest that the volume of overcutting and hence deforestation if aggregated over the decade 1945 through 1955, the two decades 1945 through 1965, or for a longer period to include the second half of the 1930s would have affected large areas of forest.

Adding peacetime deforestation to the volume of depredation experienced during World War II in many of these same regions creates the impression that major areas of the European U.S.S.R. have experienced deforestation. Its exact regional volume and form, as argued below, although hard to measure from a distance and from the silence on the subject in Soviet publications, probably have resulted in extensive uneconomic areas of forest characterized by dominant stands of poor quality poplar, birch, and alder and sparse or irregular stands of conifers.[24]

In his assessment of the Soviet pulp and paper industry in 1955, Rodgers parenthetically commented on wartime devastation of forests in the Baltics, the northwest and the center, and on the forest depredation committed by the Japanese during their tenure on Sakhalin Island from 1905 to 1945; in view of the battles waged during the war on territory of the Ukraine and Belorussia, we can assume that numerous forests must have suffered extensive damage—forests certainly were used as part of the military strategy of both belligerents.[25] Aggregate production of commercial roundwood during the four Soviet war years, however, was only 230 million cubic meters, or twice the figure for 1940 alone; production of fuelwood was approximately 450 million cubic meters.[26] Military consumption of fuelwood, especially in areas related to heavy concentrations of military forces on both sides of the front lines, must have placed great demands on local timber supplies. Vorobyev and associates report that approximately 20 million hectares of forest were felled or destroyed in areas occupied by the Nazis or subject to military activities.[27] They note that, where logging intensified elsewhere in the U.S.S.R. to offset the loss of timber production in western areas, actual cutting—especially in central regions—far exceeded the levels of established annual allowable cut. Given that allowable cut prior to the 1960s was itself exaggerated by a factor of at least 2.5 to 3, then the resulting deforestation must have been pronounced, particularly in central regions which had already undergone centuries of forest depredation.

Although World War II is the most recent military event to affect Soviet forests directly, earlier struggles in the twentieth century also must have had a profound influence on local and regional deforestation, particularly in western regions of the Russian Empire during World War I, along key transportation routes such as the Trans-Siberian Railway during the Civil War of 1918–1921, and in areas of the maritime region of the Soviet far east occupied by Japan in the early 1920s. Further analysis of these events in the present study, however, is precluded by space and by the limited extent to which fragmentary information on associated deforestation can be synthesized into a coherent account of the general problem in the U.S.S.R. Until Soviet foresters and historians themselves provide detailed estimates of the effect of war and foreign intervention on Soviet forest reserves, other analysts must by necessity take a broadly based or aggregated regional approach to twentieth-century deforestation in that country.

Lacking a coherent Soviet literature, access to relevant archives, and consistent rather than fugitive or fragmentary published data, therefore, our historical perspective on the general problem of deforestation in the U.S.S.R. must accept that, of the two major developmental time periods comprising Soviet twentieth-century experience, the second (commencing in the late 1950s) offers a better opportunity for meaningful current systemic analysis, estimation of likely future courses of development, and tentative suggestions of deforestation behavior if normal developmental problems alone had occurred between 1917 and the mid-1950s. Seen in this way, events in the earlier period are far too erratic to provide a meaningful analysis of deforestation as a process or general condition in comparison to that experienced by other countries in the first three-fifths of the twentieth century.

The first four decades of Soviet rule were associated with consolidation of power, development of central planning, years of war and destruction, and periods of reconstruction, recovery, and reassessment. The country's industrial economy was engaged in a struggle for survival, rather than in a balanced and managed use of such elements of the natural environment as the extensive forests in European Russia, the Urals, and accessible Siberia. Analysis of Soviet deforestation, therefore, especially in the worldwide context of the problem, appears to be most appropriately addressed by evaluation of events and processes that have characterized the past 2½ decades of Soviet industrial activity, in which economic conditions hopefully more indicative of normal Soviet practice seem to have prevailed. The major problem in

Soviet deforestation consequently appears to be the extent to which the national central administration and political structure is prepared to establish prices, plans, and programs essential for sustained yield management of the resource, and to monitor effectively at all levels the activities of those responsible for their implementation.

The twentieth-century transformation of the Soviet forest industry from a state of chronic underdevelopment into a modern commercial industrial sector has reached a point in the 1980s where further progress and enhanced resource utilization probably cannot be achieved without important modifications to existing administrative and operating practices and without major revisions to central planners' understanding of the geographical dimensions underlying the national space economy.[28] The improvement of management in forestry and wood-processing, the allocation of capital for thorough inventories of the forest resource, and the provision of significant amounts of capital for reforestation and silviculture all suggest that the U.S.S.R. to this point in time has reached only a minimal level of effective utilization of its forest resource. Future increases in the utilization and productivity of timber resources and attendant reductions in forest depredation can probably best be achieved by more intensive management and better utilization of forests in accessible but often significantly deforested regions—such as central European Russia, the Baltics, and the Urals—which have superior locations in the national space economy and which generate higher rates of annual tree growth.

Recent Changes in Forest Volume Area

The area and volume of timber related to forested land has apparently been increasing in the past quarter century (see table 11.4). The sizes of forested land and primary growing stock shown in the four reliable postwar inventories (there have been five major inventories in the Soviet period, all since World War II) suggest that the U.S.S.R. has been making considerable strides in its reforestation programs. As Holowacz has observed, the increase in forest area between 1966 and 1973 "is equivalent to more than the entire forested area of Sweden."[29] Given, however, that the U.S.S.R. cuts approximately 400 million cubic meters annually (about half its mean annual increment)

Table 11.4 U.S.S.R. Forest Resource (forested area and related growing stock, from inventory, as of 1 January)

	1961	1966	1973	1978
Total forested area (ha. × 10^6)	738	747	769	792
Collective farm forest	32	26	20	20
Industrial forest	706	721	749	772
Forest managed by central state forestry agencies	647 (est.)	661	675	694
Total growing stock (cu. m. × 10^9)	80	80	82	84
Collective farm forest	3	2	2	2
Industrial forest	77	78	80	82
Forest managed by central state forestry agencies	71 (est.)[a]	73	75	77

Sources: Narodnoye Khozyaystvo SSSR. 1922–1982 (Moscow, 1982), 389; personal communication from J. H. Holowacz.
a. Estimated by author.

from approximately 2 million hectares, loses timber on a significant further area to fire and disease, and claims to regenerate approximately 2.4 million hectares (table 11.5) annually, the forested area and the volume of forest should be declining.[30] Yet Holowacz notes that the reforestation program tends to focus on forest in the sparsely forested regions but that increases in forest volume are recorded in the taiga forest. Given that the area of conifers is increasing, Holowacz concludes that "most of the additional forested area of 22 million hectares included in the 1973 inventory, but not accounted for during that of 1966, originated in the taiga forest region where coniferous species predominate."[31] A similar increase occurred between 1973 and 1978. Presumably such increases "occurred not as a result of regeneration operations, but mainly due to the inclusion of new areas not having been accounted for during the preceding inventories."[32]

The U.S.S.R. is increasing the area of artificially established forests, however, and Holowacz reports an increase between 1966 and 1973 of 5.1 million hectares, which brought the national total of such forests to 15.9 million hectares in 1973.[33] He further notes that the mean annual increment reached 1.39 cubic meters/hectare in 1973 from

Table 11.5 U.S.S.R. Forest Management by Central State Forestry Agencies

Early Soviet periods	Forest restoration and afforestation		
	Total (ha. × 10^3)	Seeding and planting (ha. × 10^3)	Assisted natural regeneration (ha. × 10^3)
1917–37	1,835	1,326	509
1938–41	1,054	964	90
1942–45	245	165	80
1946–49	2,182	1,373	809
Selected years			
1950	1,088	729	359
1955	1,188	582	606
1960	1,577	824	753
1965	2,011	1,223	788
1970	2,354	1,290	1,064
1975	2,363	1,275	1,088
1976	2,329	1,238	1,091
1977	2,354	1,247	1,107
1978	2,395	1,291	1,104
1979	2,421	1,323	1,098
1980	2,454	1,358	1,096
1981	2,427	1,332	1,095
Total 1917–81	64,817	36,561	28,256

1.35 cubic meters/hectare in 1966. In the 1978 inventory this figure was reported as 1.38 cubic meters/hectare for all lands administered by central state forestry management; the mean annual increment on European-Uralian forested lands thus administered was 2.13 cubic meters/hectare, an increase from 2.00 cubic meters/hectare in 1973.

The mean annual increment (yearly net addition of wood fiber) in forests is a useful measurement to compare with annual allowable cut (a management term based on present age and species composition of the forest and incorporating numerous variables associated with rates of regrowth, planned future utilization of forest land, and the principles of sustained-yield forest management). The former criterion has greater value in general analyses than annual allowable cut but

Table 11.5 Continued

Selected years	Managed/ regulated forests (ha. × 10^6)	Improvement thinnings and selection fellings (ha. × 10^3)	Total volume (cu. m. × 10^6)	Marketable volume of solid timber (cu. m. × 10^6)
1950	20.7	2,523	24.2	n/a
1955	31.7	2,365	21.0	n/a
1960	41.3	2,699	22.3	n/a
1965	39.5	3,110	30.0	25.4
1970	41.6	3,823	44.1	36.7
1975	46.0	3,979	49.7	40.4
1976	46.4	4,015	51.6	41.9
1977	47.1	3,971	52.8	42.6
1978	47.0	3,952	52.1	41.9
1979	47.1	3,927	52.4	42.3
1980	46.8	3,919	52.6	42.4
1981	47.4	3,975	54.7	44.3

Sources: Early Soviet periods and 1917–81 total from J. H. Holowacz and *Lesnoye Khozyaystvo SSSR za 50 Let* (Moscow: Lesnaya Promyshlennost, 1967), 20; remainder compiled from *Narodnoye Khozyaystvo SSSR v 1977 g.* (Moscow, 1978), 221; *Narodnoye Khozyaystvo v 1980 g.* (Moscow, 1981), 220–21; *Narodnoye Khozyaystvo SSSR. 1922–1982* (Moscow, 1982), 389–90.
Note: n/a = data not available.

obviously is a cruder measurement of the amount of timber that could be cut without jeopardizing the health of a forest. At present, one-third of the mean annual increment occurs in European forests (25 percent of the coniferous increment, 74 percent of the shade tolerant hardwood increment, and 47 percent of the shade intolerant hardwood increment), and the remaining two-thirds occurs in forests east of the Urals. In the European U.S.S.R., conifers constitute 52 percent of the mean annual increment, shade tolerant hardwoods 9 percent, and shade intolerant hardwoods 39 percent. In Asian regions, these shares are 68 percent, 4 percent, and 28 percent respectively. The share of deciduous species in the annual increment of Asian forests is twice that of their proportion in forested areas due to the excessively high share of middle-aged coniferous forest with attendant low annual increments, and also due to the fact that deciduous stands occupy more favorable sites.

These observations suggest that the precision of forest inventory methods and procedures in the U.S.S.R. is likely increasing with each successive attempt to measure the forest resource. The area included in the inventory is obviously being extended into the accessible taiga areas of the European north and Siberia, and the areas subject to rough estimates are obviously becoming relatively smaller. Official releases and assessments of inventory data claim that improvements in the forest resource stem from the effectiveness of forestry measures carried out in management of the resource.[34] Certainly the expansion of forested area and volumes of timber comply with the Soviet economic philosophy of expanded spheres of reproduction, but the main credit for such increases appears to belong to administrative and technology-related changes in the areas being measured and in the reliability of forest mensuration techniques. Further observations by Glotov and Eronen suggest that Soviet forest authorities have been revising their terminology in the post–World War II period, especially in the definition of annual allowable cut.[35] At a national level, therefore, the size of the Soviet forested area and primary growing stock appears to be expanding despite major annual incursions by loggers, fires, and insects.

Within economic regions, however, destruction of growing stock and deleterious change in forest volume and species composition are well documented. Furthermore, many critical aspects of these processes occur in the accessible forests of European Russia and Siberia. Greater felling of mature (adult trees with decreasing rates of annual growth; minimum size for normal commercial exploitation) and overmature (adult trees nearing or at maximum size, increasingly decadent and vulnerable to disease and insect depredation) timber in accessible or exploitable European-Uralian forests between 1973 and 1978 reduced the volume of coniferous growing stock by 450 million cubic meters (slightly more than one year's total national timber harvest). The commercial reserve of coniferous timber declined by 5 percent in the Northwest, 16 percent in the Urals, 17 percent along the Volga, and 12 percent in the Volgo-Vyatka districts. The significant reduction of mature timber and the limited availability of stands approaching maturity have prompted calls for greater precision in determination of allowable cut in these regions in order to ensure sustained yields.[36] The problem of definition is compounded by considerable species and geographical variation in the size of trees at any given age.

The growing consumption of spruce pulpwood by the pulp and paper industry and fuller utilization of pine for resin rather than pulp is causing the felling of spruce in the European-Uralian forests to exceed the allowable cut. When coupled with the length of time required for regeneration of spruce, overutilization is reducing the forested area of European-Uralian spruce forests; improved regeneration and forest management practices are urgently needed.

Despite Soviet efforts to improve the species composition of coniferous forests (particularly by planting pine and spruce) to ensure continual improvement in the economic utility of the growing stock, much of the 6.7 million hectare increase of first-class coniferous juvenile stands between 1973 and 1978 comprises larch forest, the utilization of which is weakly developed in the U.S.S.R.[37] During this period, mature stands of conifers declined by 5.1 million hectares.

In the European-Uralian forests a customary but undesirable replacement of conifers by shade intolerant species during natural regeneration appears to have been arrested generally, although in some regions of the north and the Urals it continues, albeit at a reduced rate. Consequently, and assisted by the weak utilization of conifers, their area increased by 1.9 million hectares between 1973 and 1978; 0.9 million hectares of this increase were in European-Uralian forests. Perhaps due to their growing environment, aspen forests have been replaced more successfully than birch by such economically desirable species as pine and spruce.[38]

The process of replacing mature and overmature timber (in European forests primarily) by juvenile and maturing stands appears to be occurring in a ratio of nearly 3 to 1. This replacement is highly significant for the long-term management of forests and for effective definition of annual allowable cut. As Eronen notes, the Soviet "allowable cut remains much smaller than that in similar conditions in the Nordic countries, where it is based on total growth (provided the age class distribution is even)."[39] The Soviet Union seems to use five methods for determining annual allowable cut but periodically levels undergo drastic change (between 1951 and 1973, for example, the limit was reduced from 1,752 million to 620 million cubic meters).[40] Only mature stands are taken into account in determination of annual allowable cut. Inclusion of younger stands in thoroughly managed forests, for example, would enhance regional productivity, improve the general utility of accessible—especially European-Uralian—forest, and would reduce the need for expensive development of Siberian accessible and currently inaccessible forests, including those in the Baikal-Amur Mainline (BAM) service areas.

Such remote, unpopular, and inhospitable areas are "being constrained by a labour shortage . . . it is unlikely that BAM will open up forestry in the short run."[41] Expectation of future harvesting in this area of not more than 27 million cubic meters annually out of 6 billion cubic meters—or 10 percent of the country's resource—however, allegedly comprises one of the major justifications for construction of this second Trans-Siberian railway because it "will provide the most likely route for East Siberian timber" to Pacific ports.[42] The total planned future volume from the BAM service area equals the maximum allowable volume annually recoverable from thinnings in European reforested areas (of which only two-thirds is presently harvested).

Nationally, mature and overmature stands comprise two-thirds of the growing stock. Conifers represent three-fifths and shade intolerant species most of the remainder. Mature stands are less significant in the European-Uralian forests but their importance diverges among developed regions. Mature timber is predominant in European-Uralian regions only in the Northwest and the Urals; elsewhere, juvenile stands predominate—for example mature stands comprise 5.6 percent of the forested area in the Central Black Earth region, 2.6 percent in Belorussia, 6.3 percent in the Ukraine, and 6.2 percent in Lithuania.[43] Thus between 1973 and 1978, the general volume of timber resources in the U.S.S.R. increased by 2.2 billion cubic meters but the volume of mature and coniferous forests declined by 860 million cubic meters.[44]

Europe-Uralia versus Siberia and the Far East

Given the distance of many forests from domestic and international markets and the priorities for investment outside Soviet forestry (in such sectors as agriculture, aerospace, defense, computers, and programs to improve efficiency in consumption of fuel and most industrial materials), national increases in volumes of growing stock and forested area mask the true economic and ecological significance of overexploitation of the European-Uralian forest. Significant opportunity costs arise when European forests especially are not thoroughly managed and effectively reforested (the importance of regional forest to the timber and wood-processing industry, for example, is further explored in various publications by the author).[45]

Eastern forests (particularly those east of the Baikal Meridian, which defines the usual eastern economic limit for areas tributary to Soviet Europe) have significance to Asian markets, but the volume of roundwood exported from Siberian and far eastern forests to Pacific Rim countries (including China) in 1981 was less than 2 percent of the Soviet national timber harvest.[46] The dominant market is Japan, with which the Soviet Union has an overall trade deficit. While Japan takes approximately two-fifths of Soviet roundwood exports (primarily from adjacent areas in the U.S.S.R.), its purchases in 1981 comprised only 17 percent of far eastern roundwood production for 1975, and 5.5 percent of combined eastern Siberian and far eastern production for that year (the last year for which regional production of roundwood was published).[47] Although occasional sales of roundwood, such as the sale of one million cubic meters of roundwood to China announced early in 1983, receive justified publicity, the level of significance of Soviet eastern forests is determined mainly by their marginal utility to domestic processors and European markets.[48]

This general, low regional utility not only reflects a spatial bias in the location of processing toward the European U.S.S.R. and the commitment of that country to many eastern and western European markets, but it also clearly reflects the heavy bias in eastern forest toward larch, whose utility is still much more problematic than that of any other major coniferous species. Discussions relevant to species composition of regional forests and demand by major geographical areas, however, lie beyond the present analysis. Although these subjects have been mentioned in numerous other studies, they deserve major consideration in future evaluations of the BAM railway's service zone and in analyses of Soviet trade opportunities in Pacific Asia.

Relative location and harsh environment together restrict the potential economic utility of most projects envisioned for all eastern forest. Mean annual increment per hectare is approximately three times greater in accessible Europe than in Siberia. Many of the forests of the European north, Siberia, and the far east have severe climates and excessively moist or poorly drained soils. In these districts, for example, on the five-point Soviet site quality scale, stands on the sites in two best categories comprise only 11 percent of the forested area, those on the next two sites about 51 percent, and those on the worst site, 38 percent.[49] State forests contain 73.6 percent of the country's swamp, all of its ponded stands, and over 26 percent of the area of its fresh water bodies.[50] Approximately one quarter of these areas are located in the European U.S.S.R. Active amelioration is recognized as a major solution to these poor environmental conditions but to date the scale of operations has been insufficient to affect national levels of productivity. In the period from 1973 to 1978, drainage of 300,000 hectares of swamps in the southern provinces of the northwest seems to have been the major accomplishment in reducing the size of swampy nonforest forest land; the associated land uses, however, appear to be pasture and natural hayfields, not forest plantations.[51]

The relative superiority of European forest environments is also noticeable in the utilization practices pursued among the three official groups of Soviet forests.[52] Thinning and improvement fellings contribute approximately 10 percent of Soviet annual timber harvest, 47 percent occurs in "protection forests" (Group 1), 48 percent originates in the "industrial forests in densely populated areas" (Group 2), and the remainder is found in the "general industrial forest" (Group 3).[53] Groups 1 and 2 comprise half of the Soviet European state forest.

Thinnings comprise 44 percent of the timber obtained from Group 1 forests, 17 percent of that from Group 2 forests, and 1.4 percent of that obtained from Group 3 forests. Unfortunately, the large relative share of thinnings in some regions is unwarranted, and is due to the needless thinning out of stands and the underutilization of more valuable mature timber. The greatest prospect for increasing the volume of thinnings exists in the Asian U.S.S.R., although the most comprehensive system for obtaining thinnings relies on a well-developed infrastructure, logging system, and supply of labor.[54] Thus despite the excessively large portion of mature and overmature timber in Siberian and far eastern forests, thinnings and improvement fellings have considerable importance in European forests, where sustained-yield management is perceived as a cost-effective alternative to deeper penetration into the taiga. Most of the intensive forest management practices represented by thinnings and improvement fellings occur in the European U.S.S.R. and serve as a constant reminder of the real alternative that effective management and timber utilization in Europe represent to peripheral northern and Siberian logging.

Group 3 is the largest category and comprises forests suitable for full commerical exploitation. It accounts for some 45 percent of the state forest in the European U.S.S.R. and for 84 percent of that in the Asian U.S.S.R. The groups are further broken down into operational (accessible and potentially accessible) forest open to harvesting, reserve forest, and inaccessible forest. Approximately 55 percent of Soviet mature timber volume is found in accessible or potentially accessible forests (and only about half of this is being utilized); the accessible share of mature stands is 87 percent in European forests and 48 percent in Asian forests.[55] Group 2 forests are the most accessible; even in Asia, 80 percent of their mature timber is accessible, although 96 percent of their mature timber is accessible in the European U.S.S.R. The annual allowable cut allocated is 75 percent to Group 3 forests, 17 percent to Group 2 forests, and 8 percent to Group 1 forests.[56]

Deforestation through Deleterious Management

Despite comprehensive environmental planning and legislation and extensive and thorough resource assessments by professionals and academics, environmental abuses including destruction of the forest resource continue to draw the ire of responsible observers in the U.S.S.R.[57] The litany of trangressions continues relentlessly to suggest that implementation of environmentally and economically sound solutions to the problems does not receive priority attention by those ultimately responsible for economic activity and resource management—the Council of Ministers and the CPSU Politburo—and from which all other organizations take their cue.

Notwithstanding the inherent wealth of the Soviet forest resource, the U.S.S.R. actually encounters timber shortages in its economy. *Izvestia* asked in 1981, “Why is there a shortage of timber?—despite the great wealth of our green ocean,” and suggested that it was the authors' "profound conviction that the answer must be sought above all in the rational utilization of the timber that we earmark for cutting and that we actually cut."[58] In the Ministry of Timber, Pulp and Paper, and Wood-Processing Industry (the principal Soviet timber consumer), for example, the best stands are cut, the rest are left in a confusing array of cut and uncut stands unprofitable for further exploitation, trimmings and trunks are ignored, scrap suitable for processing is burned needlessly, and logging enterprises are prematurely shutdown. Of the 400 million cubic meters harvested annually, 160 million cubic meters are lost in various stages of production; only 51 percent reaches the consumer and only 8 percent of the remainder (scrap) is utilized anywhere.[59] Many of the processes causing the Soviet form of deforestation can be illustrated from such penetrating outcries in the Soviet press because they echo and supersede myriad other exposés carried in the press and scientific literature in the thirty years since Stalin's demise.

The damage to the resource and the poor record of product utilization are caused partially by the activities of 10,000 enterprises working in the forests under the jurisdiction of 34 ministries and departments. The U.S.S.R. State Planning Committee (GOSPLAN) allocates assignments in terms of total timber removals undifferentiated by coniferous and deciduous species. When loggers then seek permission to obtain the specified, planned allocations, they must pass through several other agencies before organizations of the State Forestry Committee, which supervises forest resources and allocates timber stands. Often the required timber stands are not allocated; the 220 million cubic meters harvested annually by the major consumer actually is about 6 million cubic meters short of the requirement planned by GOSPLAN; of the remaining 180 million cubic meters harvested, 80 million cubic meters are cut by enterprises of the U.S.S.R. State Forestry Committee, and most of the remainder is harvested by "so-called unorganized forest users," which are enterprises of various ministries and departments cutting timber for their own needs.[60] These entrepreneurs are not part of the official pulp, paper, and wood-processing industry. Utilizing timber as poles, planks, boards, and so forth in myriad aspects of construction and repair, these enterprises subject logs to little, if any, value added through manufacturing and are not induced by royalty (stumpage) fees or wholesale prices to attach scarcity to their timber supplies. Their activities appear most pernicious in areas relatively remote from general central (Moscow) administration but that are integral components in the spatial economy of the expanding pulp, paper, and wood-processing industry.

The current Five-Year Plan and guidelines up to 1990 stipulate that greater use must be made of European forests without harm to the environment; about 80 percent of the country's timber requirements are found in Europe but the eastern region operating costs that include "normative depreciation" (*privedyonniye zatraty*) (i.e., capital investments, outlays on current operations, transportation expenses) are 1.5 to 2 times greater than in Europe. The solution is to make greater comprehensive use of roundwood in the east before products are shipped to Europe and to make greater use of allowable cut in Europe, where some 23.3 million hectares of forest or 19 percent of the total stand are not being utilized. Apparent confusion generating earlier curtailment of European cutting has caused unwarranted migration of logging to the east although the European forests could sustain 27 million cubic meters of cutting during forest management, thinning, and improvement felling.

Such practice currently yields 18 to 20 million cubic meters in the European-Uralian regions and thus could yield another 50 percent without violating regulations. Enterprises of the State Forestry Committee engage in this type of cutting and nationally produce approximately 40 million cubic meters annually from cleaning and improvement cutting; the remainder of their harvest unfortunately is mature timber that increasingly comes from stands allocated to the principal consumer (a specialist term referring to all the official enterprises subordinate to the U.S.S.R. Ministry of Timber, Pulp and Paper, and Wood-Processing Industry). In fact, the State Forestry Committee should establish sustained-yield practices in core (but currently sparsely or poorly forested) areas and leave the cutting of mature timber to the planned, comprehensive activities of the principal consumer in forest surplus regions.

To fulfill their own objectives, agencies other than the principal consumer are harvesting timber supplies earmarked for planned cutting near major facilities and thus undermine the resource base of the main consumer and the period of future viability of its loggers. This "autonomous" logging apparently is necessary because the principal consumer cannot meet the country's timber requirements and has not been organized to do so. The principal consumer carries out only about one-third of the annual reforestation, but the significance of even this modest effort for the long-run material spatial planning of production facilities is negated by spatially concurrent logging activities of other agencies.[61] After enterprises are built, settlements are established, and roads and railroads are constructed, the principal consumer frequently finds that allocated cutting areas have had their timber pilfered by autonomous or nomadic operators whose unit costs and wage expenditures are two to three times greater than those of the principal consumer.

The principal consumer (Ministry of Timber, Pulp and Paper, and Wood-Processing Industry) found that in major areas of recent expansion in Siberia such as forests adjacent to the Ivdel-Ob railroad, the Bratsk timber complex, and the western portions of the BAM, over

40 percent of the time reserves had been allocated to independent loggers. In Tyumen oblast, for example, which has over 8 percent of Soviet mature and overmature timber, all of it coniferous, "the very utilization of the forest is not sensible. Thus intensive cutting adjacent to the Ivdel-Ob and Tavda-Sotnik Railroads exceeds the allowable by more than one and a half million cubic meters. As a result, the natural climatic and hydrological regimens in the area have been noticeably disturbed."[62] Planning for the region has been so poor that, in 1976, this timber-rich area imported 200,000 cubic meters from other provinces to satisfy its growing need for industrial timber.

In the early 1970s in Irkutsk oblast 140 independent logging camps were operated by various construction enterprises—twice the number operated by the principal consumer. Of the 7 million cubic meters required annually by the Bratsk wood-processing complex, the largest in the U.S.S.R., only 2.5 million cubic meters could be obtained from the forested land previously allocated to ensure its supply, although a million cubic meters were annually harvested on that land. The difference was taken by independent loggers.[63] The annual allowable cut from the land in question was 6.5 million cubic meters; thus 1.5 million cubic meters were being cut in excess of permissible levels by all loggers. Throughout Irkutsk oblast, independent loggers were overharvesting the allowable cut by 5 million cubic meters annually and made a large contribution to the accumulated overcut of 50 million cubic meters recorded by 1972. By 1972, 52 million cubic meters previously allocated to official consumers of timber in Irkutsk province had been recently transferred to the independent loggers. This overcutting did not apply to areas cleared for hydroelectric establishments in the province but to those areas necessary to sustain the timber supply for large, capital-intensive processing facilities built in Irkutsk province since the 1960s. *Pravda* concluded by noting "that if the cutting in the zone allotted to the wood chemistry giants continues at the same rate, all the timber will have been cut down within 40 years. There are already some sad lessons: the raw materials resources of the Zima, Birusa, Yurty, Tulun and Chunsky combines, whose construction cost many millions of rubles, are exhausted. Timber is hauled hundreds of miles to them by railroad."[64] Reports published subsequently confirm that independent logging continues to plague the forest resources of Irkutsk province.

Thus, throughout the major forest regions of the northwest, Volgo-Vyatka, the Urals, western and eastern Siberia, and the far east, enterprises of different ministries and departments operate side by side, duplicate processing, transportation, maintenance, and social facilities, and erode the long-term stability and viability of the forest resource. The answer to this confusion is clearly apparent to Soviet management, in such forms as comprehensive, integrated large-scale timber operations, but those guiding the operational structure of the economy have been unable or unwilling to act.

Conclusion

The foregoing observations suggest that the Soviet Union has a form of deforestation in which major regional changes in forested land and species composition are being affected by conflicts in central planning and institutional management. These practices act to the detriment of the entire resource base although national forest land, forested area, and volumes of standing timber in aggregate are expanding through improved mensuration and inventory technology. Unlike the forests of many nations, Soviet timberlands are not under attack by landless peasants, unemployed urban dwellers, or plantation agriculture. Although nominally a type of federal state, the U.S.S.R. does not have the kind of problems causing deforestation in such societies as Canada and Malaysia, where provincial ownership of the resource is at odds with federal or national planning objectives. Nevertheless, the apparently unwieldy Soviet central planning apparatus, the size and multiplicity of forest-related institutions, and the divergence, both short-run and long-term, among the objectives of forest users create a situation in the U.S.S.R. that is clearly contrary to the ideological and economic goals of the state and party. Many of the shortcomings in forest management that create forms of deforestation can be masked both by secrecy and by clever adjustments to successive inventories, but eventually, as is evident in the cries of indignation by responsible foresters, the state itself reaches the point where the confusion in management and paralysis of production cause economic stagnation that the authorities "can no longer tolerate."[65] In this case, the Soviet form of deforestation clearly has become inimical to the stated economic and social goals of that society.

NOTES

1
Deforestation in the Araucaria Zone of Southern Brazil, 1900–1983

1 Brazil. Instituto Nacional do Pinho, *Anuário Brasileiro de Economia Florestal* (1948): 416.
2 Food and Agriculture Organization of the United Nations (hereinafter cited as FAO), *Proyecto de evaluación de los recursos forestales tropicales. Los recursos forestales de la America tropical* (Rome, 1981), 58; Dora A. Romariz, "Mapa de vegetação original do estado do Paraná," *Revista Brasileira de Geografia* 15 (1953): 597–612; Beneval de Oliveira, "As regiõas de occorência normal de Araucaria," *Anuário Brasileiro de Economia Florestal* 1 (1948): 185–99.
3 A. Aubreville, "A Floresta do pinho do Brasil," *Boletím Geográfico* 12 (1954): 164–73; Edgar Kuhlmann, "A Vegetação original do Rio Grande do Sul," *Boletím Geográfico* 11, no. 1 (1952): 157–63.
4 Roberto M. Klein, "Fitosonomia e notas sobre a vegetação para accompanhar a planta fitogeográfica de partes dos Municípios do Rio Branco do Sul, Bocaiuva do Sul, Almirante, Jamandaré e Colombo (Paraná)," *Boletím da Universidade do Paraná Instituto de Geologia* 3 (1962): 1–33; Roberto M. Klein, "Fitosonomia e notas sobre a vegetação para accompanhar a planta fitogeográfica do Município de Curitiba e arredores (Paraná)," *Boletím da Universidade do Paraná Instituto de Geologia* 3, no. 2 (1962): 1–29; Aubreville, "A Floresta do pinho," 165.
5 Auguste de Sainte-Hilaire, *Viagem à Comarca de Curitiba (1820)* (São Paulo, 1964), 10–12; Roberto M. Klein, "O Aspecto dinâmico do pinheiro brasileiro," *Sellowia* 12 (1960): 34; *Anuário Brasileiro de Economia Florestal* (1949): 339–40; R. L.

Rogers, "Problemas silviculturais da Araucaria angustifolia," *Anuário Brasileiro de Economia Florestal* 6 (1953): 308–70; Aubreville, "A Floresta do pinho," 165–67.

6 Arturo Dias, *Brazil of To-Day*, trans. Louis Raposo (Nivelles, Belgium, 1907), 496.

7 C. M. Delgado de Carvalho, *Le Brésil méridional* (Rio de Janeiro, 1910), 361; Rio de Janeiro Instituto de Expansão Commercial, *O Brasil atual* (Rio de Janeiro, 1930), 78–80.

8 Pierre Denis, *Brazil* (London, 1911), 276.

9 Reinhard Maack, "O Aspecto fitogeográfico atual do Paraná," *Anuário Brasileiro de Economia Florestal* 6 (1953): 38; Orlando Valverde, *Geografia agrária do Brasil* (Rio de Janeiro, 1964), 351; Temístocles Linhares, *Paraná vivo* (Rio de Janeiro, 1953), 111–12; Joe Foweraker, *The Struggle for Land. A Political Economy of the Pioneer Frontier in Brazil from 1930 to the Present Day* (Cambridge, 1981), 34; Ditmar Brepohl, "A Contribução econômica da exploração da araucaria angustifolia a economia paranaense," in *Forestry Problems of the Genus Araucaria* (Curitiba, 1979), 349–50; FAO, *Recursos forestales tropicales*, 38.

10 Maack, "O Aspecto fitogeográfico do Paraná," 34; R. Andrew Nickson, "Brazilian Colonization of the Eastern Border Region of Paraguay," *Journal of Latin American Studies* 13 (1981): 111–31.

11 This aspect of frontier history is by no means unusual; see Alfred Crosby, *Ecological Imperialism* (New York: Cambridge University Press, 1986), chap. 2, "Pangaea Revisited, The Neolithic Considered," 8–41.

12 Carvalho, *Brésil méridional*, 364; Valverde, *Geografia agrária*, 349–50; Preston James, *Brazil* (New York, 1946), 200.

13 Denis, *Brazil*, 285.

14 Luiz F. G. Laboriau, "Notas preliminares sôbre a região da Araucaria," *Anuário Brasileiro de Economia Florestal* 1 (1948): 216; Harry A. Franck, *Working North from Patagonia* (New York, 1921); Raphael Zon and William N. Sparhawk, *Forest Resources of the World*, 2 vols. (New York, 1923), 2:704; Merle S. Lowden, "Paraná, Brazil Fire Team Report" (1963), unpublished paper shown to me through the courtesy of J. L. Whitmore, U.S.F.S.

15 Cecilia Maria Westphalen and Altiva Pilatti Balhana, "Nota prévia ao estudo da expansão agricola no Paraná moderno," *Boletím do Departemento de História da Universidade do Paraná* 25 (1977): 7–10; James Henderson, *A History of the Brazil, Comprising Its Geography, Commerce, Colonisation, Aboriginal Inhabitants, etc., etc.* (London, 1821), 141; Centro Indústrial do Brasil, *O Brasil. III. Indústria de transportes. Indústria fabril* (Rio de Janeiro, 1909), 2:248.

16 Julio Pompeu, *Vier staaten brasiliens. Four Brazilian States* (Rio de Janeiro, 1910), 34–40; Dias, *Brazil of To-Day*, 502–3; Nadia A. Cancian, "Romario Martins e 'O Livro das arvores do Paraná,'" *Boletím do Departamento da História da Universidade do Paraná* 21 (1974): 174; Ernani Silva Bruno, *História do Brasil. V. São Paulo e o Sul* (São Paulo, 1967), 196–97; Centro Indústrial do Brasil, *O Brasil*, 2:77.

17 Cecilia Maria Westphalen, Brasil Pinheiro Machado, and Altiva Pilatti Balhana, "Nota prévia ao estudo da occupação da terra no Paraná moderno," *Boletím do Departamento da História da Universidade do Paraná* 7 (1968): 8; Linhares, *Paraná vive*, 99–118; Zon and Sparhawk, *Forest Resources*, 2:780; Charles S. Sargeant, *The Spatial Evolution of Greater Buenos Aires, 1870–1930* (Tempe, Ariz., 1974), 29–31,

64–66; Institute of Inter-American Affairs, *Forest Resources of Paraguay* (Washington, D.C., 1946), 58. I owe a debt here to Jack C. Westoby, former senior director of the FAO Department of Forestry, for alerting me to my muddled interpretation of official statistics. Of all plantation softwoods in Brazil, the southern states had about 24% and the southeast 54% (1985 unpublished data from IBDF).

18 Brepohl, "A Contribução econômica," 347.

19 Franck, *Working North from Patagonia*, 140–55; M. G. Mulhall and E. T. Mulhall, *Handbook of Brazil* (Buenos Aires, 1877), 17; Nickson, "Brazilian Colonization," 111–31.

20 The figure of 65 cubic meters per hectare derives from data cited above in note 1. It is, if anything, low. Brepohl offers data that suggest yields of between 104 and 166 cubic meters per hectare. His figures derive from the practices of the 1970s, which may have been more efficient than those in use earlier. Thus I have stuck with the lower figure, which comes from the 1950s. If one prefers Brepohl's data, the area deforested by timber operations amounts to roughly 50% less. Brepohl, "A Contribução econômica," 349.

21 Westphalen and Balhana, "Nota prévia ao estudo da expansão agricola," 10–16; Nilo Bernardes, "Expansão do povoamento no estado do Paraná," *Revista Brasileira de Geografia* 14 (1952): 438–40. On settlement in Paraná generally, see Romário Martins, *Quanto somos e quem somos?* (Curitiba, 1941); Lydia M. C. Bernardes, "Crescimento da população do estado do Paraná," *Revista Brasileira de Geografia* 12 (1951): 265–74; Lydia M. C. Bernardes, "O Problema das 'frentes pioneiras' no estado do Paraná," *Revista Brasileira de Geografia* 15 (1953): 335–84. On agriculture see Brasil Pinheiro Machado, *Campos Gerais—estruturas agrárias* (Curitiba, 1968); Fundação Instituto Agronômico do Paraná, *Manual agropecuário para o Paraná* (Londrina, 1976).

22 Denis, *Brazil*, 270–308; Carvalho, *Brésil méridional*, 291; Pompeu, *Vier staaten brasiliens*, 48. For a definition of caboclo see James B. Watson, "Way Station of Westernization: The Brazilian Caboclo," in *Brazil: Papers Presented in the Institute for Brazilian Studies*, (Nashville, 1953).

23 Westphalen, Machado, and Balhana, "Nota prévia ao estudo da occupação da terra"; Lydia M. C. Bernardes, "Distribução da população no estado do Paraná," *Revista Brasileira de Geografia* 12 (1950): 565–86; Denis, *Brazil*, 270–71; Foweraker, *Struggle for Land*, 64–65.

24 Craig L. Dozier, "Northern Paraná, Brazil: An Example of Organized Regional Development," *Geographical Review* 46 (1956): 318–33; Westphalen et al., "Nota prévia ao estudo da occupação da terra"; Westphalen and Balhana, "Nota prévia ao estudo da expansão agricola."

25 Foweraker, *Struggle for Land*, 106–49.

26 Valverde, *Geografia agrária*, 353–54. Dozier, "Northern Paraná," 318–33. Foweraker, *Struggle for Land*, 86.

27 Brepohl, "A Contribução econômica." Carvalho, *Brésil méridional*, 364. Newton Carneiro, "O Pioneirismo florestal do Romário Martins," *Boletím do Departamento da História da Universidade do Paraná* 11 (1964). Embaixada do Brasil, *Documents on Brazil. IV. Survey of the Brazilian Economy* (Washington, D.C., 1966), 108–10.

28 Personal communication from Antonio Carlos do Prado, 30 July 1986. The fiscal

incentives program is discussed in a white paper done for the Instituto de Planejamento Economico e Social (Brasilia) by Antonio Carlos do Prado, "Uma avaliação dos Incentivos fiscais do FISET-Florestamento/Reflorestamento" (February, 1986).

29 All figures on reforestation since 1967 came from unpublished IBDF data kindly provided to me by Antonio Carlos do Prado of IPLAN/IPEA and William Beattie of the World Bank. Some figures are in Brepohl, "A Contribução econômica," 350.

30 Foweraker, *Struggle for Land*, 33–35, 64–65, et passim.

31 FAO, *Proyecto de evaluación*, 50–58.

2
Southern Mount Kenya and Colonial Forest Conflicts

This research was funded by grants from the National Science Foundation, the University of California at Santa Barbara, and the Inter-Cultural Studies Foundation of New York City. Their support is gratefully acknowledged. Opinions and findings expressed here do not reflect the views of any of the institutions named above.

Abbreviations used in citations include the following:

A.C.F.	Assistant Conservator of Forests
A.O.	Agricultural Officer
Ag. P.C.	Acting Provincial Commissioner
C.L.G.L.S.	Commissioner of Local Government, Land, and Settlements
C.L.S.	Commissioner of Land and Settlements
D.C.	District Commissioner
D.F.O.	District Forest Officer
D.O.	District Officer
E-	File located in the Embu District Archives
EAPR	East African Protectorate Annual Report
EDAR	Embu District Annual Report
ELNC	Embu Local Native Council, Minutes of Meetings (As of 12–16 Feb. 1951, the council was known as the African District Council. Since independence it is known as the Embu County Council).
E-LO	Embu District Archives, Law and Order, Mau Mau Emergency Prohibited Areas file
E-LND	Land Commission
E.O.E.C.	Executive Officer, Emergency Committee
E.O.P.E.C.	Executive Officer, Provincial Emergency Committee
FDAR	Kenya Forest Department Annual Report
Hs.A.Ds.	Heads of All Departments
KLC	Kenya Land Commission Report
KLC-E	Kenya Land Commission. Evidence (3 vols.)

KPAR Kikuyu Province Annual Report
NDAR Nyeri District Annual Report

1 Recent works include two reports by the Food and Agriculture Organization of the United Nations (hereinafter cited as FAO): "Intensive Multiple-Use Forest Management in Kerala" (Rome: FAO Forestry Department, 1984); and "Intensive Multiple-Use Forest Management in the Tropics" (Rome: FAO Forestry Department, 1985).
2 For example, see M. Sorrenson, *Origins of European Settlement in Kenya* (Nairobi: Oxford University Press, 1967); Frank Furedi, "The Social Composition of the Mau Mau Movement in the White Highlands," *Journal of Peasant Studies* 1 (1974): 486–505; R. Tignor, *The Colonial Transformation of Kenya* (Princeton: Princeton University Press, 1976).
3 See R. Troup, *Colonial Forest Administration* (Oxford: Oxford University Press, 1940); J. Logie and W. Dyson, *Forestry in Kenya* (Nairobi: Government Printer, 1962).
4 Troup, *Colonial Forest Administration*, 120.
5 For further background on this area, see A. Castro, "Facing Kirinyaga: A Socio-Economic History of Resource Use and Forestry Interventions in Southern Mount Kenya" (Ph.D. diss., University of California at Santa Barbara, 1987).
6 See J. Ojiambo, *The Trees of Kenya* (Nairobi: Kenya Literature Bureau); and Republic of Kenya, *Statistical Abstract* (Nairobi: Ministry of Finance and Economic Planning, 1981).
7 S. Wimbush, "The Forests of Mount Kenya," in E. Dutton, *Kenia Mountain* (London: Cape, 1930).
8 Kenya Land Commission, *Evidence*, 3 vols. (hereinafter cited as KLC-E) (Nairobi, 1934), 1:96.
9 G. Stigand, *The Land of Zinj*, (London: Cass, 1913 and 1966); KLC-E, vol. 1.
10 See J. Thomson, *Through Masailand* (London: Sampson, 1885); C. Peters, *New Light on Dark Africa* (London: Ward and Lock, 1891); E. Gedge, "A Recent Exploration of the River Tana to Mount Kenya," *Geographical Society, Proceedings* 14 (1892): 513–33; F. Lugard, *The Rise of Our East African Empire*, vol. 1 (Edinburgh: Blackwood, 1893); L. von Hohnel, *Discoveries of Lakes Rudolf and Stephanie*, vol. 1. (London: Cass, 1898 and 1968); J. Gregory, *The Great Rift Valley* (London: Nelson, 1896); H. Mackinder, "A Journey to the Summit of Mount Kenya," *Geographical Journal* 15 (1900): 453–86; A. Arkell-Hardwick, *An Ivory Trader in North Kenia* (London: Longmans, Green, 1903); B. Dickson, "The Eastern Borderland of Kikuyu," *Geographical Journal* 22 (1903): 36–39; G. Stigand, *Land of Zinj*.
11 Arkell-Hardwick, *Ivory Trader*, 59.
12 Peters, *New Light*, 215–16.
13 G. Muriuki, *A History of the Kikuyu, 1500–1900* (Nairobi: Oxford University Press, 1974), 159.
14 R. Meinertzhagen, *Kenya Diary, 1902–1906* (London: Oliver and Boyd, 1957), 138–54; G. Mungeam, *British Rule in Kenya, 1895–1912* (Oxford: Clarendon Press, 1966), 146–47.
15 For historical background, see Sorrenson, *Origins of European Settlement*; R. Wolff, *Britain and Kenya, 1870–1930* (Nairobi: Transafrica, 1974); Tignor, *Colonial Transformation of Kenya*; G. Kitching, *Class and Economic Change in Kenya* (New Haven,

Conn.: Yale University Press, 1980); S. Stichter, *Migrant Labour in Kenya* (Harlow: Longman, 1982).

16 Sir Charles Eliot, *The East African Protectorate* (London: Arnold, 1905), 74–75, 194–96.

17 Ibid., 74–75, 103–104.

18 Mungeam, *British Rule in Kenya*, 161–64.

19 See Logie and Dyson, *Forestry in Kenya*.

20 D. Hutchins, *Report on the Forests of Kenya* (London: HMSO, 1907), 5–6, 31–34. See also East African Protectorate, *Annual Report* (London: HMSO, 1906–7), 9 (hereinafter cited as EAPR).

21 Hutchins, *Report* (1907); Mungeam, *British Rule in Kenya*, 166.

22 Hutchins, *Report* (1907), 24, 34–38.

23 Ibid., and D. Hutchins, *Report on the Forests of British East Africa* (London: HMSO, 1909).

24 Hutchins, *Report* (1909), 16–17, 66–68.

25 Hutchins, *Report* (1907), 30.

26 Hutchins, *Report* (1909), 69.

27 Mungeam, *British Rule in Kenya*, 165–66.

28 Lord Cranworth, *A Colony in the Making* (London: Macmillan, 1912), 160–61.

29 Cranworth, *Profit and Sport in British East Africa*, 207–8.

30 Cranworth, *Colony in the Making*, 156.

31 Ibid., 155–56.

32 Ibid., 162.

33 Cranworth, *Profit and Sport*, 207–8.

34 Cranworth, *Colony in the Making*, 157; idem, *Profit and Sport*, 209.

35 Idem, *Colony in the Making*, 47.

36 W. Routledge and K. Routledge, *With a Prehistoric People*, (London: Cass, 1910), 68.

37 KLC-E (1934), 3362–65.

38 Stigand, *Land of Zinj*, 253–58.

39 Ibid., 238.

40 Ibid., 258.

41 Ibid., 259.

42 Major G. Orde-Browne, *The Vanishing Tribes of Kenya*, 19–20; Hutchins, *Report* (1909), 66–67.

43 EAPR (1934), 32.

44 KLC-E (1934), 3:3182.

45 Ibid.

46 *East African Standard* (1930), 173.

47 Hutchins, *Report* (1909), 66–67.

48 Land Commission (Embu District Archives) (hereinafter cited as E-LND) 16/17/1, D.C. Embu to P.C. Kikuyu, 9 Dec. 1930.

49 For a standard history of the department, see Logie and Dyson, *Forestry in Kenya*; and Troup, *Colonial Forest Administration*.

50 *East African Standard*, 163.

51 Embu Local Native Council Archives (available in Embu and at Nairobi Archives; hereinafter cited as ELNC) (Aug. 1940), 7; ELNC (23–24 Oct. 1940), 2.

52 Nyeri District *Annual Report*, Nairobi (hereinafter cited as NDAR) (1916), appendix C.
53 Troup, *Colonial Forest Administration*, 174.
54 Kenya, Forest Department *Annual Report*, Nairobi (hereinafter cited as FDAR) (1948–1950), 35.
55 NDAR (1916), appendix C.
56 Kikuyu Province *Annual Report*, Nairobi (1915–16); and Wimbush, "Forests of Mount Kenya," 196. For excellent descriptions of indigenous bee-keeping practices, see H. Mwaniki, "Bee-Keeping: The Dead Industry among the Embu," *Mila* 1 (1970): 34–41; and L. Leakey, *The Southern Kikuyu before 1903* (New York: Academic Press, 1977), 1:250–54.
57 Embu District Archives, Forest Offenses 13/3/1 (hereinafter cited as E-FOR 13/3/1): Ag. P.C. Central to D.C. Embu, 10 May 1935.
58 E-FOR 13/3/1: A.C.F. Nyeri to D.C. Embu, 30 June 1944.
59 KPAR (1917–1918), 30.
60 ELNC (12 July 1925), 14.
61 ELNC (17 July 1925), 24.
62 ELNC (12 Feb. 1926), 66.
63 ELNC (25 Aug. 1936), 72; ELNC (10 Oct. 1927), 138; ELNC (15–16 Aug. 1933), 2.
64 ELNC (6–7 Aug. 1940), 7; ELNC (23–24 Oct. 1940), 2.
65 See Logie and Dyson, *Forestry in Kenya*, 6.
66 ELNC (10 Oct. 1927), 138–39.
67 E-FOR 13/3/1, A.C.P. Nyeri to D.C. Embu, 14 July 1931.
68 See C. Rosberg and J. Nottingham, *The Myth of "Mau Mau"* (New York: Praeger, 1966); and Tignor, *Colonial Transformation of Kenya*.
69 J. Huxley, *African View* (London: Chatto and Windus, 1931), 176.
70 E-LND 16/17/1, D.C. Embu to P.C. Nyeri, 9 Dec. 1932.
71 E-FOR 13/3/1, A.C.F. Nyeri to D.C. Embu, 20 Sept. 1932.
72 E-FOR 13/3/1, A.C.F. Nyeri to D.C. Embu, 18 July 1933.
73 KLC-E (1934), 1:257–58.
74 KLC-E (1934), 1:258.
75 Ibid.
76 Kenya Land Commission, *Report* (Nairobi) (hereinafter cited as KLC) (1934), 12–28.
77 KLC-E (1934), 1:250.
78 E-LND 16/17/1, D.C. Embu to P.C. Kikuyu, 17 Oct. 1932.
79 Rosberg and Nottingham, *Myth of "Mau Mau,"* 86, 115–16.
80 KLC-E (1934), 96.
81 KLC-E (1934), 1:96.
82 Ibid., 98.
83 KLC-E, (1934), 1:515–18.
84 KLC-E (1934), 1:518.
85 E-LND 16/17/1, D.C. Embu to P.C. Kikuyu, 31 Oct. 1932: 1.
86 E-LND 16/17/1, D.C. Embu to P.C. Kikuyu, 9 Dec. 1932: 1.
87 Ibid., 2.
88 E-LND 16/17/1, D.C. Embu to P.C. Kikuyu, 31 Oct. 1932: 2; also D.C. Embu to P.C. Kikuyu, 9 Dec. 1932.

89 E-LND 16/17/1, D.C. Embu to P.C. Kikuyu, 9 Dec. 1932: 2.
90 E-LND 16/17/1, D.C. Embu to P.C. Kikuyu, 31 Oct. 1932: 2.
91 KLC 1934: 67.
92 KLC-E (1934) 3:3181–83.
93 KLC (1934), 104–5.
94 KLC (1934), 132–35.
95 E-LND 16/17/1, The Secretariat, Nairobi, circular letter, 16 July 1934.
96 E-LND 16/17/1, D.C. Embu to P.C. Central, 13 Aug. 1934, ELNC (8–9 Aug. 1934), 11–12.
97 E-LND 16/17/1, "*Baraza* at Kerugoya, Minutes," 28 Jan. 1935.
98 E-LND 16/17/1, Ag. P.C. Central to G.L.G.L.S., Nairobi, 15 Oct. 1935; a copy is in the Nairobi archives, PC/CP/9/10/1; E-LND 16/17/1, The Secretariat, Nairobi to P.C. Central, 2 Dec. 1935.
99 E-LND 16/17/1, D.C. Embu to P.C. Central, 12 Feb. 1936.
100 E-LND 16/17/1, P.C. Central to D.C. Central, 30 May 1936.
101 E-LND 16/17/1, P.C. Central to Colonial Secretary, Nairobi, 2 June 1936. Also P.C. Central to Colonial Secretary, Nairobi, 3 Sept. 1936.
102 E-LND 16/17/1, Ag. P.C. Central to Colonial Secretary, Nairobi, 29 Nov. 1938, and other correspondence in the same file.
103 E-LND 16/17/1, C.L.S., Nairobi, to P.C. Central, 19 Dec. 1938.
104 E-LND 16/17/1, P.C. Central to C.L.S.; Nairobi, 29 Dec. 1938.
105 E-LND 16/17/1, "Notes of a Meeting, D.C.'s office, Embu," 23 Dec. 1938; also P.C. Central to C.L.S. Nairobi, 29 Dec. 1938.
106 E-LND 16/17/1, D.C. Embu to P.C. Central, 12 Jan. 1939.
107 E-LND 16/17/1, P.C. Central to C.L.S. Nairobi, 14 Jan. 1939.
108 E-LND 16/17/1, P.C. Central to D.C. Embu, 15 March 1939.
109 Embu District Annual Report (Kenya National Archives) (hereinafter cited as EDAR) (1940), 4.
110 E-LND 16/17/1, P.C. Central to D.C. Embu, 15 March 1939; EDAR (1939), 1–2; E-LND 16/17/1, "Minutes of a Meeting between Kiambu Kikuyu Elders and the Central Province Commissioner," 27 May 1936.
111 E-LND 16/17/1, A.O. Embu to D.C. Embu, 1 May 1939: 4; and C. Maher, *Soil Erosion and Land Utilization in the Embu Reserve*, 2 vols. (Nairobi, 1938).
112 Hutchins, *Report* (1909), 77.
113 EAPR (1909–10), 20.
114 Logie and Dyson, *Forestry in Kenya*, 13.
115 Logie and Dyson, *Forestry in Kenya*.
116 ELNC (10 Feb. 1943), 6.
117 EDAR (1945), 39.
118 ELNC (6–8 Nov. 1945), 4.
119 EDAR (1946), 44.
120 EDAR (1947), 11.
121 EDAR (1949), 12.
122 Ibid.
123 ELNC (11–12 Aug. 1950).
124 EDAR (1950), 8.
125 Embu District, "Handing Over Report," 1946.

126 ELNC (6 Feb. 1946), 9.
127 ELNC (21–24 Aug. 1951), 3.
128 Embu District Archives, Sawmills and Pitsawing files (E-FOR 13/1), various correspondence.
129 ELNC (21–24 Aug. 1951), 3.
130 ELNC (8 Dec. 1951), 1.
131 ELNC (8 Dec. 1951), 1–2.
132 E-FOR 13/1.
133 ELNC (21–24 Aug. 1951), 3.
134 ELNC (8 Dec. 1951), 1.
135 ELNC (21–24 Aug. 1951), 3.
136 E-LO 15/23/3, D.F.O. Nyori to A.C.F. Embu, 14 Nov. 1952.
137 E-LO 15/23/2, E.O.E.C., Embu to D.O.s, Embu, 5 Sept. 1953. Also E.O.E.C. Embu to E.O.E.C. Central, 22 July 1953.
138 E-LO 15/23/9, D.C. Embu to E.O.P.E.C. Central, 10 Oct. 1953.
139 See Castro, "Social Aspects of Deforestation and Reforestation in Kirinyaga."
140 EDAR (1952), 1.
141 African Affairs Department, *Annual Report* for 1952: 45.
142 L. Leakey, *Mau Mau and the Kikuyu* (London: Methuen, 1952), 72–73.
143 Embu District Archives, Forest Offenses (E-FOR 13/3/1), D.F.O. Embu to H.s A.Ds. Embu, 7 Nov. 1955.
144 See Castro, "Social Aspects," 1984.
145 Ibid.

3
The Impact of German Colonial Rule on the Forests of Togo

1 R. J. Harrison Church, *West Africa: A Study of Environment and Man's Use of It* (London, 1963); R. W. J. Keay, *Vegetation Map of Africa* (Oxford, 1959); Yema E. Gu-Konu, *Atlas du Togo* (Paris, 1981).
2 Merrick Posnansky, "An Archaeological Reconnaissance of Togo," *Archaeology at UCLA* 2 (1980).
3 Sharon E. Nicholson, "The Methodology of Historical Climate Reconstruction and Its Application of Africa," *Journal of African History* 20 (1979): 31–49.
4 Robert Cornevin, *Histoire du Togo* (Paris, 1962), 39–40.
5 Susan Keech McIntosh and Roderick J. McIntosh, "West African Prehistory," *American Scientist* 69 (1981): 602–13.
6 Merrick Posnansky, "Notse Town Wall Survey," *Nyame Akuma* 18 (1981): 56–57.
7 *Rural Energy Systems in the Humid Tropics* (Tokyo: United Nations University, 1980).
8 Candice L. Goucher, "Iron is Iron 'Til It is Rust: Trade and Ecology in the Decline of West African Iron Smelting," *Journal of African History* 22 (1981): 179–89.
9 Walter Busse, "Die periodischen Grasbrande im trospischen Afrika, ihr Einfluss

auf die Vegetation und ihre Bedeutung fur die Landeskultur," *Mitteilungen aus den deutsche Schutzgeieten (hereinafter cited as MDSG)* 21 (1908): 113–39.

10 Karl Gaisser, "Production der Eingeboren des Bezirkes Sokode-Bassari," *MDSG* 25 (1912): 242–44; Michael Darkoh, "The Economic Life of Buem: 1884–1914," *Ghana Geographical Association Bulletin* 9 (1964): 40–54.

11 R. S. Rattray, "The Iron Workers of Akpafu," *Journal of the Royal Anthropological Institute* n.s. 19 (1916): 431–35.

12 Friedrich Hupfeld, "Die Eisenindustrie in Togo," *MDSG* 12 (1899): 1751–94.

13 Darkoh, "Economic Life"; Posnansky, "Archaeological Reconnaissance."

14 Interviews at Dimouri, Binajouba, and Banjeli, 11–16 February 1982.

15 Hupfeld, "Die Eisenindustrie."

16 Museum für Volkerkunde, Berlin-Dahlem: Kersting 197/08 and catalog nos. 22210–22211. The species were identified as *Combretum* spp., *Detarium senegalense,* and *Crossopteryx africana.* No doubt the reluctance on the part of the smelters to speak about the fuel stemmed from their opinion as to its value owing to scarcity. Only men may have been interviewed and charcoal making was done by women.

17 Candice L. Goucher, *The Iron Industry of Bassar, Togo: An Interdisciplinary Investigation of Technological Change* (Ph.D. diss., University of California at Los Angeles, 1983).

18 Ivor Wilks, "Land, Labour, Capital and the Forest Kingdom of Asante: A Model of Early Change," in *The Evolution of Social Systems*, ed. J. Fridman and M. J. Rowlands (London, 1977).

19 M. B. K. Darkoh, "A Note on the Peopling of the Forest Hills of the Volta Region of Ghana," *Ghana Notes and Queries* 11 (1970): 8–13; *Deutsche Koloniale Zeitung (hereinafter cited as DKZ)* (1899), 123; Peter Buhler, *The Volta Region of Ghana: Economic Change in Togoland, 1850–1914* (Ph.D. diss., University of California at San Diego, 1975).

20 Heinrich Klose, *Togo unter deutsche Flagge* (Berlin, 1899).

21 Willi Koert, "Das Eisenerzlager von Banjeli in Togo," *MDSG* 19 (1906): 124.

22 Busse, "Die periodischen Grasbrande."

23 O. F. Metzger, *Unsere alte Kolonie Togo* (Neudamm, 1941), 5.

24 Busse, "Die periodischen Grasbrande"; Klose, *Togo*; Hupfeld, "Die Eisenindustrie."

25 For a general discussion of German colonial endeavors in Africa, consult Prosser Gifford and William Roger Louis, *Britain and Germany in Africa* (New Haven, Conn., 1967), and L. H. Gann and Peter Duignan, *The Rulers of German Africa, 1884–1914* (Princeton, N.J., 1979). On early trade and German commercial relations, H. Muller, "Bremen und Westafrika," *Jahrbuch der Wittheit zu Bremen* (Sonderdruck) 15 (1971): 45–92.

26 Ernest Henrici, *Das deutsche Togogebiet und meine Afrikareise* (Leipzig, 1887).

27 *DKZ* (1888): 82.

28 *DKZ* (1900): 177.

29 Cotton produced across the Volta in Danish settlements was sent back to Europe as early as 1798. See Georg Norregard, *Danish Settlements in West Africa: 1658–1850* (Boston, 1966), 180.

30 Metzger, *Unsere alte Kolonie.*

31 Ibid.

32 *Der Tropenpflanzer* Beiheft 10 (1909): 138–39; Beiheft 11 (1910): 138.

33 I am indebted to colleague David Feeny for an elaboration of this point. The period in Togo is characterized by an uneven concern regarding technology. At the outbreak of World War I, the German troops at Kamina facing the British were equipped only with rifles dating back to the Franco-Prussian War of 1870.

34 See also the German experience in Cameroon: S. H. Bederman, "Plantation Agriculture in Victoria Division, West Cameroon," *Geography* 51 (1966): 349–60.

35 Raymond Dumett, "The Rubber Trade of the Gold Coast and Asante in the Nineteenth Century: African Innovation and Market Responsiveness," *Journal of African History* 12 (1971): 79–101.

36 Henrici, *Das deutsche Togogebiet*, 122.

37 "Bericht von Premierlieutenant Kling" *MDSG* 2 (1889): 103.

38 Norman Dean Haskett, *Kete Krakye and the Middle Volta Basin, 1700–1914* (Ph.D. diss., University of California at Los Angeles, 1981), 625–26.

39 "Bericht von Premierlieutenant v. Doering aus den Jahren 1893 bis 1895," *MDSG* 8 (1895): 236, 249.

40 K. Boahen, "Commercial Agriculture in Asante," *Ghana Geographical Association Bulletin* 15 (1973): 46.

41 *Jahresbericht* 1901–1902, p. 183.

42 August Full, *Funfzig Jahre Togo* (Berlin, 1935).

43 *DKZ* (1900): 177.

44 Klose, *Togo*, 76.

45 *Jahresbericht*.

46 Heinrich Schnee, ed., *Das Buch der deutscher Kolonien* (Leipzig, 1937), 426–28.

47 Archives Nationales du Togo, FA1/394: 24 (Letter from Kersting to Lome, 23 September 1907).

48 *Tropenpflanzer*, Beiheft 1 (1900): 213.

49 A. H. Unwin, *Report on the Afforestation of Togo with Teak and African Timber Trees* (London, 1912). I am grateful to Mr. Anthony Harvey, British Museum (Natural History), for his assistance in obtaining this source.

50 *Jahrbuch für Bremische Statistik* (1875).

51 Arthur J. Knoll, *Togo under Imperial Germany, 1884–1914* (Stanford, Calif., 1978), 92.

52 Full, *Funfzig Jahre Togo*, 198.

53 Hupfeld prepared a pamphlet extolling the potential of Banjeli mines and the need for railroad transport, but he later opposed the extension as a threat to the labor supplied by northern regions to his southern planting interests.

54 National Archives of Ghana, Tamale: C 21, Informal Diary, Navrongo Section, 17 April 1934. I am grateful to Mr. Timothy Garrard for bringing this source to my attention.

55 Archives Nationales du Togo: FA1/19: 180–90 (Bezirksleiter Gaisser to Lome, May–June 1911).

56 D. C. Dorward and A. I. Payne, "Deforestation, the Decline of the Horse, and the Spread of the Tsetse Fly and Trypanosomiasis (*nagana*) in Nineteenth-Century Sierra Leone," *Journal of African History* 16 (1975): 239–56.

57 Metzger, *Unsere alte Kolonie*.

58 Archives Nationales du Togo, FA1/307: 124–25 (Letter Verein Deutschen Zellstoff-Fabrikanten, Breslau, to Lome, 23 August 1910).

59 Archives Nationales du Togo, FA1/317.
60 Archives Nationales du Togo, FA1/307: 62, 117, 124–25, 133–41.
61 The forestry program concentrated on species that included *Khaya senegalensis, Khaya grandifoliola, Anogeissus leiocarpus, Chlorophora excelsa, Erythrophleum guineense, Prosopis africana,* and *Detarium senegalense*. See Metzger, *Unsere alte Kolonie*.
62 Full, *Funfzig Jahre Togo*.
63 Unwin, *Report*.
64 A. Aubreville, "Les Forêts du Dahomey et du Togo," *Bulletin du Comité d'Etudes Historiques et Scientifiques de l'A.O.F.* 20 (1937): 78.
65 Jean Meniaud, *Nos Bois Coloniaux* (Paris, 1931), 365.
66 Meniaud, *Nos Bois Coloniaux*.
67 D. Kinloch and W. A. Miller, *Gold Coast Timbers* (London, 1949).

4
Deforestation and Desertification in Twentieth-Century Arid Sahelien Africa

1 The Club du Sahel unites a number of OECD countries and the member states of C.I.L.S.S. (Comité inter-états de lutte contre la secheresse au sahel) [Permanent Interstate Committee for Drought Control in the Sahel] in an effort to identify promising development/anti-desertification projects and to channel donor funds into priority areas in the Sahel.
2 James T. Thomson, "The Precolonial Woodstock in Sahelien West Africa: The Example of Central Niger (Damagaram, Damergu, Air)," in *Global Deforestation and the Nineteenth-Century World Economy*, ed. Richard P. Tucker and J. F. Richards (Duke Press Policy Studies; Durham, N.C.: Duke University Press, 1983), 176–77. However, deforestation was not uncommon in precolonial times around urban centers. Jean Gorse, forester at the World Bank, has brought to my attention the following passages:

"About noon we saw at a distance the capital of Kaarta [present-day Mali], situated in the middle of an open plain, the country of two miles round being cleared of wood by the great consumption of that article for building and fuel, and we entered the town about two o'clock in the afternoon." *Mungo Park's Travels in the Interior of Africa: The First Journey (1795)* (London: J. M. Dent and Sons, 1954), 70.

"The majority of the residents [in Bambara, Mali] burn nothing but millet stalks; wood is so rare that those who have some go to market to sell it." Rene Caillie, *Voyage à Tombouctou, 1824–28* (Paris: FM/La Decouverte, 1979), 115; similar comments are made concerning Brakna, in contemporary Mauretania (pp. 86–87) and Jenne, Mali (p. 145).

Other sources suggest that men caused various forms of environmental degradation, particularly on the desert edge, as early as 450 B.C., through bush fires, expansion of agriculture and cattle herding, gum arabic trade, etc. National Research Council, *Environmental Change in the West African Sahel* (Washington, D.C.: National Academy Press, 1983), 24–42.

3 Marcel Chailley, *Histoire de l'Afrique occidentale française, 1638–1959* (Paris: Editions Berger-Levrault, 1968), 346–92. Niger, the last colony to be added to the West African Empire, was effectively occupied in 1899. Several minor revolts occurred during the first two decades of the century, but the French presence was never effectively contested during the 60 years of occupation. All the West African Sahelien states regained independence in 1960.

4 For an overview of the situation in commercial logging at mid-century, see Lord Hailey, *An African Survey, Revised 1956: A Study of Problems Arising in Africa South of the Sahara* (London: Oxford University Press, 1957), 951–53.

5 Jean Rouch, "Migrations au Ghana (Gold Coast), "*Journal de la societe des africainistes* 26, nos. 1–2 (1956): 33–196; Jean-Loup Amselle, "Aspects et significations du phénomène migratoire en Afrique," in *Les Migrations africaines*, ed. Jean-Loup Amselle (Paris: François Maspero, 1976), 9–39; Jean-Marie Kohler, *Les Migrations des Mosi de l'ouest* (Paris: ORSTOM, 1972); Claude Raynaut, "Lessons of a Crisis," in *Drought in Africa/Secheresse en afrique,* 2, ed. David Dalby, R. J. Harrison Church, and Fatima Bezzaz (London: International African Institute, 1977), 21–22; extensive further references are cited by Michel Aghassian, "Les Migrations en Afrique au sud du Sahara; bibliographie selective," in *Les Migrations africaines*, ed. Jean-Loup Amselle (Paris: François Maspero, 1976), 107–14.

6 Cf. Jean-Yves Marchal, *Société, espace et desertification dans le Yatenga (Haute-Volta), ou la dynamique de l'espace soudano-sahelien* (Paris: ORSTOM, 1982), 2:337–57, 433–49.

7 Georges Yameogo, Issoufou Ouedraogo, and Sam Baldwin, "Lab Tests of Fired Clay Stoves, The Economics of Improved Stoves, and Steady State Heat Loss from Massive Stoves," (Ouagadougou, Upper Volta: CILSS/VITS, 1982), 1–30.

8 F. Fournier, "The Soils of Africa," in *A Review of the Natural Resources of the African Continent* (Paris: UNESCO, 1963), 228–29.

9 Jean Suret-Canale, *Afrique noire occidentale et centrale: l'ère coloniale (1900–45)* (Paris: Editions sociales, 1964), 289–91. When the opportunity offered, administrators happily "made a profit" for the metropole by raising taxes well in excess of expenses. See Jean-Yves Marchal, *Chronique d'un cercle de l'AOF: Ouahigouya (Haute Volta) 1908–41*, Travaux et documents de l'ORSTOM, no. 125 (Paris: ORSTOM, 1980), 98. It should be noted that most public works during the first 40 years of colonialism in Upper Volta (1896–1936) were realized by forced labor contingents.

10 Herve Derriennic, *Famines et dominations en Afrique noire: Paysans et éleveurs du Sahel sous le joug* (Paris: l'Harmattan, 1977), 30–32.

11 Marchal, *Chronique d'un cercle*, 23–90; James T. Thomson, "Capitation and Post-Colonial Niger: Analysis of the Effects of an Imposed Head-Tax System on Rural Political Organization," in *The Imposition of Law*, by Sandra B. Burman and Barbara E. Harrell-Bond (New York: Academic Press, 1979), 201–205.

12 Stephen Baier, *An Economic History of Central Niger*, Oxford Studies in African Affairs (Oxford: Clarendon Press, 1980), 105–10, esp. 108, provides a detailed illustration of the type of device some colonial officers engaged to profit illegally from the situation.

13 See reference above, note 5.
14 Marchal, *Chronique d'un cercle*, 154.
15 Rouch, "Migrations au Ghana," 51–52.
16 Amselle, "Aspects et significations du phénomène migratoire," 17; Elliott P. Skinne, "Intergenerational Conflict among the Mossi: Father and Son," *Journal of Conflict Resolution* 5 (March 1961): 55–60; Marchal, *Société, espace et desertification*, 2:331–34.
17 Richard W. Franke and Barbara H. Chasin, *Seeds of Famine: Ecological Destruction and the Development Dilemma in the West African Sahel*, LandMark Studies (Montclair, N.J.: Allanheld, Osmun, 1980), 68–78.
18 Sheldon Gellar, *Structural Changes and Colonial Dependency: Senegal 1885–1945*, Studies in Comparative Modernization Series, no. 90-036 (Beverly Hills/London: Sage Publications, 1976), 49–56; John D. Collins, "Government and Groundnut Marketing in Rural Hausa Niger: The 1930s to the 1970s in Magaria" (Ph.D. diss., Johns Hopkins University, School of Advanced International Studies, 1975), 23–44. A. G. Hopkins, *An Economic History of West Africa* (New York: Columbia University Press, 1973), 187–236, presents a broad overview.
19 Marchal, *Chronique d'un cercle*, 99.
20 Robert Delavignette, "Le Dunamisme de l'A.O.F.: Une nouvelle colonie; d'Abidjan a Ouagadougou," *Afrique française* 42 (1932): 539–40; Derriennic, *Famines et dominations*, 35–37.
21 Collins, "Government and Groundnut Marketing," 17–30; Hopkins, *Economic History*, 220.
22 Peter B. Hammond, *Yatenga: Technology in the Culture of a West African Kingdom* (New York: Free Press, 1966), 163–79; Guy Le Moal, *Les Bobo: Nature et fonction des masques* (Paris: ORSTOM, 1980), 91–159; Guy Nicolas, *Dynamique sociale et apprehension du mode au sein d'une société Hausa* (Paris: Institut d'ethnologie, 1975), 223–308.
23 *1979 World Bank Atlas: Population, Per Capita Production and Growth Rates* (Washington, D.C.: World Bank, n.d. [1980?]), 8.
24 Leonard Berry, David J. Campbell, and Ingemar Emker, "Trends in Man-Land Interaction in the West African Sahel," in *Drought in Africa/Secheresse en Afrique*, 2, ed. David Dalby, R. J. Harrison Church, and Fatima Vezzaz (London: International African Institute, 1977), 83–91.
25 Raynaut, "Lessons of a Crisis"; Marchal, *Société, espace et desertification*, 2:503–24; Berry, Campbell, and Emker, "Trends in Man-Land Interaction"; James T. Thomson, "The Politics of Sahelien Desertification: Centralization, Non-Participation, Inaction," in *Divesting Nature's Capital*, ed. H. Jeffrey Leonard (New York: Holmes and Meier, in press), ch. 7.
26 Suzanne Lallemand, "La Secheresse dans un village Mossi de Haute-Volta," in *Secheresses et famines du sahel* (Paris: François Maspero, 1975), 2:55–56, records a situation in which mutual assistance remains a reality, at least in hard times.
27 Marchal, *Société, espace et desertification*, 2:359.
28 Ibid., 518–22.
29 Ibid., 433–40, 460.
30 Joel S. Midgal, *Peasants, Politics and Revolution: Pressures towards Political and Social Change in the Third World* (Princeton, N.J.: Princeton University Press,

1974), 113–16, demonstrates this is a common response. See also Samir Amin, *Neo-Colonialism in West Africa*, trans. from the French by Francis McDonagh (New York/London; Monthly Review Press, 1973).

31 Marchal, *Société, espace et desertification*, 2:330–35; Kohler, *Les Migrations des Mosi*, 37–44.

32 Richard Rathbone, "Ghana," in *West African States: Failure and Promise: A Study in Comparative Politics*, ed. John Dunn (Cambridge: Cambridge University Press, 1978), 28; Lexie Verdon, "Nigeria Expels Illegal Aliens; Thousands Clog Port, Roads," *Washington Post*, 31 January 1983, p. 3. In 1981 Ivory Coast officials began rejecting Voltaic immigrants who had no permanent position in the country and refused to naturalize themselves Ivoirien.

33 James T. Thomson, "Law, Legal Process and Development at the Local Level in Hausa-Speaking Niger: A Trouble Case Analysis of Rural Institutional Inertia" (Ph.D. diss., Indiana University, Bloomington, 1976), 235–80; Henri Raulin, *Techniques et bases socio-économiques des sociétés rurales nigeriennes*, Etudes nigeriennes, no. 12 (Niamey, Niger/Paris: IFAN/CNRS, n.d. [1962?]), 80–99, 103–106, 133–35.

34 Michel Keita, "Analyse du secteur forestier; volet sociologique," in *Analyse du secteur forestier et propositions: Niger* (SAHEL D [81] 132; Paris: Club du Sahel, 1981), 2:61–62; James T. Thomson, "Le Processus juridique, les droits fonciers et l'amenagement de l'environnement dans un canton Hausaphone du Niger," in *Enjeux fonciers en Afrique noire*, ed. Emile Le Bris, Etienne Le Roy, and François Leimdorfer (Paris: ORSTOM/Karthala, 1982), 174–75.

35 Marchal, *Société, espace et desertification*, 2:363.

36 Emmanuel Gregoire, "Un Système de production agro-pastoral en crise: le terroir de Gourjae (Niger)," in *Enjeux fonciers en Afrique noire*, ed. Emile Le Bris, Etienne Le Roy, and François Leimdorfer (Paris: ORSTOM/Karthala, 1982), 207; Mamadou M. Niang, "Reflexions sur la reforme foncière sénégalaise de 1964," in *Enjeux fonciers en Afrique noire*, ed. Emile Le Bris, Etienne Le Roy, and François Leimdorfer (Paris: ORSTOM/Karthala, 1982), 222–24.

37 Interview with S. S., Silmi-Mossi village chief, Bendogo (pseudonym), Seguenega Subprefecture, Department of the North (Yatenga Region), Upper Volta, 6 January 1982; interviews with D. T., I. Z., and S. G., research assistants, Inuwa Canton (pseudonym), Mirriah Arrondissement, Zinder Department, Niger, June–July 1981.

38 Marchal, *Société, espace et desertification*, 2:368, tableau 46; Niang, "Reflexions sur la réforme foncière senegalaise," 237.

39 Ibrahim Najada, Directeur General du Service des Eaux et Forets, Niger, "Analyse du secteur forestier," in *Analyse du secteur forestier et propositions: le Niger* (SAHEL D [81] 132; Paris: Club du Sahel, 1981), 107–40, for Niger; generally for the Sahel, James T. Thomson, "Participation, Local Organization, Land and Tree Tenure: Future Directions for Sahelien Forestry: (SAHEL D [83] 190; Paris: Club du Sahel, 1983), p. 2, par. 1.3., and references there cited.

40 James T. Thomson, "Ecological Deterioration: Local-Level Rule-Making and Enforcement Problems in Niger," in *Desertification: Environmental Degradation in and around Arid Lands*, ed. Michael H. Glantz (Boulder, Colo.: Westview Press, 1977), 63–71.

41 Interviews with informants in Inuwa Canton (pseudonym), Mirriah Arrondisse-

ment, Zinder Department, Niger, February–May 1979, June–July 1981, and in Kokokwase and Bendogo (pseudonyms), Seguenega Subprefecture, Yatenga Region, Upper Volta, October–November 1979, November 1981–January 1982.

42 Thomson, "Participation, Local Organization," 21–31.

43 During the period 1975–1980, of $7.5 billion in public assistance funds made available to CILSS Sahelien states, only 1.4 percent or slightly more than $100 million have been devoted to forestry and related activities. Fred Weber, "Review of CILSS Forestry Sector Program Analysis Papers" USDA/OICD/TAD/WW, AID/USDA/USFS (Washington, D.C.: United States Agency for International Development, 1982), 10.

44 French West Africa (Dakar, Senegal), Decree of 4 July 1935.

45 Hailey, *An African Survey*, 954.

46 See note 1, above.

47 The three documents are, respectively, *Etude du secteur forestier au Mali* (SAHEL D [82] 165; Paris: Club du Sahel, 1982); *Analyse du sector forestier et propositions: la Haute Volta* (SAHEL D [82] 159; Paris: Club du Sahel, 1982); and *Analyse du secteur forestier . . . le Niger*.

48 For further information on this point, see: Richard W. Katz and Michael H. Glantz, "Rainfall Statistics, Droughts and Desertification in the Sahel," in *Desertification: Environmental Degradation in and around Arid Lands*, ed. Michael H. Glantz (Boulder, Colo.: Westview Press, 1977), 81–102, presenting rainfall data for the region during the 35 years beginning in 1940 and noting, as have most other writers, the impact numerous droughts have on human survival strategies. See also Sharon Nicholson, "The Climatology of Sub-Saharan Africa," appendix B, in *Environmental Change in the West African Sahel*, National Research Council (Washington, D.C.: National Academy Press, 1983), 71–92; and Marchal, *Société, espace et desertification*, 1:40–43, for rainfall data for the Yatenga region of northern Upper Volta during the years 1921–73.

49 A. Aubreville, "Les Forêts de la colonie du Niger," *Bulletin de la comité d'études historiques et scientifiques de l'A.O.F.* 19, no. 1 (March 1936): 70–77.

50 Ibid., 68.

51 Ibid., 59.

52 Ibid., 62–65.

53 Thomson, "The Precolonial Woodstock," 170.

54 Heinrich Barth, *Travels and Discoveries in North and Central Africa* (London: Frank Cass, 1965), 3:96–101.

55 Aubreville, "Les Forêts," 62; see Marchal, *Société, espace et desertification*, 3:768; Erik P. Eckholm, *Losing Ground: Environmental Stress and World Food Prospects* (New York: W. W. Norton, 1976), 104. Eckholm has effectively publicized the Sahelien wood crisis, though as anyone can attest who has frequented the region in the 1980s, he engages in poetic license in asserting that "virtually all trees within seventy kilometers of Ouagadougou [capital of Upper Volta] have been consumed by the city's inhabitants."

56 Aubreville, "Les Forêts," 62.

57 Ibid., 63–64 (my translation).

58 Ibid., 63.

59 Ibid., 62.

60 Marchal, *Chronique d'un cercle*, 101 (my translation).
61 Ibid., footnote 12 (my translation).
62 P. Foury, "Politique forestière au Senegal," *Bois et forêts des tropiques*, no. 30 (July–August 1953): 12.
63 Ibid., 12.
64 Ibid., 14–16.
65 Ibid., 15.
66 Hailey, *An African Survey*, 952.
67 *Analyse du secteur forestier . . . le Niger*, 1:50–55.
68 Warren J. Enger, "The Government of Niger's Agricultural Strategy and the Potential for Meeting Long-Term Goals," in *Niger Agricultural Sector Assessment*, ed. Warren J. Enger (Niamey, Niger: USAID, 1979), vol. 2, B, pp. 8–9.
69 Gregoire, "Un Système de production," 203, 210–11. For a broader-scale analysis of the entire Maradi region, which reached the same general conclusions for villages in three different ecological zones (based on rainfall and land use), see Claude Raynaut, "Programme de recherche multidisciplinaire dans la region de Maradi (Niger): méthodes et premiers resultats," in *Maitrise de l'espace agraire et développement en Afrique tropicale: logique paysanne et rationalité technique, Actes du colloque de Ouagadougou, 4–8 décembre 1978* (Paris: ORSTOM, 1979), 427–35.
70 *Analyse du secteur forestier . . . la Haute Volta*, 77–88, provides source materials for calculations presented in the text.
71 Marchal, *Société, espace et desertification*, 7:222.
72 For references and brief discussion of similar studies and results elsewhere in the Sahel, ibid., 1:240–42.
73 Ibid., 3:763–74.
74 Ibid., 3:768–69.
75 Ibid., 3:767.
76 *Etude du secteur forestier au Mali*, 25–44, esp. 25, 29, 34, 37, 40–44, and 183.
77 Personal communication, Moumouni Ouedraogo, extension agent in Seguenega Subprefecture, Department of the North, Upper Volta, commenting on the spread, in Ouahigouya, Upper Volta, of improved stoves devised by local blacksmiths; Claude Freud, "Analyse du secteur forestier; evaluation economique," in *Analyse du secteur forestier . . . le Niger*, 2:25, reporting on a similar development in Niamey, Niger. An enormous assistance effort is underway, motivated by Club du Sahel officials (Sahelien and from OECD member states) and by private voluntary associations, to perfect cheap improved stoves that can be mass-produced and widely distributed. In Upper Volta, it has been proposed that distribution of such stoves be subsidized if necessary, or even that the first one be given away free to each Voltaic family (*Analyse du secteur forestier . . . la Haute Volta*, 103); see also Yameogo, Ouedraogo, and Baldwin, "Lab Tests," 1–30.
78 *Etude du secteur forestier au Mali*, 62.
79 Marchal, *Société, espace et desertification*, 3:793–97; Thomson, "The Politics of Sahelien Desertification."

5
The British Empire and India's Forest Resources

1 Richard P. Tucker, "The British Colonial System and the Forests of the Western Himalayas, 1815–1914," in *Global Deforestation and the Nineteenth-Century World Economy*, ed. Richard P. Tucker and J. F. Richards (Durham, N.C.: Duke University Press, 1983), 146–66.

2 For convenience I shall use the term Kumaon, though the forests under consideration include both the broadleaf belt at the foot of the hills (the Western Division Forests in colonial terminology) and the coniferous forests of the Kumaon division, which encompassed the pre-1947 civil districts of Nainital, Almora, and Pauri Garhwal. The people of Garhwal would properly object to this distortion.

3 *India's Contribution to the Great War* (Calcutta, 1923), 124–26; De Witt Ellinwood, ed., *India and World War I* (New Delhi, 1978), 141–76.

4 *Working Plans* of those years from each forest division give full details.

5 Richard P. Tucker, "The Historical Context of Social Forestry in Kumaon, Western Himalayas," *Journal of Developing Areas* 18 (April 1984): 341–56.

6 Ibid., 346–48.

7 E. A. Smythies, *India's Forest Wealth* (London, 1925).

8 B. R. Tomlinson, *The Political Economy of the Raj, 1914–47* (Cambridge, 1979), chaps. 3–4.

9 R. G. Marriott, *Resin Industry of Kumaon*, United Provinces Forest Department Bulletin no. 9 (Allahabad, 1937); United Provinces Forest Department, *Annual Progress Reports on the Resin Industry in the Kumaon and Utilization Circles* (Allahabad, annual from 1918).

10 United Provinces, *Legislative Council Debates*, 1923.

11 E. P. Stebbing, *The Forests of India*, vol. 4, ed. H. Champion and F. C. Osmaston (London, 1962), chap. 8; A. C. Chatterjee, *Notes on the Industries of the United Provinces* (Allahabad, 1908); C. G. Trevor and E. A. Smythies, *Practical Forestry Management in the United Provinces* (Allahabad, 1920); and [Anonymous], *Forest Research and Indian Industry* (Delhi, 1936).

12 See Government of India, *Annual Return of Statistics Relating to Forest Administration in British India*, for full statistical tables (hereinafter cited as SRFABI).

13 Tomlinson, *Political Economy of the Raj*, chap. 4.

14 Stebbing, *Forests of India*, vol. 4, chap. 10; for detailed statistics, see *Progress Reports on Forest Administration in the United Provinces*, annual (hereinafter cited as PRFAUP).

15 P. S. Narayana Prasad, "The World Depression and India," *Indian Journal of Economics* (October 1935): 121–44; C. H. Lee, "The Effects of the Depression on Primary Producing Countries," *Journal of Contemporary History* (1969): 139–55.

16 See SRFABI, annually, for details.

17 PRFAUP, for annual statistics; Stebbing, *Forests of India*, vol. 4, chap. 21.

18 See the figures for the hill districts of U.P. in the Government of India's decennial censuses.

19 Reconstruction Committee of Council (Government of India), *Second Report on Reconstruction Planning* (New Delhi, 1944); Stebbing, *Forests of India*, vol. 4, chap. 9.

20 Herbert Howard, *Post-War Forest Policy for India* (New Delhi, 1944); R. D. Richmond, "Post-War Forest Policy for India," *Empire Forestry Journal* 23, 2 (1944): 103–9, and 24, 1 (1945): 52–55.

21 For a survey of the results, see: *Report of the Tarai and Bhabar Development Committee* (Allahabad, 1947).

22 See several discussions in PRFAUP, 1945–49.

23 P. D. Strachey, "The Development of Forestry in Assam in the Last Fifty Years," *Indian Forester* (December 1956): 619–23; H. P. Smith and C. Purkayastha, *Assam: A Short History of the Assam Forest Service, 1850–1945* (Shillong, India, 1946).

24 M. C. Jacob, *The Forest Resources of Assam* (Shillong, India, 1940).

25 For the region's general history, see E. A. Gait, *A History of Assam*, 2d ed. (Calcutta and Simla, 1926), for a British perspective; and the fuller study by Amalendu Guha, *From Planter Raj to Swaraj: Freedom Struggle and Electoral Politics in Assam, 1826–1947*. (New Delhi: Indian Council for Historical Research, 1977).

26 Myron Weiner, *Sons of the Soil* (Princeton, 1978), chap. 2.

27 Ibid.; for the context in land tenure law, see J. N. Das, *An Introduction to the Land Laws of Assam* (Gauhati, India, 1968).

28 Sunil K. Sharma, "Origin and Growth of the Tea Industry in Assam," in *Contributions to Indian Economic History*, ed. T. Raychaudhuri (Calcutta, 1963); also Guha, *Raj to Swaraj*, 14; Weiner, *Sons of the Soil*, 88–92.

29 Percival Griffiths, *The History of the Indian Tea Industry* (London, 1967), 35–76.

30 For details, see Government of Assam, *Progress Reports on Forest Administration in Assam*, annual, 1919–25.

31 This principle had been supported from the beginning of forest management in Assam, as elsewhere in India. See Dietrich Brandis, *Suggestions Regarding Forest Administration in Assam* (Calcutta, 1879); and B. Ribbentrop, *Notes on an Inspection of the Forests of Assam during January to April 1889* (Simla, India, 1889).

32 Guha, *Raj to Swaraj*, 261–62.

33 Smith and Purkayastha, *Short History*, 42.

34 Bertram Hughes Farmer, *Agricultural Colonization in India Since Independence* (Oxford, 1974), 13–17, 28–31, 56–59.

6

Agricultural Expansion and Forest Depletion in Thailand, 1900–1975

The author wishes to acknowledge the helpful comments of Richard Barichello, Robert Evenson, Jan Laarman, François Mergen, Ansil Ramsay, John F. Richards, Suthad Setboonsarng, and Richard P. Tucker on an earlier version of this paper.

1 Jere R. Behrman, *Supply Response in Underdeveloped Agriculture: A Case Study of Four Major Annual Crops in Thailand, 1937–1963*, (Amsterdam: North-Holland Publishing Co., 1968), 108.

2 Thailand, Ministry of Commerce (1968), 27–28; Dusit Banijbatana, *The Management of Forests in Thailand* (Bangkok: Royal Forest Department).

3 See Lucien M. Hanks, *Rice and Man: Agricultural Ecology in Southeast Asia* (Chicago: Adline and Atherton, 1972); and David Feeny, *The Political Economy of Productivity: Thai Agricultural Development, 1880–1975* (Vancouver: University of British Columbia Press, 1982), for descriptions of the settlement patterns that accompanied the expansion of the area under cultivation.

4 Because cloud cover is at its minimum in the dry season, aerial photography and satellite image surveys are generally conducted in that season. Given that much of the cropping in Thailand is rain fed, it is difficult in the dry season to distinguish between areas that were cropped in the previous wet season and areas that were fallow; thus the tendency for these surveys to overestimate the area under cultivation.

5 Pat Tasanasanta Damrongsak, "Sources of Agricultural Output Growth in Thailand, 1950–76" (Ph.D. diss., Washington State University, 1978).

6 While the more complete coverage should bias our estimate of the rate of increase of the area under cultivation upward, a downward bias is also imported by using harvested rather than planted area for the later period.

7 According to Ministry of Agriculture data, as a percent of the area under major crops in the whole kingdom, the Central Plain major crop area was 61% in 1911–1912, 56% in 1921–1922, 54% in 1931–1932, 50% in 1938–1939, 45% in 1950–1952, 44% in 1958–1960, and 46% in 1965–1967. See Feeny, *Political Economy of Productivity*, 128, 141; and James C. Ingram, *Economic Change in Thailand: 1850–1970*, 2d ed. (Stanford, Calif.: Stanford University Press, 1971), 238.

8 Dusit Banijbatana, "Forest Policy in Northern Thailand," in *Farmers in the Forest: Economic Development and Marginal Agriculture in Northern Thailand*, ed. Peter Kunstadter, E. C. Chapman, and Sanga Sabhasri (Honolulu: The University of Hawaii Press, 1978), 56.

9 See *Area Handbook for Thailand* (Washington, D.C.: U.S. Government Printing Office, 1971), 15; and Wold Donner, *Five Faces of Thailand: An Economic Geography* (New York: St. Martin's Press, 1978), 141.

10 For the 1900–1930 period teak logs floated via the Chao Phraya River represented 75% of total extractions, logs floated via the Salween represented 17.5%, and logs floated via the Mekong accounted for 7.6% of the total; see Ministry of Commerce (1930), 130. The Salween route was probably relatively more important in the late 19th-century period than in the 1900 to 1930 period.

11 For general discussions of the Thai teak industry and its history, see Ingram, *Economic Change in Thailand*; A. J. C. Dickson, "The Teak Industry," in *Twentieth Century Impressions of Siam: Its History, People, Commerce, Industries, and Resources*, ed. Arnold Wright and Oliver T. Breakspear (London: Lloyd's Greater Britian Publishing Co., Ltd., 1908); Dusit, *Management of Forests*; Virginia Thompson, *Thailand: The New Siam, Second Edition* (New York: Paragon Book Reprint Corp., 1967); W. A. Graham, *Siam*, 3d ed., 2 vols. (London: The De La More Press, 1924); Asian Development Bank (hereinafter ADB), *Asian Agricultural Survey* (Seattle: University of Washington Press, 1969); Ministry of Commerce (1930); S. Mahaphol, *Teak in Thailand* (Bangkok: Royal Forest Department, 1954); Con-

stance M. Wilson, *Thailand: A Handbook of Historical Statistics* (Boston: G. K. Hall, 1983); and Peter Kunstadter, E. C. Chapman, and Sanga Sabhasri, eds., *Farmers in the Forest: Economic Development and Marginal Agriculture in Northern Thailand* (Honolulu: University of Hawaii Press, 1978).

12 The ban was partially lifted later to allow for the export of teak planks and boards.

13 See Thompson, *Thailand: Area Handbook*; and Ingram, *Economic Change in Thailand*.

14 Krit Samapuddhi, *Forestry Development in Thailand* (Bangkok: Royal Forest Department, 1966). Ministry of Commerce (1930); and Sukhum Thirawat, *Brief Information on Forestry Information in Thailand* (Bangkok: Royal Forest Department, 1955) list the major species found in the Thai forests.

15 The integration of northern Thailand into the Bangkok-dominated government and the role of forestry policy in that integration are carefully analyzed in James Ansil Ramsay, "The Development of a Bureaucratic Policy: The Case of Northern Siam" (Ph.D. diss., Cornell University, 1971). Catthip Nartsupha and Suthy Prasartset, eds., *Socio-Economic Institutions and Cultural Change in Siam, 1851–1910: A Documentary Survey* (Singapore: Southeast Asian Perspective Series no. 4), 4–9, provide a translated copy of a 1903 report found in the Thai National Archives describing the changes in forestry policy instituted in northern Thailand in the late nineteenth- and early twentieth-century period.

16 Ministry of Commerce (1930), 128. Under Thai law permits are required for the harvesting of all timber on public property, which includes virtually all forests except for the coastal mangrove forests; for a discussion of the mangrove forests see: F. J. Taylor, "The Utilization of Mangrove Areas in Thailand and Peninsular Malaysia," *Journal of Southeast Asian Studies* 13 (March 1982): 1–9. Timber cutting on private land is legal unless the species is a reserved one (such as teak or yang), in which case a permit is required.

17 Robert L. Pendleton, *Thailand: Aspects of Landscape and Life* (New York: Duell, Sloan and Pearce, 1962), 217.

18 Charlermrath Khambanonda, *Thailand's Public Law and Policy for Conservation and Protection of Land with Special Attention to Forests and Natural Areas* (Bangkok: National Institute of Development Administration, 1972), 50.

19 Graham, *Siam*; idem, "Note on Economic Development and Conservancy" (memorandum, 10 October 1925).

20 Sukhum, *Brief Information on Forestry*, 34.

21 See Sukhum, *Brief Information on Forestry*; Thailand, *Thailand, FAO Progress Report in Forestry, 1952–54* (Bangkok: Royal Forest Department, 1954); Donald F. Gienty, *Thailand's Forest Development and Its Effect on Rural Peoples: A Survey and Evaluation of Thailand's Forest Legislation and Policy in Light of Increased Demands on the Forest by Its Citizens, with Emphasis on Northeast Thailand* (Bangkok: United States Operations Mission, 1967); Kunstadter, Chapman, and Sabhasri, *Farmers in the Forest*; Charlermrath, *Thailand's Public Law and Policy*; Thompson, *Thailand*.

22 Thompson, *Thailand*, 341.

23 Sukhum, *Brief Information on Forestry*, 24. Krit, *Samapuddhi, The Forests of Thailand and Forestry Programs* (Bangkok: Royal Forest Department, 1957), 23, reports

that 27 million hectares was the goal set by a Food and Agriculture Organization (hereafter FAO) expert in his 1948 survey in Thailand. See also Charlermrath, *Thailand's Public Law and Policy*.

24 See International Bank for Reconstruction and Development (hereafter cited as IBRD), *A Public Development Program for Thailand* (Baltimore: Johns Hopkins Press, 1959), 82–84, and Krit, *Forestry Development*, 9. The ADB suggests a goal of 25 million hectares in forest reserves in *Asian Agricultural Survey* (Seattle: University of Washington Press, 1969), 465.

25 Thailand, Office of the Prime Minister, *The Fourth National Economic and Social Development Plan* (Bangkok: National Economic and Social Development Board [NESDB], 1977), 151.

26 Chote Suvatti and Dheb Menasvetts, "Threatened Species of Thailand's Aquatic Fauna and Preservation Problems," in *Conservation in Tropical Southeast Asia: Proceedings of the Conference on Conservation of Nature and Natural Resources in Tropical South East Asia*, ed. Lee M. Talbot and Martha H. Talbot (Morges, Switzerland: International Union for Conservation of Nature and Natural Resources, 1968), 332; Charlermrath, *Thailand's Public Law and Policy*, 51.

27 Boonsong Lekagul, "Threatened Species of Fauna of Thailand," in *Conservation in Tropical South East Asia: Proceedings of the Conference on Conservation of Nature and Natural Resources in Tropical South East Asia*, ed. Lee M. Talbot and Martha H. Talbot (Morges, Switzerland: International Union for Conservation of Nature and Natural Resources, 1968), 267; Krit, *Forestry Development*; Gienty, *Thailand's Forest Development*; and Charlermrath, *Thailand's Public Law and Policy*.

28 See Gienty, *Thailand's Forest Development*; and James A. Hafner, "Man and Environment in Rural Thailand," *Journal of the Siam Society* 61 (July, 1973): 129–38.

29 Charlermrath, *Thailand's Public Law and Policy*, 147.

30 See Sukhun, *Brief Information on Forestry*; and Gienty, *Thailand's Forest Development*.

31 See ADB, *Asian Survey*, 476–980; Tomoo Hattori and Kazutake Kyuma, "The Soil and Rice Growing," in *Thailand: A Rice Growing Society*, ed. Yoneo Ishii (Honolulu: University of Hawaii Press, 1978), 199; Charlermrath, *Thailand's Public Law and Policy*; and Kunstadter, Chapman, and Sabhasri, *Farmers in the Forest*.

32 Peter Kunstadter, "Alternatives for the Development of Upland Areas," in *Farmers in the Forest: Economic Development and Marginal Agriculture in Northern Thailand*, ed. Peter Kundstadter, E. C. Chapman, and Sanga Sabhasri (Honolulu: University of Hawaii Press, 1978), 306; Charlermrath, *Thailand's Public Law and Policy*; Dusit, "Forest Policy," 55; Thailand, *NESDB, The Fifth National Economic and Social Development Plan (1982–1986)* (Bangkok: Office of the Prime Minister, 1981).

33 IBRD, *Public Development Program*, 82–84; Charlermrath, *Thailand's Public Law and Policy*; Donner, *Five Faces of Thailand*; T. H. Silcock, ed., *Thailand: Social and Economic Studies in Development* (Canberra: Australian National University Press, 1967), 297–98; ADB, *Asian Survey*; Hiroshi Tsujii, "A Multidisciplinary Theory of Postwar Agricultural and Economic Change in Thai Village, the Agricultural Sector, and the Whole Country," in *A Comparative Study of Paddy Growing Communities in Southeast Asia and Japan*, ed. M. Kuchiba and L. E. Bauzon (Kyoto: Ryukoku University Press, 1979); Kunstadter, Chapman, and Sabhasri, *Farmers in the Forest*.

34 ADB, *Asian Survey*, 465.

35 See Kunstadter, Chapman, and Sabhasri, *Farmers in the Forest.*

36 ADB, *Asian Survey*, 479.

37 Thiem Komkris, "Forestry Aspects of Land Use in Areas of Swidden Cultivation," in *Farmers in the Forest: Economic Development and Marginal Agriculture in Northern Thailand*, ed. Peter Kunstadter, E. C. Chapman, and Sanga Sabhasri (Honolulu: University of Hawaii Press, 1978), 66–69.

38 Thompson, *Thailand*, 472–73; See also: Robert L. Pendleton, "Some Interrelations between Agriculture and Forestry, Particularly in Thailand," *Journal of the Thailand Research Society* (formerly Siam Society Natural History Supplement) 12 (December 1939); idem, "Land use in Northeastern Thailand," *The Geographical Review* 33 (January 1943): 15–41.

39 David Feeny, "Post–World War II Thai Agricultural Development," Policy: Continuity or Change?" in *Poverty and Social Change in Southeast Asia*, ed. Ozay Mehmet (Ottawa: University of Ottawa Press, 1979), 61–82; idem, *Political Economy of Productivity*.

40 The informal mechanisms for evading forest regulations and the legal loopholes have already been briefly discussed. Administrative inefficiencies in forest regulation arise in part because the responsibility for that regulation is shared by both the Ministry of the Interior (through its provincial and district officials) and the Royal Forest Department (a part of the Ministry of Agriculture since 1956. While the RFD is in charge of technical matters, the provincial and district level forest officers are under the command of local administration officials and are reluctant to enforce forestry policies that are opposed by those officials. The Forest Police Division (created in 1960) is also under the control of the Ministry of the Interior, although for budgetary purposes it is under the RFD. For more on these issues see Gienty, *Thailand's Forest Development*; and Charlermrath, *Thailand's Public Law and Policy*. Numerous observers have suggested administrative reforms.

41 Issues of optimal land use in the context of tropical development are discussed in Robert E. Evenson, "Tropical Forests in Economic Development," in *International Symposium on Tropical Forests: Utilization and Conservation*, ed. François Mergen (New Haven, Conn.: Yale University School of Forestry and Environmental Studies, 1981), 125–42.

42 Charlermrath, *Thailand's Public Law and Policy*, 128.

43 See Kunstadter, Chapman, and Sabhasri, *Farmers in the Forest*; Charlermrath, *Thailand's Public Law and Policy*; Gienty, *Thailand's Forest Development*; Talbot and Talbot, *Conservation in Tropical South East Asia*; Sukhum, *Brief Information on Forestry*; Krit, *Forestry Development*; Mahaphol, *Teak in Thailand*; Erik P. Eckholm, *Losing Ground: Environmental Stress and World Food Prospects* (New York: W. W. Norton, 1976); Tsujii, "Multidisciplinary Theory of Postwar Change"; Thailand, *Fourth National Development Plan* (NESDB, 1977, 1981); and Jack D. Ives, Sanga Sabhasri, and Pisit Varaurai, eds., *Conservation and Development in Northern Thailand: Proceedings of a Programmatic Workshop on Agro-Forestry and Highland-Lowland Interactive Systems, Held at Chiang Mai, Thailand, 13–17 November 1978* (Tokyo: United Nations University, 1980).

44 See Behrman, *Supply Response in Underdeveloped Agriculture*, 200–218. For additional detail see Ingram, *Economic Change in Thailand*, 240–43; and ADB, *Asian Survey*, 691.

45 For the 1950 to 1955 period paddy accounted for more than 85% of the total area in a more comprehensive list of crops in which rubber, kenaf, oil seeds, cassava, garden crops, and other fruits were also included; see T. H. Silcock, *The Economic Development of Thai Agriculture* (Ithaca, N.Y.: Cornell University Press, 1970), 54.

46 Carle C. Zimmerman, "Some Phases of Land Utilization in Siam," *Geographical Review* 27 (July 1937): 378–93. Recent data from the Division of Agricultural Economics in Thailand for 1975 indicate that for the whole kingdom 63.1% of the area of farm holdings is accounted for by paddy, 18.2% by other major field crops, and 18.7% by homesteads, gardens, fruit trees, tree crops, woodlots, and other. Relative to the 1930s, nonpaddy field crops have become more important, but paddy has basically retained its primacy.

47 See Feeny, *Political Economy of Productivity*.

48 Controls on the export of rice were instituted during both World War II and in the 1945 to 1949 postwar period. From 1947 through 1955 rice exporting was subject to an implicit tax through the operation of a multiple exchange rate system and after 1955 to a rice export tax and the rice premium (a specific export tax). During the 1945 to 1949 period there was a significant amount of rice smuggled out of the country. In the period since 1955 rice smuggling may have been important in periods when the rice premium was set at very high levels and world rice prices were also high. Thus official export data for the 1945 to 1949 period and some portions of the 1949 to 1955 period may understate the true level of exports. Overall the results for the rice balance sheet analysis of the 1906 to 1955 period are unlikely to be sensitive to these data problems.

49 See Feeny, *Political Economy of Productivity*.

50 See ibid.; and Supanee Milindankura and Melvin M. Wagner, *The Demand for Thai Agricultural Products: A Nutritional Approach*, Kasetsart Economic Report no. 30 (Bangkok: Kasetsart University, 1969).

51 See G. William Skinner, *Chinese Society in Thailand: An Analytical History* (Ithaca, N.Y.: Cornell University, 1957); Larry Sternstein, "A Critique of Thai Population Data," *Pacific Viewpoint* 6 (May 1965): 15–38; Jean Bourgeois-Pichat, "An Attempt to Appraise the Accuracy of Demographic Statistics for an Under-Developed Country: Thailand," paper presented at the United Nations Seminar on Evaluation and Utilization of Population Census Data in Latin America (Siantiago, Chile, 30 November–18 December 1959); and J. C. Caldwell, "The Demographic Structure," in *Thailand: Social and Economic Studies in Development*, ed. T. H. Silcock (Canberra: Australian National University Press, 1967), 27–64. Thai population statistics are further discussed in Feeny, *Political Economy of Productivity*; and Ralph Tomlinson, *Thailand's Population: Facts, Trends, Problems, and Policies*, Wake Forest University Developing Nations Monograph Series, no. 5 (Winston-Salem, N.C., 1972).

52 Bourgeois-Pichat presents estimates for the 1919 through 1956 period; the rate of growth of the Thai population taken from census data for the 1911 through 1919 period was used to interpolate over the 1906 to 1919 period. For Sternstein the 1906 estimate was interpolated from a graph, because the figure was not given in a table.

53 See Feeny, *Political Economy of Productivity*, tables A1–A19, p. 150.

54 Thailand, *National Income of Thailand*, Office of the Prime Minister, B.E. 2511–

2512 (Bangkok: Office of the National Economic Development Board, 1968–1969 edition).

55 Ingram, *Economic Change in Thailand*, 241–43. Ingram assumed 150 kilograms per person for rice consumption, a milling rate of 66%, and a seeding rate of 60 kilograms per hectare. Behrman, *Supply Response in Underdeveloped Agriculture*, made similar assumptions about domestic nonhuman consumption and computed human consumption as a residual after accounting for exports and other uses. For the 1947 to 1962 period, consumption averaged 181 kilograms per person per year, a figure that he did not consider to be excessive.

56 One additional caveat to this conclusion needs to be discussed, but given the sensitivity tests that have been performed and the congruence of the results with those arrived at by others, the validity of the overall conclusion still stands. A better method of constructing the rice balance sheets would be to multiply age-sex specific per capita consumption rates by the number of persons in each of the categories. The method used here may marginally overestimate consumption over time in that as the population age structure became younger (through the rapid rate of natural increase), the national average per capita consumption rate should have fallen as an increasing share of the population was in younger age groups with lower rates of per capita consumption. Unfortunately we lack the precise information on the age-sex specific consumption rates needed to construct these more accurate consumption estimates.

57 Similarly using the Bourgeios-Pichat population series, a consumption rate of 170 kilograms and a seeding rate of 125 kilograms per hectare, the mean of the ratio would fall from 1.3835 to 1.1804 over a thirty-eight-year period.

58 The underestimation by the Rice Department that was apparent when rice balance sheet estimates were constructed and the discrepancies between Rice Department estimates and census crop-cutting survey figures led the Rice Department to abandon their traditional system in 1968.

7
Export of Tropical Hardwoods in the Twentieth Century

1 Jack C. Westoby, "Halting Tropical Deforestation: The Role of Technology," paper commissioned by Congress of the United States, Office of Technology Assessment, panel on Sustaining Tropical Forest Resources (Washington, D.C., 1982), 1.

2 Ibid.; Christopher J. N. Gibbs, "Causes of Deforestation in Tropical Asia," paper commissioned by Congress of the United States, Office of Technology Assessment, panel on Sustaining Tropical Forest Resources (Washington, D.C., 1982).

3 Norman Myers, *Conversion of Tropical Moist Forests* (Washington, D.C.: National Academy of Sciences, 1980); Jean-Paul Lanly, "Tropical Forest Resources," FAO Forestry Paper No. 30 (Rome, 1982); Office of Technology Assessment, Congress of the United States (hereafter cited as OTA), *Technologies to Sustain Tropical Forest Resources* (Washington, D.C., 1984), 89–102.

4 Norman Myers, "Whose Hand on the Axe?" in *Tropical Deforestation: Hearings before the Subcommittee on International Organizations of the Committee on Foreign Affairs*, House of Representatives, 96th Congress (Washington, D.C., 1981), 138–43.
5 Food and Agriculture Organization (hereafter cited as FAO), *Yearbook of Forest Products, 1983* (Rome, 1985).
6 FAO, *Yearbook, 1983*, 22–23.
7 Raphael Zon and William N. Sparhawk, *Forest Resources of the World* 1, 2 (New York: McGraw-Hill Book Company, 1923).
8 Ibid., 536.
9 Ibid., 475.
10 Ibid., 559, 575, 648.
11 Ibid., 536.
12 Ibid., 672–73.
13 Ibid., 827, 832, 866, 869, 884.
14 Ibid., 355.
15 Ibid., 386.
16 International Institute of Agriculture (hereinafter cited as IIA), *International Yearbook of Forestry Statistics, 1932* (Rome, 1933), 30–39.
17 IIA, *International Yearbook of Forestry Statistics, Africa* (Rome, 1942), 212–13.
18 Union pour le Commerce des Bois Tropicaux dans la CEE (hereinafter cited as UCBT), "Characteristic Features of Individual Markets for Tropical Hardwoods in Europe," in *Tropical Hardwood Utilization: Practice and Prospects*, ed. R. A. A. Oldeman (The Hague: Martinus Nijhoff/D. W. Junk Publishers, 1982), 267, 278.
19 FAO, *Forestry and Forest Products, World Situation 1937–1946*, (Stockholm, 1946), 15.
20 Harvey A. Price, "Philippine Forestry and the Japanese Occupation," *Journal of Forestry* 44 (1946): 272–73.
21 Carlisle P. Winslow, "Wood and War," *Journal of Forestry* 40 (1942): 920–22.
22 FAO, *World Forest Products Statistics, a Ten-Year Summary 1946–1955* (Rome, 1958), 3.
23 A discussion of the growth factors is found in: FAO, "Wood: World Trends and Prospects," *Unasylva* 20, nos. 80–81 (1966): 1–135; Economic Commission for Europe/Food and Agriculture Organization (hereafter cited as ECE/FAO), "Consumption of Tropical Hardwoods in Europe," *Unasylva* 21, no. 84 (1967): 31–38; Kenji Takeuchi, "Hardwood Trade in the Asia-Pacific Region," World Bank Occasional Papers No. 17 (Baltimore: Johns Hopkins University Press, 1974); S. L. Pringle, "Tropical Moist Forests in World Demand," *Unasylva* 28, nos. 112–13 (1976): 106–18; idem, "The Outlook for Tropical Wood Imports," *Unasylva* 31, no. 125 (1979): 10–18.
24 Myers, *Conversion of Forests*, 41.
25 William C. Siegel and Clark Row, "U.S. Hardwood Imports Grow as World Supplies Expand," U.S. Forest Service Research Paper SO-17, U.S. Department of Agriculture, Southern Forest Experiment Station (New Orleans, 1965), 8–9.
26 Santiago Berbano, "Some Economic Aspects of Lumbering in Northern Negros," *Philippine Journal of Forestry* 21 (1938): 13–26.
27 Takeuchi, "Tropical Hardwood Trade," 4.

28 Jan G. Laarman, "Timber Exports from Southeast Asia: Away from Logs towards Processed Wood," *Columbia Journal of World Business* 19 (1984): 77–82.
29 Takeuchi, "Tropical Hardwood Trade."
30 Pringle, "Outlook for Tropical Wood Imports," 15.
31 Benjamin F. Sanvictores, "Moving Away from Log Exports," *Unasylva* 27, no. 108 (1975): 10–14; Gerard F. Schreuder and Richard P. Vlosky, "Indonesia as an Exporter and Importer of Forest Products," *World Trade in Forest Products* 2, ed. G. F. Schreuder (Seattle: University of Washington Press, 1975), 168–90.
32 S. Kolade Adeyoju, *Forestry and the Nigerian Economy* (Ibadan, Nigeria: Ibadan University Press, 1975), 118–30.
33 FAO, "Wood: World Trends," 106–107.
34 Pringle, "Tropical Moist Forests," 110.
35 Lanly, "Tropical Forest Resources," 42.
36 Food and Agriculture Organization/ Economic Commission for Latin America/ United Nations Industrial Development Organization (hereafter cited as FAO/ ECLA/ UNIDO), *Development of the Sawmilling Industry in Latin America*, Regional Consultation on Forest Industries in Latin America (Mexico, 1970).
37 FAO, "Wood: World Trends," 106.
38 Lanly, "Tropical Forest Resources," 73.
39 Myers, *Conversion of Forests*, 7–8; William M. Denevan, "Causes of Deforestation and Forest and Woodland Degradation in Tropical Latin America," paper commissioned by Congress of the United States, Office of Technology Assessment, Panel on Sustaining Tropical Forest Resources (Washington, D.C., 1982), 5–7; Lanly, "Tropical Forest Resources," 73–77.
40 Roger A. Sedjo and Marion Clawson, "How Serious Is Tropical Deforestation?" *Journal of Forestry* 81 (1983): 792–94.
41 Lanly, "Tropical Forest Resources," 74.
42 James C. Riddell, "Causes of Deforestation and Forest and Woodland Degradation in Tropical Africa," paper commissioned by Congress of the United States, Office of Technology Assessment, Panel on Sustaining Tropical Forest Resources (Washington, D.C., 1982), 29.
43 Westoby, "Halting Tropical Deforestation," 1–4; John W. Gray, "Forest Revenue Systems in Developing Countries," FAO Forestry Paper No. 43 (Rome, 1983), 8–9.
44 Lanly, "Tropical Forest Resources," vii.
45 Ibid., 75–76.
46 Ibid., 76–77.
47 Westoby, "Halting Tropical Deforestation," 8.
48 Sedjo and Clawson, "How Serious Is Deforestation?" 792–94.
49 Zon and Sparhawk, *Forest Resources*, 559–60, 608–9, 648–49, 672–74, 780–83.
50 Westoby, "Halting Tropical Deforestation," 4–5.
51 Pringle, "Outlook for Tropical Wood Imports," 17.
52 Zon and Sparhawk, *Forest Resources*, 441–42.
53 Tom Gill, *Tropical Forests of the Caribbean*, for the Tropical Plant Research Foundation (Baltimore: Read-Taylor Co., 1931), 271.
54 Westoby, "Halting Tropical Deforestation," 45–46.
55 Ibid., 34.
56 For example, see Myers, *Conversion of Forests*, 176.

57 Pringle, "Outlook for Tropical Wood Imports," 17–18.
58 J. I. Gammie, *World Timber to the Year 2000*, The Economist Intelligence Unit Special Report No. 98 (London, 1981), 72–73.
59 Pringle, "Outlook for Tropical Wood Imports," 17.
60 Isamu Nomura, "Demand and Supply Outlook of Forest Products in Japan," in *World Trade in Forest Products* 2, ed. G. F. Schreuder (Seattle: University of Washington Press), 57–74.
61 Laarman, "Timber Exports from Southeast Asia," 77–82; Schreuder and Vlosky, "Indonesia as an Exporter," 175–76.
62 Pringle, "Tropical Moist Forests," 114–18; T. Erfurth and H. Rusche, "The Marketing of Tropical Wood in South America," FAO Forestry Paper No. 5 (Rome, 1978), 4–5.
63 United Nations, *International Tropical Timber Agreement, 1983* (New York, 1984).
64 World Resources Institute, World Bank, and United Nations Development Program, "Tropical Forests: A Call for Action" (part 1), (Washington, D.C., 1985).
65 Myers, *Conversion of Forests*, 41–42.
66 Gray, "Forest Revenue Systems," 7–10.
67 Westoby, "Halting Tropical Deforestation," 30–33; Riddell, "Causes of Deforestation," 26–29.
68 George M. Guess, "Technical and Financial Policy Options for Development Forestry," *Natural Resources Journal* 21 (1981): 37–55.

8
The North American-Japanese Timber Trade

Research on this essay was supported in part by a grant from the San Diego State University Foundation. For consistency, all Japanese names are rendered below in Western order.

1 On the earlier trade, see Tamotsu Watanabe, *Gaizai yunyū no jōsei to sono taisaku* [Foreign timber importation situation and its countermeasures] (Tokyo: Teikoku Shinrinkai, 1925); E. A. Selfridge, *American Lumber in Japan*, U.S. Bureau of Foreign and Domestic Commerce, Trade Promotion Series no. 59 (Washington: GPO, 1928); M. Arikawa, "American Lumber in Japan," *Timberman* 29 (March 1928): 188–90; Ivan M. Elchibegoff, *United States International Timber Trade in the Pacific Area* (Stanford, Calif.: Stanford University Press, 1949); Walter A. Radius, *United States Shipping in Transpacific Trade, 1922–1928* (Stanford, Calif.: Stanford University Press, 1944), 88–98; Frank Langdon, *The Politics of Canadian-Japanese Economic Relations, 1952–1983* (Vancouver: University of British Columbia Press, 1983), 10.
2 On Japan's Southeast Asian timber trade, see Hikojiro Katsuhisa, "Forest Products Trade of Japan," in Ryoichi Handa et al., *The Current State of Japanese Forestry (III): Its Problems and Future* (Tokyo: Japanese Forest Economic Society, 1984): 10–16, 19–22; Michio Tsutsui, *Tenkanki no nan'yōzai mondai* [The turning point of the South Pacific timber trade] (Tokyo: Ringyō Chōsakai, 1978).

3 Imperial Japanese Commission to the Louisiana Purchase Exposition, *Japan in the Beginning of the 20th Century* (Tokyo: For the Commission by the *Japan Times*, 1904), 224–25, 237, 241, 250–51, 274, 282; Tsuneaki Sato, "Japanese Industries: Agriculture and Forestry," in *Fifty Years of New Japan*, ed. Shigenobu Okuma (London: Smith, Elder, 1909), 583–93; Minoru Kumazaki, "The Role of Forestry in Japanese Industrialization," in Handa, *Current State of Japanese Forestry (III)*, 84–96; Yasunori Fukushima and Shin Nagata, "Japanese Forestry before World War II," in *The Current State of Japanese Forestry (IV): Its Problems and Future*, ed. Ryoichi Handa et al. (Tokyo: Japanese Forest Economic Society, 1985): 76–86; William N. Sparhawk and Raphael Zon, *Forest Resources of the World*, 2 vols. (New York: McGraw-Hill, 1923), 1:442–49; Yutaka Ishii, "Development Process of Forestry in Hokkaido," ibid., 89–95; Thomas C. Smith, *Political Change and Industrial Development in Japan* (Stanford, Calif.: Stanford University Press, 1955), 113–14; John A. Harrison, *Japan's Northern Frontier* (Gainesville: University of Florida Press, 1953), 106–8; Tamotsu Watanabe and Ushimaro Hayao, *Nihon no ringyō* [Japanese forestry] (Tokyo: Teikoku Shinrinkai, 1930).

4 Fukushima and Nagata, "Japanese Forestry before World War II," 86–88; Ishii, "Development of Forestry in Hokkaido," 95–96; Edward A. Ackerman, *Japan's Natural Resources and their Relation to Japan's Economic Future* (Chicago: University of Chicago Press, 1953), 250–55, 332–41; Natural Resources Section, General Headquarters, Supreme Commander for the Allied Powers, *Reforestation in Japan*, report no. 113 (Tokyo: SCAP, 1948), 19–22; Supreme Commander for the Allied Powers, *Japanese Natural Resources: A Comprehensive Survey* (Tokyo: SCAP, 1949), 1:275–76; Ministry of Agriculture and Forestry [hereafter cited as MAF], *White Paper on Japan's Forestry Industry, 1965*, trans. Japan Lumber Journal (Tokyo: Japan Lumber Journal, 1966), 19.

5 Japan turned to foreign sources for wood pulp and chips as well as logs and lumber. That trade possesses characteristics of its own that place it beyond the scope of this study.

6 Interview, Shigeru Tanaka [chief, organization department, National Federation of Forest Owners Cooperatives], Tokyo, March 7, 1984; interview, Tsutomu Sasaki [managing director, Sakai Minato Timber Importers Association], Sakai Minato, May 8, 1984; Katsuhisa, "Forest Products Trade of Japan," 23–24; Isamu Nomura, *Shin Gaizai Tokuhon* [New foreign timber reader] (Tokyo: Ringyō Shimbunsha, 1978), 364ff. So separate is the demand for lumber from Soviet timber from that for lumber from American logs that one operator in Gumma prefecture actually owns mills for each and finds that they do not compete with one another for sales at all. Interview, Keikichi Ukiji [president, Ukiji Lumber Company], Numata, March 21, 1984.

7 Tanaka interview; interview, Isamu Enokido [president, Enokido Lumber Company], Tokyo, March 15, 1984; interview, Kazunari Tomita [manager, Matsue Mutual Lumber Market], Matsue, April 28, 1984; interview, Kazuo Yokoji [general manager, Nita-cho Forest Owners Cooperative], Minari, May 21, 1984 (quotation). See also Kiyoshi Yukutake, "Simulation Analysis of Log and Lumber: Econometric Model for Japan," in Handa, *Current State of Japanese Forestry (IV)*, 2–3; Michio Tsutsui, *Gendai ringyō kōgi*, vol. 3, *rinseigaku* [Discourses on present forestry, vol. 3, Government forestry studies] (Tokyo: Chikyūsha, 1983), 179–87.

8 Forestry Agency, *Ringyō tōkei yōran* [Forestry statistics survey] (Tokyo: Forestry Agency, 1973), 136–41. On forest-dependent mountain villages, see: Ushiomi Toshitaka, *Forestry and Mountain Village Communities in Japan: A Study of Human Relations* (Tokyo: Kokusai Bunka Shinkōkai, 1968); Hisayoshi Mitsuda, "Impacts of Rapid Economic Growth on the Structure of Mountain Communities: Depopulation and Disorganization in Charcoal-Producing Mountain Villages of Kyoto Prefecture in Japan," in *History of Sustained-Yield Forestry: A Symposium*, ed. Harold K. Steen (Santa Cruz, Calif.: Forest History Society, 1984), 251–69; Norio Takahashi, "Production Activities of Forestry Households Not Engaged in Farming in Japan," in *The Current State of Japanese Forestry (II): Its Problems and Future*, ed. Ryoichi Handa, Isamu Nomura, and Minoru Kumazaki (Tokyo: Japanese Forest Economic Society, 1982), 64–74; Shigeru Shimotori, "Hokkaidō ni okeru sanson to ringyō mondai" [Woodlands and forestry problems in Hokkaido], *Hoppō-ringyō* [Northern Forestry] 22 (1970): 38–41. See also: Tosihiko Yagi, "Ringyō rōdō kenkyū no konnichiteki kadai" [On forestry labor studies today], *Tottori Daigaku Nōgakubu Enshūrin Hōkoku* [Bulletin of the Tottori University Forests] 11 (Dec. 1979): 141–67; MAF, *White Paper*, 114–22.

9 MAF, *White Paper*, 129–83; Forestry Agency, *Ringyō tōkei yōran*, 161–78.

10 Interview, Ryoichi Takano [president, Canada House Company], Tokyo, March 7, 1984; interview, Masao Nakane, [executive director, National Federation of Forest Owners Cooperatives], Tokyo, April 5, 1984; Tanaka interview; *Japan Lumber Journal* 3 (Sept. 5, 1962): 1; ibid. 4 (Oct. 5, 1963): 1; ibid. 4 (Oct. 21, 1963): 2; ibid. 5 (June 5, 1964): 1; ibid. 6 (Dec. 6, 1965): 4; ibid. 13 (Aug. 10, 1972): 1; Minoru Kumazaki, "Long Term Trends in Wood Consumption in Japan," *Rural Economic Problems* 4 (1967): 56–74; Isamu Nomura, "Long Range Demand and Supply Prospects for Forest Products in Japan, and Some Problems," paper read at North American Conference on Forest Sector Models, Williamsburg, Va., Dec. 4, 1981.

11 Occupation authorities encouraged Forest Owners' Associations—cooperatives—in an attempt to do for forest areas what agricultural cooperatives did for farming districts. On the FOA situation, see: MAF, *White Paper*, 98–100, 155–56, 179–81; Forestry Agency, *Ringyō tōkei yōran*, 181–88; Shigeru Tanaka, *Nihon ringyō no hatten to shinrin kumiai* [The development of Japanese forestry and forest owners associations] (Tokyo: Nihon Ringyō Chōsakai, 1982); John W. Bennett, "Economic Aspects of a Boss-Henchman System in the Japanese Forest Industry," *Economic Development and Cultural Change* 7 (Oct. 1958): 13–30; Manabu Morita, "The Forest Owners' Association as an Organizer of Regional Forestry by Small Forest Owners," in *The Current State of Japanese Forestry: Its Problems and Future*, ed. Ryoichi Handa and Isamu Nomura ([Tsukuba: Forestry and Forest Products Research Institute], 1981), 136–40; Narao Ebata, "Shinrin kumiai no bunseki" [An analysis of forest associations], *Ringyō Shikenjō Kenkyū Hōkoku* [Bulletin of the Government Forest Experiment Station; the provided English title varies] 121 (March 1960): 1–110; Heihachi Kanamaru, "Ringyōshi kenkyū (II): Shinrin kumiai no seikaku to sono seika ni tsuite" [A study of forestry history (II): the character and results of forest associations], *Mita Gakkai Zasshi* [Mita Journal of Economics] 47 (1954): 64–82; and *Shinrin Kumiai*, the monthly journal of the National Federation of Forest Owners Cooperatives.

12 Ryoichi Handa, "Structure of Forest Ownership," in Handa and Nomura, *Current State of Japanese Forestry*, 96–104; MAF, *White Paper*, 82–113; Mitsuma Matsui, "Forest Administration in Japan," *Journal of Forestry* 58 (1980): 417–19; interview, Masahisa Nishizawa [professor of forestry, Kyushu University], Fukuoka, May 25, 1976.

13 Yoshio Utsuki, "Timber Demand and Supply in Japan, Recent Situation and Prospect," in Handa and Nomura, *Current State of Japanese Forestry*, 3; MAF, *White Paper*, 94; Shinji Kamino, "System of Japanese Forestry Extension," in Handa, *Current State of Japanese Forestry (IV)*, 64–68; MAF, Forestry Agency, *Forestry in Japan* ([Tokyo]: MAF, Forestry Agency, 1981), 70–71; Shoji Matsuda, comp., *Forestry Technology in Japan* (Tokyo: Japan FAO, 1976), 45–47; Nishizawa interview. Cf. Minoru Kumazaki, "Trends in Small Private Forests of Japan, 1960–1980," in Handa and Nomura, *Current State of Japanese Forestry*, 77–86.

14 Takashi Kato, "A Regional Comparison of Forest Productivity, Stumpage Prices, Logging and Regeneration Costs Among Japan, Canada and the United States," in Handa and Nomura, *Current State of Japanese Forestry*, 29–40; Takashi Kato, "Comparison of Softwood Lumber Manufacturing and Selling Costs between the Pacific Coast of North America and Japan," in Handa, Nomura, and Kumazaki, *Current State of Japanese Forestry (II)*, 30–39; Keiichi Hasegawa and Kiji Hisada, "Zōrin tōshi no saisansei no hikaku" [A comparison of the profitability of silvicultural investment], *Ringyō Shikenjō Kenkyū Hōkoku* 311 (Aug. 1980): 1–16; MAF, *White Paper*, 43; Tsuneta Yano Memorial Society, ed., *Nippon: A Charted Survey of Japan, 1973* (Tokyo: Kokusei-sha, 1973), 261–63, 267–72. See also: Bruce A. McCarl and Richard W. Haynes, "Exchange Rates Influence Softwood Lumber Trade," *Journal of Forestry* 83 (1985): 368–70.

15 MAF, *White Paper*, 53–55; Matsuda, *Forestry Technology in Japan*, 22–23; Forestry Agency, *Forestry Technology in Japan*, 22–24, 39–40; Forestry Agency, *Forestry in Japan*, 42, 46–47, 86; Mitsuma Matsui, "Forest Resources in Japan," *Journal of Forestry* 128 (1980): 98–99; Boston Consulting Group and Boston Consulting Group of Japan, *Penetrating the Japanese Market for Softwood Lumber and Plywood Products* (Boston: Boston Consulting Group [c. 1967]), 8–9. Identified in Japan in 1972 and named *Bursaphelenchus lignicolus*, the nematode was subsequently found to be the same as *A. xylophilus*, first identified in the United States in 1929. This suggests that the Japanese belief that the nematode was introduced into Japan on pine from the United States may be true. American species seem to be much more resistant to the effects of the nematode than are Japanese species. See: Fujio Kobayashi, "Matsukure taisaku no tembo" [Observations on pine wilt disease counterplans], *Shinrin Bunka Kenkyū* [Forest Culture Studies] 2 (1981): 49–53; Michael J. Wingfield, Robert A. Blanchette, and Thomas H. Nichols, "Is the Pine Wood Nematode an Important Pathogen in the United States?" *Journal of Forestry* 82 (1984): 232–35.

16 Forestry Agency, *Forestry Technology in Japan*, 25, 42–43; MAF, *White Paper*, 75–77; Forestry Agency, *Ringyō tōkei yōran*, 39; Mitsuda, "Depopulation and Disorganization," 5–11, 16–18; Takeshi Akaha, "The Coppice Forest and Its Utilization in Japan," Handa, Nomura, and Kumazaki, *Current State of Japanese Forestry (II)*, 87–98; Shigeru Shimotori, "Hokkaidō ni okeru mokuzai shijō no henbō to kōyōju seizaigyō no dōkō" [The transfiguration of the timber market and trends of the lum-

ber industry using broad-leaved trees in Hokkaido], *Hōkkaido Daigaku Nōgakubu Enshūrin Hōkoku* [Research bulletin of the Hokkaido University Experimental Forest] 34 (1975): 1–32. Most of what potential for lumber that these species might have had was negated by the former management of stands as coppice rather than high forest.

17 Matsui, "Forest Resources in Japan," 99; MAF, *White Paper*, 139–46, 169–78; Forestry Agency, *Forestry in Japan*, 20–21; Sukeharu Tsuru, "Financial Assistance for Silvicultural Practices in Japan," in Handa and Nomura, *Current State of Japanese Forestry*, 118–26; Shigeru Iida, "Roles of Public Corporation on Afforestation Policy," in ibid., 127–35; Katsuya Fukuoka, "Financing the Forestry in Japan," in ibid., 141–52.

18 Interview, Michio Tsutsui [professor of forestry Tokyo University], Tokyo, April 5, 1984; Forestry Agency, *Forestry in Japan*, 26–30, 50–57; Japan Institute of International Affairs, *White Papers of Japan, 1972–73: Annual Abstract of Official Reports and Statistics of the Japanese Government* (Tokyo: Japan Institute of International Affairs, 1974), 87; Forestry Agency, *Ringyō tōkei yōran*, 148–56; Ayakira Okazaki, "Forests and Recreation in Japan," *American Forests* 78 (1972): 20–21, 60–62; Katsuya Fukuoka, "Multiplicative Estimation of Forest and Economic Equilibrium," in Handa, *Current State of Japanese Forestry (IV)*, 109–17; Ichiro Kato, Nobuo Kumamoto, and William H. Matthews, eds., *Environmental Law and Policy in the Pacific Basin Area* (Tokyo: University of Tokyo Press, 1981), 96, 145–51. For a fuller treatment, see: Michio Tsutsui et al., *Kankyō hozen to shinrin kisei* [Environmental preservation and forest regulation] (Tokyo: Nōrin Shuppan, 1976).

19 MAF, *White Paper*, 89–100 and passim; Kumazaki, "Trends in Small Private Forests"; Morita, "Forest Owners' Association"; Forestry Agency, *Forestry Technology in Japan*, 50–58; Matsuda, *Forestry Technology in Japan*, 17–21; Shimotori, "Mokuzai shijō no henbō to kōyōju seizaigyō no dōkō"; Tsutsui, *Gendai ringyō kōgi*, 3:145–94. See also n. 10 above.

20 Enokido interview; Tomita interview; interview, Mineji Inoue [manager, Higashi Agano Forest Owners Cooperative], Higashi Aganao, March 4, 1984; interview, Isamu Nomura [professor of economics, Nihon University], Tokyo, March 16, 1984; interview, Katsuya Fukuoka [professor of economics Rissho University], Tokyo, March 17, 1984; Minaji Ito [sales manager, National Federation of Forest Owners Cooperatives], Tokyo, April 5, 1984; Shizuo Shigesawa, "Timber Market and Timber-Processing Industries in Japan," in *World Trade in Forest Products*, ed. James S. Bethel (Seattle: University of Washington Press, 1983), 194–230.

21 Environmental concern has largely revolved around the question of pollution and its effect on humans, but ecological and other considerations are beginning to receive attention. Tsutsui interview; Margaret A. McKean, *Environmental Protest and Citizen Politics in Japan* (Berkeley: University of California Press, 1981); Julian Gresser et al., *Environmental Law in Japan* (Cambridge, Mass.: MIT Press, 1981); Kato, Kumamoto, and Matthews, *Environmental Law and Policy*, 77–156. The broadened outlook is reflected in articles in *Shinrin Bunka Kenkyu* and such recent work of Michio Tsutsui as *Nihon to yama to nihonjin* [Japan, the mountains, and the Japanese] (Tokyo: Asahi Shimbunsha, 1982).

22 On the rise of plantation forestry in Japan, see: Conrad Totman, *The Origins of Japan's Modern Forests: The Case of Akita* (Honolulu: University of Hawaii Press,

1984); Masako Osako, "Forest Preservation in Tokugawa Japan," in *Global Deforestation and the Nineteenth-Century World Economy*, ed. Richard P. Tucker and J. F. Richards (Durham, N.C.: Duke University Press, 1983), 129–45. Japanese work on the subject is fragmented, but extensive. For examples, see: Kitami Tanaka, "Kinsei shoki eirin seisaku no ichi kōsatsu" [A study of the policy of forest management in the beginning of modern times], *Iwate Shigaku Kenkyū* [Iwate Historical Studies] 24 (1958): 11–18; Michio Tsutsui, "Akita-han ni okeru ringyō ikusei seisan shoshiki to gyōsei no hōkō" [The development of the organization, production, and the administration of forestry in Akita-han], *Tōkyō Daigaku Nōgakubu Enshūrin Hōkoku* [Bulletin of the Tokyo University Forests] 53 (July 1957): 1–26; Rurio Motoyoshi, "Kyōto-fu Maizuru, Miyazu chihō ni okeru rin'ya shoyū no keisei to zōrin no shiteki hatten" [The formation of forest land ownership and the historical development of artificial reforestation in Maizuru and Miyazu districts of Kyoto prefecture], *Kyōto Furitsu Daigaku Nōgakubu Enshūrin Hōkoku* [Bulletin of the Kyoto Prefectural University Forests] 15 (Oct. 1970): 1–49; Motoyoshi, "Okumikawa chihō ni okeru ringyō no shiteki hatten ni kansuru kenkyū" [Studies in the historical development of forestry in Okumikawa district], ibid., 20 (1976): 1–33; and numerous articles in *Tokugawa Rinseishi Kenkyūjo Kenkyū Kiyō*, the bulletin of the Tokugawa Institute for the History of Forestry. For discussions of recent shifts in silvicultural practices in artificial forests and of their impact, see: Matsuda, *Forestry Technology in Japan*, 17–20; Forestry Agency, *Forestry Technology in Japan*, 50–59, 116–18; Yuji Uozumi, "Kokuyū rin'ya shigyohō no kenkyū" [Studies on the management system in national forestry], *Tottori Daigaku Nōgakubu Enshūrin Hōkoku* 10 (Dec. 1977), 61–72.

23 Evidence on changing smallholder attitudes and their impact is, at this point, fragmentary. The above rests primarily on John B. Cornell, "Three Decades of Matsunagi: Changing Patterns of Forest Land Use in an Okayama Mountain Village," in Steen, *History of Sustained-Yield Forestry*, 237–50, and on a series of interviews conducted in various parts of Japan in 1984. See also: *Western Timber Industry* 16 (Nov. 1965): 11; *Japan Lumber Journal* 6 (Dec. 6, 1965): 4; Kumazaki, "Trends in Small Private Forests," 79–86; Handa, "Structure of Forest Ownership," 99–104; Forestry Agency, *Ringyō tōkei yōran*, 94–95; MAF, *White Paper*, 80–107; Matsuda, *Forestry Technology in Japan*, 17–20; Forestry Agency, *Forestry Technology in Japan*, 50–58; Shigeru Shimotori, "Trends and Problems in the Logging Industry in Japan," in Handa, Nomura, and Kumazaki, *The Current State of Japanese Forestry (II)*, esp. 56–58; Shimotori, "Hokkaidō ni okeru rinchi ryūdōka no taiyō to kōzō" [The pattern and economic structure of forest land movement in Hokkaido], *Hokkaidō Daigaku Nōgakubu Enshūrin Hōkoku* 34 (1977): 1–42; Asia Foundation, trans., "Japan and the Timber Trade: An Analysis from that Side," *Forest Industries* (July 1981): 37–38. Cf. "Broad-Leaved Evergreens and Pines Disappearing on the Japanese Archipelago," *The East* 18 (1982): 43.

24 On the diversity of stands and conditions, see: Forestry Agency, *Forestry in Japan*, 8–17; Forestry Agency, *Forestry Technology in Japan*, 1–27; Kato, "A Regional Comparison"; Hasegawa and Hisada, "Zōrin tōshi"; Kiji Hisada and Akira Ouchi, "Zōrin no saisan no chiku hikaku ni tsuite" [Researches on the economical comparison of silviculture in the various areas of Japan], *Ringyō Shikenjō Kenkyū Hōkoku* 121 (March 1960): 189–206; Robert B. Forster, *Japanese Forestry: The Resources, Indus-*

tries and Markets, Information Report E-X-30 (Ottawa: Canada, Department of the Environment, 1978), 50–71. For unusually successful local adaptation to the new conditions inaugurated by imports and related changes, see: Tamutsu Ogi, "Katsuyama shijō no hatten katei to kongo no kadai" [Development process and problems of the timber producing district in Katsuyama], *Tōkyō Daigaku Nōgakubu Enshūrin Hōkoku* 69 (Dec. 1979): 169–77.

25 *Japan Lumber Journal* 1 (Oct. 29, 1960): 2; ibid., 3 (Nov. 20, 1962): 4; ibid., 4 (March 5, 1963): 3; ibid., 5 (Jan. 20, 1964): 3; ibid., 5 (June 20, 1964): 1; ibid., 6 (March 5, 1965): 2; ibid., 7 (Nov. 10, 1966): 1; ibid., 7 (Dec. 10, 1966): 1; *Western Timber Industry* 13 (May 1962): 1; ibid., 15 (Jan. 1964): 1; ibid., 16 (May 1965): 4; ibid., 17 (Oct. 1966): 9; ibid., 18 (April 1967): 24; ibid., 18 (Aug. 1967): 4, 7; ibid., 19 (Feb. 1968): 5; Warren S. Hunsberger, *Japan and the United States in World Trade* (New York: Harper & Row for the Council on Foreign Relations, 1964), 100–101; Yoshinao Murashima, "Problems of Foreign Timber Trade," in Handa and Nomura, *Current State of Japanese Forestry*, 11–16; Boston Consulting Group, *Penetrating the Japanese Market*, 12, 43–55; Forster, *Japanese Forestry*, 34–38, 77; MAF, *White Paper*, 159–60, 278–79; Akira Takahashi, Yozo Shiota, Chiaki Tanaka, and Arno P. Schneiwind, "Recent Trends in the Wood Industry of Japan," *Forest Products Journal* 30 (1980): 24–25; Asia Foundation, "Japan and the Timber Trade," 37. For an example of large-scale import business activity, see: obituary for H. Yamasaki, *Japan Lumber Journal* 21 (Jan. 31, 1980): 4–5.

26 There are numerous works on the log export trade and its regulation, several suggesting that the effects of log exports and their subsequent restriction have been exaggerated. These studies tend to focus exclusively on economic factors. See: A. A. Wiener, "Export of Forest Products: Would Cutting Off Log Exports Lower Prices of Wood Products?" *Journal of Forestry* 71 (1973): 215–16; Darius Darr, *Softwood Log Exports and the Value and Employment Issues*, USDA, Forest Service, Research Paper PNW-200 (Portland: Pacific Northwest Forest and Range Experiment Station, 1975); Richard W. Haynes, *Price Impacts of Log Restrictions under Alternative Assumptions*, USDA, Forest Service, Research Paper PNW-212 (Portland: Pacific Northwest Forest and Range Experiment Station, 1976); R. A. Oliviera and G. W. Whittaker, *An Examination of Dynamic Relationships—and the Lack Thereof—Among U.S. Lumber Prices, U.S. Housing Starts, U.S. Log Exports to Japan, and Japanese Housing Starts*, Oregon State University, Agricultural Experiment Station, Special Report 565 (Corvallis: Oregon State University, 1979); David R. Darr, "Softwood Log Export Policy: The Key Question," *Journal of Forestry* 78 (1980): 138–40, 151; Roger A. Sedjo and A. Clark Wiseman, "Log Export Restrictions: Some Findings," *Journal of Forestry*, 78:738; David R. Darr, Richard W. Haynes, and Darius M. Adams, *The Impact of the Export and Import of Raw Logs on Domestic Timber Supplies and Prices*, USDA, Forest Service, Research Paper PNW-277 (Portland: Pacific Northwest Forest and Range Experiment Station, 1980); William E. Penoyer, "The Japanese Market for Solid Wood Products, *Forest Products Review* 36 (1980): 17–20; Roger A. Sedjo, ed., *Issues in U.S. International Forest Products Trade: Proceedings of a Workshop* (Washington, D.C.: Resources for the Future, 1981), 187–263; Roger A. Sedjo and A. Clark Wiseman, "The Effectiveness of an Export Restriction on Logs," *American Journal of Agricultural Economics* 65 (1983):

113–16; A. Clark Wiseman and Roger A. Sedjo, "Have Controls Reduced Log Exports in the Pacific Northwest?" *Journal of Forestry* 83 (1985): 680–82.

27 See Thomas R. Cox, *Mills and Markets: A History of the Pacific Coast Lumber Industry to 1900* (Seattle: University of Washington Press, 1974); Edwin T. Coman, Jr., and Helen M. Gibbs, *Time, Tide and Timber: A Century of Pope & Talbot* (Stanford, Calif.: Stanford University Press, 1949); Edmond S. Meany, Jr., "History of the Lumber Industry of the Pacific Northwest to 1917" (Ph.D. diss., Harvard University, 1935).

28 Selfridge, *American Lumber in Japan*; C. B. Rathbun, "Seattle Exports Double—Japan Relief Starts," *Export and Shipping Journal* 4 (Sept. 1923): 6, 20; editorial, *Export and Shipping Journal*, 4:12; Harold Crary, "Seattle Feels Trade Stimulus," ibid. (Oct. 1923): 6, 15; D. H. Blake, "Future of Japanese Trade," ibid. (Dec. 1923): 3–4, 15, 17; T. C. Williams, "1923 Banner Portland Shipping Year," ibid., 5–6, 13; A. F. Haines, "Pacific Shipping Outlook," ibid. (Feb. 1924), 5:3–4, 12; editorial, ibid., 10–11; ibid. (April 1924): 10, 12; ibid. (May 1924): 9, 10; *Timberman* 26 (June 1925): 76.

29 Business Executives Research Committee, *The Forest Products Industry of Oregon: A Report* (Portland: Business Executives Research Committee, 1954), 30–33; Stephen Haden-Guest, John K. Wright, and Eileen M. Teclaff, *A World Geography of Forest Resources* (New York: The Ronald Press, 1956), 558, show a temporary decline in sawnwood and log imports in 1925 and a steady decline commencing in 1928. The shipments at this time were largely made up of "squares" because logs could not be readily shipped with the technology and facilities then available. Nevertheless, these shipments involved so little manufacturing that they were essentially log exports.

30 Radius, *United States Shipping in Transpacific Trade*, 332–34; Coman and Gibbs, *Time, Tide and Timber*, 297–303, 321–24; *Timberman* 39 (June 1938): 11–13; *Timberman* 40 (Dec. 1938): 40–41.

31 Portland *Oregonian*, Oct. 3, 1936; Donald R. Zobel, "Port-Orford-Cedar: A Forgotten Species," *Journal of Forest History* 30 (1986): 32–33, 35.

32 *Timberman* 40 (Feb. 1939): 9–10; Franklin D. Roosevelt to Clarence D. Martin, March 19, 1939, reprinted in *Western Equipment and Timber News* 12 (June 1961): 11; Roosevelt to Martin, March 18, 1939, in *Franklin D. Roosevelt & Conservation, 1911–1945*, ed. Edgar B. Nixon (New York: Franklin D. Roosevelt Library, 1957), 2:309–10; *Congressional Record*, 76th Cong., 1st sess. (1939), 998, 6,742, 7,091–92, 8,365–67, 9,398, 9,643; "Senate Report No. 563", 76th Cong., 1st sess. (1939), *Senate Executive Reports* 7, pp. 10,294–10,304. Debates over the exportation of logs were also taking place in British Columbia. See: *Timberman* 39 (April 1938): 12.

33 Sources on the log export controversy of the 1960s are voluminous. Among the more valuable are runs of *Western Timber Industry* (under various titles) and *Japan Lumber Journal*, together with U.S. Senate, Select Committee on Small Business, *Log-Exporting Problems: Hearings before the Subcommittee on Retailing, Distribution, and Marketing Practices*, 4 vols. (Washington: GPO, 1968); U.S. Senate, Select Committee on Small Business, *Log Exporting Problems: Report . . . on the Impact of Increasing Log Exports on the Economy of the Pacific Northwest* (Washington, D.C.: GPO, 1968); Edward P. Cliff [Chief, U. S. Forest Service], "Export of Logs from Fed-

eral Administered Forests," paper presented at meeting of Western Wood Products Association, Portland, Ore., April 24, 1968 (copy in Cliff Papers, Forest History Society, Durham, N.C., vol. 4); State of Washington, Department of Natural Resources, *Report on Log Export Problem* (Olympia: Department of Natural Resources, 1965); W. Halder Fisher, *Report on Analysis of the Relationship of Softwood Log Exports to the Economy of the State of Washington* (Columbus, Ohio: Battelle Memorial Institute, 1964). For an overview, see: John Clark Hunt, "You Go Crazy with Logs," *American Forests* 74 (July 1968): 14–17, 49; (Aug. 1968), 32–35, 58–59.

34 *Japan Lumber Journal* 15 (Nov. 30, 1974): 13; ibid., 16 (May 31, 1975): 4–5; ibid., 16 (June 16, 1975): 14; ibid., 17 (Feb. 16, 1976), 11; Yoshitomo Ando, "House Building and Lumber Distribution in Japan," in Handa, Nomura, and Kumazaki, *Current State of Japanese Forestry (II)*, 40–51; *Western Timber Industry* 18 (Jan. 1967): 1; Steve Forrester, "Rebound or Not, Forest Industry Must Build Exports," Seattle *Argus*, Sept. 24, 1982; Seattle *Post-Intelligencer*, Aug. 12 and Oct. 24, 1982; Portland *Oregonian*, Feb. 26, May 9 and 10, and Dec. 27, 1982; *Portland Journal of Commerce*, Feb. 29, 1982; Longview (Wa.) *Daily News*, May 19, 1982, and Jan. 27, 1983; Olympia (Wa.) *Daily Olympian*, Feb. 2, 1983; "America's Biggest Wood Exporter," *Timber Trades Journal*, Oct. 23, 1982; Stuart U. Rich, "Export Market/Domestic Market Relationships of Northwestern Lumber Mills," *Forest Products Journal* 31 (Aug. 1981), 18–19; Kathleen W. Weigner, "Forest Products," *Forbes Magazine*, Jan. 3, 1983. Rising acceptance did not mean that opposition to log exports vanished altogether; it did not. See: Steve Woodruff, "Logs Across the Water; Jobs Across the Sea," Portland *Willamette Week*, June 2, 1980; Steve Woodruff, "Log Exports Mean Fewer Lumber Jobs," *Forest Planning* (July 1980), 10–14; Longview (Wa.) *Daily News*, Jan. 20, 1983. For a more cautious view of the future of wood product exports, see: David R. Darr and Gary R. Lindell, "Prospects for U.S. Trade in Timber Products," *Forest Products Journal* 30 (June 1980): 16–20.

35 *Western Timber Industry* 19 (April 1968): 1; ibid. (June 1968): 1–3, 10–12; ibid. (July 1968), 4, 10; ibid. (Aug. 1968), 1, 4, 19; ibid. (Sept. 1968), 4; ibid. (Dec. 1968), 1, 2; *National Timber Industry* 20 (April 1969): 5; *Japan Lumber Journal* 9 (June 15, 1968): 2; *Japan Lumber Journal* 9 (Nov. 30, 1968), 2; U. S. Department of Agriculture, Forest Service, *U.S. Timber Production, Trade, Consumption, and Price Statistics, 1950–1980*, Miscellaneous Publication No. 1408 (Washington: GPO, 1981), 21–37; Select Committee on Small Business, *Log-Exporting Problems*, 1:5–11, 4:1,427–43, 1,619–39, and passim; Edward P. Cliff, "Forest Resources of the Pacific Northwest," address to Pacific Northwest Trade Assoc., Eugene, Ore., April 4, 1968 (Cliff Papers, vol. 4).

36 *Western Equipment and Timber News* 12 (May 1961): 1; *Western Timber Industry* 13 (Feb. 1962): 5; ibid., 15 (Oct. 1964): 6; ibid., 18 (Feb. 1967): 4; ibid., 18 (March 1967): 5; ibid., 18 (July 1967): 1; Merl Jay Arnot, "Employment Effects of Log Exports to Japan from the Oregon South Coastal Region" (M.S. thesis, University of Oregon, 1968); Ted W. Nelson and Sharlene P. Nelson, "Fire, Insects, Wind, Volcano: The History of Disaster Management by the Weyerhaeuser Company," in Steen, *History of Sustained-Yield Forestry*, 21–30; Zobel, "Port-Orford-Cedar," 34. The account of the sawing of Douglas fir peeler logs in Kyoto is based on personal observation by the author.

37 *Western Timber Industry* 14 (May 1963): 4; ibid., 15 (June 1964): 11; ibid., 15 (July 1964): 4; Walter J. Mead to Vernon S. White, Dec. 4, 1967, reprinted in ibid., 19 (Jan. 1968): 2; Eugene (Ore.) *Register-Guard*, Nov. 11, 1985. See also James O. Howard and Bruce A. Hiserote, *Oregon's Forest Products Industry, 1976*, USDA, Forest Service Research Bulletin PNW-79 (Portland: Pacific Northwest Forest and Range Experiment Station, 1978), 5–15.

38 "No More Kinzua: Way of Life Ends with Town," Portland *Sunday Oregonian*, March 12, 1978; Tom Alkire, "The Mills Run Out: This Time It's For Real," Portland *Willamette Week*, May 26, 1980; Foster Church, "In Mill Shutdowns: Fear Overwhelms Jobless," Portland *Oregonian*, Jan. 25, 1981; Eugene, *Register-Guard*, Nov. 10, 1985. Appropriately, the key legislation limiting the exportation of logs from the west coast was known as the Morse amendment. The activities of Morse and Hatfield can be followed in *Western Timber Industry*, especially: 17 (Oct. 1966): 1; (Dec. 1966): 3; 18 (May 1967): 9; 19 (Feb. 1968): 1; (Aug. 1968): 1; (Sept. 1968): 4.

39 *Western Timber Industry* 19 (June 1968): 1–3, 10–12; ibid. (July 1968): 1; *Sacramento Bee*, March 31, 1973; *San Diego Union*, April 8, 1973; *San Diego Labor Leader*, April 13, 1973.

40 The story of the Japanese factory ship was heard in a bar in Washington state by a colleague. The *Japan Lumber Journal* 5 (Nov. 20, 1964): 4, denounced earlier claims by the "Save Our Logs Committee" that Japan was flooding the United States with plywood made from American logs as demagogic; indeed, they were. On Japan's plywood industry, see: Isamu Nomura, "The Current State of the Plywood Industry and the Tropical Hardwood Import of Japan," in Handa and Nomura, *Current State of Japanese Forestry*, 21–28; Forestry Agency, *Forestry in Japan*, 62–64; Forestry Agency, *Forestry Technology in Japan*, 162–65; Forster, *Japanese Forestry*, 45–49. The plywood and woodpulp industries, of course, involve complexities of their own, complexities beyond the scope of this essay.

41 Hunsberger, *Japan and the U.S. in World Trade*, 95; *Crow's Western Equipment and Timber News* 8 (Oct. 31, 1957): 1; *Western Timber Industry* 15 (June 1964): 7; *Western Timber Industry* 18 (Nov. 1967): 16; *Forest Industry* (Jan. 1983), 7; Eugene *Register-Guard*, Nov. 15, 1985; Forster, *Japanese Forestry*, 85–86, 93–105; John Clarke, "Exporting Raw Logs—A Serious Issue," *British Columbia Lumberman* (Oct. 1982), 28–29; Langdon, *Canadian-Japanese Economic Relations*, 119–20; Rinseigaku Kyōshitsu, ed., *Saikin no mokuzai shigen to shijō* [Recent timber resources and markets] (Morioka: Iwate Daigaku Nōgakubu, 1975), 5–7, 10–14, 19–26, and passim. For the larger context, see: Robert John Francis, "An Analysis of British Columbia Lumber Shipments, 1947–1957" (M.A. thesis, University of British Columbia, 1961); J. S. Kendrick, "Consumption of Forest Products in Asia and the Pacific Area," in *British Columbia's Future in Forest Products Trade in Asia and the Pacific Area* (mimeographed; Vancouver: University of British Columbia, 1965); G. R. Lindell, *Log Export Restrictions of the Western United States and British Columbia*, USDA, Forest Service, General Technical Report PNW-63 (Portland: Pacific Northwest Forest and Range Experiment Station, 1978); G. H. Manning and C. J. Macklin, *Waterborne Exports of Forest Products from British Columbia Ports: 1985 and 1990*, Canadian Forest Service, Research Paper BC-X-

211 (Victoria, B.C.: Pacific Forest Research Centre, 1980); Lawrence W. Rakestraw, *A History of the United States Forest Service in Alaska* (Anchorage: Alaska Historical Commission and United States Forest Service, 1981), 108–13, 127–28.

42 USDA, Forest Service, *Report of the Chief of the Forest Service, 1963* (Washington, D.C.: GPO, 1964), 20. The advisability and implications of conversion from old-growth to managed second-growth stands is itself a subject of considerable controversy. For discussions, see: The Conservation Foundation, *Forest Land Use: Issues, Studies, Conferences* (Washington: The Conservation Foundation, 1980), 5–12, 37–39; Marion Clawson, *Decision Making in Timber Production, Harvest and Marketing*, Research Paper No. R-4 (Washington: Resources for the Future, 1977); *America's Renewable Resource Potential—1975: The Turning Point*, Proceedings, 1975 National Convention, Society of American Foresters (Washington: Society of American Foresters, 1976), 72–107.

43 Walter J. Mead, *Mergers and Economic Concentration in the Douglas-Fir Lumber Industry*, Research Paper PNW-9 (Portland: Pacific Northwest Forest and Range Experiment Station, 1964); Mead, *Competition and Oligopsony in the Douglas-Fir Lumber Industry* (Berkeley: University of California Press, 1966); Dennis C. LeMaster, *Mergers Among the Largest Forest Products Firms, 1950–1970*, Research Center Bulletin 854 (Pullman: Washington State University, 1977); Kendrick, "Consumption of Forest Products"; Lowell Besley, "Potential International Timber Supply," in *Transactions of the Fifth British Columbia Natural Resources Conference* (Victoria, B.C.: Government Printers, 1952); J. C. Westoby, *World Forest Development: Markets, Men and Materials*, 1965 H. R. MacMillan Lectureship (Vancouver: University of British Columbia, 1966); John A. Zivnuska, *U.S. Timber Resources in a World Economy* (Washington: Resources for the Future, 1967), esp. 103–10; Eugene *Register-Guard*, Nov. 17, 1985.

44 For discussions of these problems, see: Marion Clawson, ed., *Forest Policy for the Future: Conflict, Compromise, Consensus* (Baltimore: Johns Hopkins University Press for Resources for the Future, 1974); idem, *Forests for Whom and for What?* (Baltimore: Johns Hopkins University Press for Resources for the Future, 1975); John V. Krutilla and Anthony C. Fisher, *The Economics of Natural Environments: Studies in the Valuation of Commodity and Amenity Resources* (Baltimore: Johns Hopkins University Press for Resources for the Future, 1975); Dennis C. LeMaster and Luke Popovich, eds., *Crisis in Federal Forest Land Management* (Washington: Society of American Foresters, 1977); Thomas R. Cox, et al., *This Well-Wooded Land: Americans and Their Forests from Colonial Times to the Present* (Lincoln: University of Nebraska Press, 1985), 237–64.

45 There are exceptions to this generalization, of course. Some small firms have catered to and profited from the Japanese trader. See: Eugene *Register-Guard*, Nov. 14, 1985. Further study of the impact of the trade is needed. Not only are most of the studies to date narrow and incomplete, but also none are broadly comparative. This paper represents the present author's first step toward such a broadly cast study.

46 For discussions of the relationships of the international economy and environmental concerns, see Seymour J. Rubin and Thomas R. Graham, eds., *Environment and Trade: The Relations of International Trade and Environmental Policy* (Totowa, N.J.:

Allanheld, Osmun, 1982); John Larsen, "Environmental Issues and Their Influence on World Trade in Wood," in Bethel, *World Trade in Forest Products*, 129–36.

9
Changing Capital Structure, the State, and Tasmanian Forestry

1 The concept of the capitalist world system is developed by Immanuel Wallerstein, *The Capitalist World Economy* (Cambridge: Cambridge University Press; Paris: Editions de la Maison des Sciences de l'Homme, 1979) and *The Modern World-System*, 2 vols. (New York: Academic Press, 1974–80). The need to construe the world in *both* structuralist and pluralist terms is developed by Ralph Pettman, *State and Class: A Sociology of International Affairs* (London: Croom Helm, 1979).

2 Ernest Mandel, *Late Capitalism* (London: Verso, 1978).

3 A fuller account covering the nineteenth and twentieth centuries is given in John Dargavel, *Development of the Tasmanian Wood Industries: A Radical Analysis* (Ph.D. diss., Australian National University, 1982).

4 In 1642, Abel Tasman landed in Tasmania and named it Van Diemen's Land but sailed on during his search for a route to "known but rich places" on behalf of the Dutch East India Company.

5 G. Martin, ed., *The founding of Australia: The Argument About Australia's Origins* (Sydney: Hale & Iremonger, 1978).

6 H. Reynolds, "'That Hated Stain': The Aftermath of Transportation in Tasmania," *Historical Studies* 14 (1969): 14–31.

7 Mandel, *Late Capitalism*; Peter Cochrane, *Industrialization and Dependence: Australia's Road to Economic Development* (St. Lucia, Queensland: University of Queensland Press, 1980).

8 Tasmania, "Petition for Amendment of Land Laws," *Journals of the House of Assembly* (1875): paper 80; Tasmania, *The Waste Lands Amendment Act, 1881* (hereafter all legislation is Tasmanian).

9 *The State Forests Act*, 1885; *The Crown Lands Act*, 1890, did enable small areas of public lands to be leased for up to 14 years for tramways, wharves, jetties, and mill sites but did not allow the forest areas that the tramways served to be leased.

10 Tasmania had 3.8 percent of the Australian trade from 1898 to 1900. M. Row, *The Tasmanian Timber Trade 1820–1930: A Case Study in Spatial Interaction* (B.S. (Honors) thesis, University of Tasmania, 1977).

11 *The Crown Lands Amendment Act*, 1898.

12 *Government Notice*, 347, 25 Nov. 1898; 59, 20 Jan. 1899; 210, 21 June 1900.

13 M. Row, "The Huon Timber Company and the Crown: A Tale of Resource Development," *Papers and Proceedings of the Tasmanian Historical Research Association* 27, no. 3 (1980): 87–102.

14 *The Geeveston Tramways and Timber Leases Act*, 1901; *The Tasmanian Timber Corporation Act*, 1902.

15 See, for example, William Heyn (timber inspector, British Admiralty Harbour Works), "Present and Future Prospects of Timber in Tasmania," *Papers and Proceedings of The Royal Society of Tasmania*, 1901, 21–37; L. Rodway (Tasmanian government botanist), "Forestry for Tasmania", in ibid., 1898–99, liii–lvii; A. North (Launceston architect), "The economic aspect of Tasmania's forests," *Report of the 10th meeting of the Australian Association for Advancement of Science* (1904): 546–60.

16 D. E. Hutchins, *A Discussion of Australian Forestry* (Perth: Government Printer, 1916).

17 Peter Cochrane, *Industrialization and Dependence*.

18 Philip Ehrensaft and Warwick Armstrong, "Dominion Capitalism: A First Statement," *Australia and New Zealand Journal of Sociology* 14, no. 3 (1978): 352–63; idem, "The Foundation of Dominion Capitalism: Economic Truncation and Class Structure," in *Inequality*, ed. Allan Moscovitch and Glenn Drover (Toronto: University of Toronto Press, 1981).

19 A most useful theoretical apparatus for separating the effects of distinct but coexisting economic structures within capitalism is given by K. D. Gibson & R. J. Horvath, "Aspects of a Theory of Transition within the Capitalist Mode of Production," *Environment and Planning D: Society and Space* 1 (1983): 121–138.

20 "Report of the Inaugural Meeting of the Empire Forestry Association," *Empire Forestry* 1 (1922): 3–10.

21 *National Dictionary of Biography* 1931–40 (Oxford University Press, 1949).

22 R. W. Connell and T. M. Irving, *Class Structure in Australian History* (Melbourne: Longman Cheshire, 1980).

23 *Proceedings [5th] Interstate Conference on Forestry* (Canberra: Government Printer, 1920).

24 Australia, Development and Migration Commission, *Report on Afforestation and Reforestation in Tasmania* (Canberra: Government Printer, 1929).

25 S. W. Steane, "State Forestry in Tasmania," *Australian Forestry* 2, no. 1 (1937): 19–24.

26 F. G. Davidson and B. R. Stewardson, *Economics and Australian Industry* (Hawthorn, Victoria: Longman Cheshire, 1974).

27 The only published reference I have found to the contract with the Canadian mills is a statement by W. Bunston in the Australian Newsprint Mills' house journal, *Newsprint Log* 1, no. 1 (Nov. 1945).

28 Legislation was passed in 1924, 1926, and finally in the *Associated Pulp and Paper Mills Act*, 1936.

29 Legislation was passed in 1932, 1935, and finally in the *Florentine Valley Paper Industry Act*, 1937.

30 Conservator of Forests to Minister Controlling Forestry, 25 Nov. 1924, State Archives of Tasmania, file FC5/5459.

31 E. L. Whelwright, "The Age of Transnational Corporations," in G. Crough et al., *Australia and World Capitalism* (Ringwood, Victoria: Penguin Books, 1980).

32 K. Tsokhas & M. Simms, "The Political Economy of the United States Investment in Australia," *Politics* 13, no. 1 (1978): 65–80.

33 Tasmania, Royal Commission, "Report on Forestry Administration," *Journals and*

Printed Papers of Parliament (hereafter cited as JPPP), 1946, paper 1; P. May, *Problems in the Analysis of Political Corruption* (Ph.D. diss., University of Tasmania, 1976).

34 *Forestry Act*, 1946.

35 *Florentine Valley Paper Industry Act*, 1961. ANM was obliged to sell 30,100 cubic meters a year of sawlogs to sawmillers, but as the concession contained 35–40 percent of sawlogs, at least as much again was pulped.

36 Legislation was passed in 1954, 1956, and in the *Huon Valley Pulp and Paper Industry Act*, 1959.

37 *Wesley Vale Pulp and Paper Industry Act*, 1959.

38 Legislation was passed in 1968 and in *Pulpwood Products Industry (Eastern and Central Tasmania) Act*, 1971.

39 The pulpwood cut in Tasmania during the 1970s was supplied: 41 percent from private forests, 51 percent from public forests, and 7 percent as waste from sawmills.

40 Tasmania, Board of Inquiry, *Private Forestry Development in Tasmania*, JPPP, 1972, paper 70.

41 Tasmania, Forestry Commission, *The Sawmilling Industry in the Wesley Vale Concession Area* (Hobart: Forestry Commission, 1978).

42 More complex actions by the state in Tasmanian forestry are analyzed in Dargavel, *Development of the Tasmanian Wood Industries*. A general review of recent theories of the state is Bob Jessop, *The Capitalist State* (Oxford: Martin Robertson, 1982).

10
The Death and Rebirth of the American Forest

1 Michael Williams, "Clearing the United States Forests: Pivotal Years," *Journal of Historical Geography* 8 (1982): 12–18; M. L. Primack, "Land Clearing under Nineteenth Century Techniques: Some Preliminary Calculations," *Journal of Economic History* 22 (1962): 485–96.

2 Frederick Starr, "American Forests: Their Destruction and Preservation," *Annual Report of Commissioner of Agriculture* (Washington, 1865), 210–34; Increase A. Lapham, J. G. Knapp, and H. Crocker, "Report on the Disastrous Effects of the Destruction of Forest Trees Now Going On So Rapidly in the State of Wisconsin" (Madison, Wisc., 1867); George Perkins Marsh, *Man and Nature, or Physical Geography as Modified by Human Action* (New York, 1864); F. B. Hough, *Report upon Forestry*, vols. 1–3 (Washington, D.C.: GPO, 1878, 1880, and 1882, respectively).

3 John J. Thomas, "Culture and Management of Forest Trees," *Annual Report of Commissioner of Agriculture* (Washington, D.C.: GPO, 1864), 43–44.

4 Frederick Starr, "American Forests," 219.

5 See Harold K. Steen, *The U.S. Forest Service: A History* (Seattle, 1976), particularly pp. 47–144; and R. Nash, *Wilderness and the American Mind* (New Haven, 1967), 183–208.

6 See William B. Greeley, *Gifford Pinchot, Forester-Politician* (Princeton, 1960), 54–55; and Roy Robbins, *Our Landed Heritage: The Public Domain, 1776–1936* (Princeton, 1942), 337.

7 Theodore Roosevelt, "The Forest in the Life of the Nation," pp. 3–12 of *Proceedings of the American Forest Congress* (Washington, D.C., 1905), extract on pp. 8–9.

8 Gifford Pinchot, *The Fight for Conservation* (New York, 1910), 7, 9.

9 James E. Defebaugh, *History of the Lumber Industry of America*, 2 vols. (Chicago, 1906), 1:272.

10 Marion Clawson, "Forests in the Long Sweep of American History," *Science* 204 (1979): 1168–74, reprinted, Resources for the Future reprint 164 (Washington, D.C., 1979).

11 William B. Greeley, "Reduction of the Timber Supply through Abandonment or Clearing of Forest Lands," Report of the National Conservation Commission, Senate document no. 676. 60th Cong., 2d Sess., (Washington, D.C.: GPO, 1909), 633–44.

12 For example: ibid., 633; U.S. Department of Agriculture, Forest Service, *Timber Depletion, Lumber Prices, Lumber Exports and Concentration of Timber Ownership*, Report on Senate Resolution 311, 66th Cong., 2d Sess., (Capper Report) (Washington, D.C.: GPO, 1920), 31–32, 37–39; William B. Greeley, "The Relation of Geography to Timber Supply," *Economic Geography* 1 (1925): 3–5; and William B. Greeley, et al., "Timber: Mine or Crop?" pp. 83–180 of U.S. Department of Agriculture, *Year Book, 1922*, particularly pp. 84–93.

13 The three maps were first produced in Greeley, "Relation of Geography to Timber Supply" (1925) and were subsequently included in Charles O. Paullin, *Atlas of the Historical Geography of the United States* (Washington, 1932), plates 3A, 3B, and 3C.

14 E. A. Zeigler, "Rates of Forest Growth," *Report of the National Conservation Commission*, Senate document no. 676, 60th Cong., 2d Sess. (Washington, D.C.: GPO, 1909), 203–69.

15 Clawson, "Forests in the Long Sweep," 4.

16 Zeigler, "Rates of Forest Growth," 219, 222–23.

17 Greeley, et al., "Timber."

18 For a general appraisal of the opposing views see Donald Swain, *Federal Conservation Policy, 1921–1922* (Berkeley, Calif., 1963), 11–15; and Jenks Cameron, *The Development of Governmental Forest Control in the United States* (Baltimore, Md., 1928), 296–402. In detail, for the cooperative point of view of H. S. Graves and of Greeley afterward see H. S. Graves, "A Policy of Forestry for the Nation," *Journal of Forestry* 17 (1919): 901–10; and for the coercion argument see "Forest Devastation: A National Danger and a Plan to Meet It," Report of the Committee for the Application of Forestry. This was written largely by Pinchot and was reprinted in the *Journal of Forestry* 17 (1919): 911–45, with a preface by Gifford Pinchot, "The Lines are Drawn," pp. 899–900.

19 U.S. Department of Agriculture, Forest Service, *A National Plan for American Forestry*, Senate document no. 12, 73d Cong., 1st Sess., (Copeland Report), 2 vols. (1933), 1:222–25.

20 Ibid., 242.

21 U. S. Department of Agriculture, Forest Service, *An Analysis of the Timber Situation*

in the United States, 1952–2030, Forest Resource Report no. 23 (Washington, D.C., December 1982), p. 134, table 6.13; p. 136, table 6.14, and p. 137.

22 See also Clawson, "Forests in the Long Sweep," table 1.

23 William B. Greeley, *Forests and Men* (New York, 1951), 15–29.

24 Based on U.S. Bureau of the Census, *Historical Statistics of the United States from Colonial Times to 1970* (Washington, D.C., 1977), part 1, tables L48 and L49.

25 For a brief summary of the Clarke-McNary Act see Samuel Trask Dana and Sally K. Fairfax, *Forest and Range Policy: Its Development in the United States*, 2d ed. (New York, 1980), 126–29.

26 One of the first inquiries to establish the trend was C. I. Hendrickson, "The Agricultural Land Available for Forestry," *A National Plan for American Forestry*, Senate document no. 12, 73d Cong., 1st Sess. (Copeland Report) (1933), 1:151–69.

27 This and the following paragraphs, and the statistics of clearing in table 10.3 are based on J. Fraser Hart, "Loss and Abandonment of Cleared Farm Land in the Eastern United States," *Annals, Association of American Geographers* 58 (1968): 417–40, and is author's analysis of the 1965, 1969, 1975, and 1979 agricultural censuses and the country distribution of clearing and abandonment.

28 See J. Fraser Hart, "Land Use Change in a Piedmont County," *Annals, Association of American Geographers* 70 (1980): 492–527, especially pp. 514–16.

29 Dana and Fairfax, *Forest and Range Policy*, 130–31.

30 Based on U.S. Bureau of the Census, *Historical Statistics of the United States from Colonial Times to 1970* (Washington, D.C., 1970), tables L32–L43 and accompanying notes.

31 See Hamlin L. Williston, *A Statistical History of Tree Planting in the South, 1925–79* (Atlanta: U.S. Department of Agriculture, Forest Service, 1981).

32 For a more extensive discussion of consumption of various forms of timber see Clawson, "Forests in the Long Sweep," 5.

11 Perspectives on Deforestation in the U.S.S.R.

1 See "Intensifikatsiya i povysheniye effektivnosti proizvodstva—vazhneyshaya zadacha lesovodov," *Lesnoye Khozyaystvo* 12 (1981): 2–5; V. D. Baytala, "Vse reservy —v deystviye," *Lesnoye Khozyaystvo* 10 (1981): 6–9; P. I. Moroz, "Perspektivy razvitiya sovetskogo lesoustroystva," *Lesnoye Khozyaystvo* 6 (1979): 42–49; S. G. Sinitsyn, "Leninskiye idei—Osnova sotsiyalisticheskoy sistemy lesopolzovaniya," *Lesnoye Khozyaystvo* 4 (1980): 10–14; N. V. Timofeyev, *Lesnaya Industriya SSSR* (Moscow: Lesnaya Promyshlennost, 1980); G. I. Vorobyev, "Novye rubezhi lesnogo khozyaystva," *Lesnoye Khozyaystvo* 2 (1981): 2–5; idem, "Lesnoye khozyaystvo v odinnadtsatoy pyatiletke," *Lesnoye Khozyaystvo* 5 (1981): 2–5; idem, "Odinnadtsatoy pyatiletke—udarny trud," *Lesnoye Kozyaystvo* 1 (1982): 2–5.

2 B. I. Yunov, "O lesopolzovanii v evropeyskoy chasti SSSR," *Lesnoye Khozyaystvo* 10 (1978): 46.
3 P. R. Pryde, *Conservation in the Soviet Union* (Cambridge: The University Press, 1972); W. R. J. Sutton, "The Forest Resources of the USSR: Their Exploitation and Their Potential," *Commonwealth Forestry Review* 54, no. 2 (1975): 110–38.
4 Pryde, *Conservation*, 96–100.
5 P. R. Pryde, "The 'Decade of the Environment' in the U.S.S.R.," *Science* 220 (1983): 274–79.
6 Pryde, *Conservation*.
7 Boris Komarov, *The Destruction of Nature in the Soviet Union* (White Plains, N.Y.: M. E. Sharpe, Inc., 1980); reviewed by Brenton M. Barr in *Canadian Slavonic Papers* 23, no. 2 (1981): 217–18.
8 Sutton, "Forest Resources," 136.
9 Brenton M. Barr, *Domestic and International Implications of Regional Change in the Soviet Timber and Wood-Processing Industries, 1970–1990*, Discussion Paper no. 4, Association of American Geographers Project on Soviet Natural Resources in the World Economy (Washington, D.C.: 1978); idem, "Soviet Timber: Regional Supply and Demand, 1970–1990," *Arctic* 4 (1979): 308–28; idem, "Regional Dilemmas and International Prospects in the Soviet Timber Industry," in *Soviet Natural Resources in the World Economy*, ed. R. G. Jensen, T. Shabad, and A. W. Wright (Chicago: University of Chicago Press, 1983), 411–41; Sutton, "Forest Resources," 136; R. N. North and J. J. Solecki, "The Soviet Forest Products Industry: Its Present and Potential Exports," *Canadian Slavonic Papers* 19, no. 3 (1977): 281–311.
10 M. M. Drozhalov, "Postoyanno uluchshat lesnoy fond SSSR," *Lesnoye Khozyaystvo* 9 (1979): 41.
11 See G. I. Vorobyev et al., *Ekonomicheskaya Geografiya Lesnykh Resursov SSSR*, discussed in Brenton M. Barr, "Soviet Forest Resources: A Review and Summary," *Soviet Geography, Review and Translation* 6 (1982): 452–62, and N. V. Timofeyev, *Lesnaya Industriya*.
12 V. V. Glotov, *Razmeshecheniye Lesopromyshlennogo Proizvodstva* (Moscow: Lesnaya Promyshlennost, 1977), 7.
13 A. Rodgers, "Changing Locational Patterns in the Soviet Pulp and Paper Industry," *Annals of the Association of American Geographers* 45, no. 1 (1955): 93. The early Soviet period is reviewed by Glotov, *Razmeshcheniye*, 7–14.
14 Rodgers, "Changing Patterns," 96.
15 Ibid.
16 Reviewed in Glotov, *Razmeshcheniye*, 7–14; and in N. V. Timofeyev et al., *Les —Natsionalnoye Bogatstvo Sovetskogo Naroda* (Moscow: Lesnaya Promyshlennost, 1967), 73–87, 141–55.
17 *Vneshnyaya Torgovlya SSSR 1922–1981* (Moscow: 1982), 52–53. The early decades of development of Soviet forestry and its industrial significance are discussed in *Lesnoye Knozyaystvo SSSR za 50 Let* (Moscow: Lesnaya Promyshlennost, 1967), 16–24. The sectoral and spatial growth of the timber industry and various wood-processing industries from 1917 to 1941 and from World War II to 1980 is described in N. V. Timofeyev, *Lesnaya Industriya SSSR* (Moscow: Lesnaya Promyshlennost, 1980), 52–212. The issue of deforestation, however, is omitted from this otherwise informative volume, but prewar exports are discussed on pages 223–27.

18 Barr, "Regional Dilemmas," 420.
19 Timofeyev et al., *Les—Natsionalnoye*, 74.
20 V. P. Tseplyaev, *Lesnoye Khozyaystvo SSSR* (Moscow: Lesnaya Promyshlennost, 1965), 144.
21 G. I. Vorobyev et al., *Ekonomika Lesnogo Khozyaystva SSSR* (Moscow: Vysshaya Shkola, 1980), 25.
22 Glotov, *Razmeshcheniye*, 74.
23 *Promyshlennost SSSR. Statisticheskiy Sbornik* (Moscow: 1957), 247.
24 P. Blandon, *Soviet Forest Industries* (Boulder: Westview Press, 1983), 5.
25 Rodgers, "Changing Patterns"; J. W. Miller, Jr., "Forest Fighting on the Eastern Front in World War II," *The Geographical Review* (April 1972): 186–202.
26 Timofeyev et al., *Les—Natsionalnoye*, 154.
27 Vorobyev et al., *Ekonomika*, 35.
28 Reviewed in ibid., 26–45. Estimates of differential rent by region are derived and discussed in J. Thornton, *Soviet Resource Valuation and the Efficiency of Resource Use*, Discussion Paper no. 26, Association of American Geographers Project on Soviet Natural Resources in the World Economy (Washington, D.C., 1980): 30, 31; in Brenton M. Barr, *The Soviet Wood-Processing Industry: A Linear Programming Analysis of the Role of Transportation Costs in Locational and Flow Patterns* (Toronto: University of Toronto Press, 1970), 118–28; and in Barr, *Domestic and International Implications*, 24–37.
29 J. H. Holowacz, "USSR," *World Wood* 8 (1975): 16.
30 *Ekonomicheskaya Gazeta* 27 (1982): 2.
31 Holowacz, "USSR," 18.
32 Ibid.
33 Ibid.
34 Drozhalov, "Postoyanno uluchshat"; V. A. Nikolayuk, "Izmeneniya v lesnom fonde v resultate khozyaystvennoy deyatelnosti," *Lesnoye Khozyaystvo* 7 (1975): 2–6.
35 J. Eronen, "Soviet Pulp and Paper Industry: Factors Explaining Its Areal Expansion," *Silva Fennicam* 3 (1982): 276–77; Glotov, *Razmeshcheniye*, 20.
36 Drozhalov, "Postoyanno uluchshat," 42.
37 Ibid.; recent planting has comprised pine (57%), spruce (17%), larch (4%), oak and beech (5%), and miscellaneous species (17%); personal communication from J. H. Holowacz.
38 Ibid.
39 Eronen, "Soviet Pulp," 227.
40 Ibid.; and Glotov, *Razmeshcheniye*, 4.
41 Blandon, *Soviet Forest*, 83.
42 *Current Digest of the Soviet Press* (hereafter cited as *CDSP*) 28, no. 14 (1976): 21; *CDSP* 28, no. 19 (1976): 2; J. Eronen, "Routes of Soviet Timber to World Markets," *Geoforum* 14, no. 2 (1983): 210.
43 Drozhalov, "Postoyanno uluchshat," 42.
44 Ibid.
45 Barr, *Soviet Wood-Processing*; Barr, *Domestic and International Implications*; Barr, "Soviet Timber"; Barr, "Regional Dilemmas."
46 *Vneshnyaya Torgovlya*, 27, 68–69, 239.
47 *Narodnoye Khozyaystvo RSFSR v 1975 g.* (Moscow: 1976), 86–89.

48 Kathleen Braden (Seattle Pacific University), personal communication (April 1983).
49 Drozhalov, "Postoyanno uluchshat," 43.
50 Barr, "Soviet Forest Resources," 461.
51 Drozhalov, "Postoyanno uluchshat," 41.
52 Discussed in Barr, "Soviet Forest Resources."
53 Ibid., 460.
54 Ibid.
55 Nilolayuk, "Izmeneniya v lesnom," 4.
56 Barr, "Soviet Forest Resources," 461.
57 Environmental planning legislation is outlined at length in L. Brezhnev and M. Georgadze, "Zakon Soyuza sovetskikh sotsialisticheskikh respublik ob utverzhdeni osnov lesnogo zakonodatelstva soyuza SSR i soyuznykh respublik," *Lesnoye Khozyaystvo* 8 (1977): 40–54; see discussion in *CDSP* 29, no. 22 (1977): 14–15; *CDSP* 29, no. 24 (1977): 4–12; *CDSP* 29, no. 25 (1977): 9–16; *CDSP* 29, no. 42 (1977): 6–7; *CDSP* 31, no. 1 (1979): 1–4; *CDSP* 32, no. 48 (1980): 15; *CDSP* 34, no. 23 (1982): 1–6; *CDSP* 34, no. 12 (1982): 9–10. Critical assessments are found in V. L. Gorovoy and V. M. Shlykov, "Basic Trends in the Development of the Forest Industry along the Baykal-Amur Mainline," *Soviet Geography: Review and Translation* 2 (1978): 84–98; E. A. Medvedkova and G. I. Malykh, "Cartographic Evaluation of the Use of Timber Resources in Irkutsk Oblast," *Soviet Geography: Review and Translation* 3 (1973): 184–94; idem, "Changes in the Spatial Organization of the Timber Industry of Irkutsk Oblast," *Soviet Geography: Review and Translation* 7 (1982): 494–505. Examples of continuing abuse of the environment are found in G. I. Galziy, "The Ecosystem of Lake Baykal and Problems of Environmental Protection," *Soviet Geography: Review and Translation* 4 (1981): 217–25; B. Komarov, *Destruction of Nature*.
58 *CDSP* 33, no. 31 (1981): 10–11; *CDSP* 33, no. 41 (1981): 23.
59 *CDSP* 33, no. 41 (1981): 23.
60 *CDSP* 33, no. 31 (1981): 11.
61 *Ekonomicheskaya Gazeta* 27 (1982): 2.
62 *CDSP* 29, no. 19 (1977): 17.
63 *CDSP* 23, no. 51 (1972): 17.
64 Ibid.
65 *CDSP* 33, no. 31 (1981): 11; *CDSP* 33, no. 41 (1981).

INDEX

Abandonment of farmland to forest: in Carroll County, Georgia (U.S.), 224–25; in the United States, 218, 222–25
Acacia albida, 81, 82
Afforestation: in Australia, 199; in Togo, 274 n.61; in the United States, 225–27, 229; in the U.S.S.R., 250. *See also* Forest management; Reforestation; Regeneration
Africa. *See* Kenya; Niger; Togo; West African Sahel
Agricultural Conservation Program (U.S.), 225. *See also* Conservation
Agricultural expansion: during pre-colonial times, 274 n.2; impact on forests: in Assam, 92, 105; in Kenya, 38, 40; in Kumaon, 92–93; in the Sahel, 71, 83; in Thailand, 112–13, 117, 126–29, 282 n.6, 286 n.45; in Togo, 58, 60, 61; in the U.S.S.R., 231, 235. *See also* Fallowing; Shifting cultivation
Agricultural statistics: in French West Africa, 11; in Thailand, 131–43
Agriculture, Ministry of (Thailand), 285 n.40
Akamatsu (*Pinus densiflora*), 171; nematode infestation, 293 n.15
Alabama, 223. *See also* United States: regional forests in the South
Almora, 280 n.2; *See also* India
Amazonia, 15. *See also* Brazil
American Federation of Labor, 178
American Forestry Congress, 212
Araucaria forests: deforestation in, 17–18; description of, 15–17; as frontier zone, 18–19; settlement patterns in, 25–27. *See also* Brazil
Argentina, 150, 198
Arkansas, 227
Armenia, 236, 237
Assam, 5, 6; agriculture in, 92–93, 104–6; area of, 104; export agriculture, 92, 106–7, 110; *jhum* system in, 104; reserved forests, 106–10; Revenue Department in, 107; timber-cutting industry, 106–8. *See also* Shifting cultivation; Timber and wood products trade
Associated Pulp and Paper Mills (Australia), 202, 205, 208, 209
Aubreville, A.: on deforestation and desertification in West Africa, 80–82

Australia: economic conditions in, 193–95, 198, 204–5, 207; forestry legislation, 301 n.9; general history of, 190–93, 301 n.4, 301 n.10; geographic description of, 190; history of forestry, 6, 195–200, 201–4, 206, 210; industry, impact on forests, 199. *See also* Tasmania
Australian Forest League, 199
Australian Newsprint Mills, 202, 205
Australian Paper Manufacturers, 202, 204, 205
Australian Paper Mills, 199
Azerbaydzhan, 236, 237, 238

Baikal Meridian (U.S.S.R.), 255. *See also* U.S.S.R.
Baltics, 236, 237, 240, 241, 248
Bamboo (*Arundinaria alpina*), 34
Baykal-Amur Mainline (U.S.S.R.), 231, 253, 255, 259. *See also* Railroads
Begar labor: in India, 96–97
Belorussia, 236, 237, 238, 240, 241, 243, 253
Birch, 233, 253
Birusa combine (U.S.S.R.), 260
Black Earth region (U.S.S.R.), 236, 237, 238, 240, 241, 253. *See also* U.S.S.R.
Brazil: Amazonia, 15; deforestation, 15, 17–18, 19; forest policy, 26, 27–32; species composition, Araucaria forests, 15–19. *See also* Colonialism; Forest management; Forestry agencies; Timber and wood products trade
Brazilian Institute for Forest Development, 60
British Transvaal Company, 195–96. *See also* Private forests
Broad-leaved species: in India, 280 n.2; in Japan, 171
Burkina Faso. *See* Upper Volta

Cabinetwoods. *See* Tropical hardwoods; Tropical softwoods
Caboclos, 26. *See also* Brazil; Shifting cultivation
Caltex, 208. *See also* Timber and wood products trade: in United States
Camphor (*Cordia abyssinica*), 42
Canada, 198, 243, 261. *See also* Timber and wood products trade
Cargo mills, 176–77. *See also* Sawmills; Timber and wood products trade: in United States
Caribbean islands: export trade, of cabinetwoods, 150; export trade, of softwoods, 148
Carroll County, Georgia (U.S.), 224–25. *See also* Abandonment of farmland to forest; United States: regional forests in the South
Cassipourea malosaria, 34
Central America: export trade, of softwoods, 148
Central Asia (U.S.S.R.), 235, 238, 239, 256. *See also* U.S.S.R.
Central Industrial Region (U.S.S.R.), 243, 256. *See also* U.S.S.R.
Central U.S.S.R., 236, 237, 240, 241. *See also* U.S.S.R.
Chao Phraya River (Thailand), 282 n.10
Charcoal industry: in Togo, 59–60
China, 231
Chir pine (*Pinus longifolia*), 93, 97–98; reserves of, 94
Chunsky combine (U.S.S.R.), 260. *See also* Timber and wood products trade: in U.S.S.R.
Civil War (Russian), 247
Clarke-McNary Act of 1924, 220, 225. *See also* Forest legislation
Clear-felling: in Tasmania, 209; in the United States, 210. *See also* Forest management
Climate: impact on crops, in Thailand, 282 n.4; impact on forest growth, in Tasmania, 193; impact on forest growth, in the U.S.S.R., 255; impact on timber trade, in Japan, 170
Club du Sahel, 80, 84, 86, 274 n.1, 279 n.77. *See also* Niger; Togo
Collins House, 199, 201, 202. *See also*

Timber and wood products trade: in Tasmania
Colonialism: impact of British economic policy, in Tasmania, 198; impact of British forest policy, in Kenya, 37–38, 44; impact of British forest policy, in Tasmania, 194–95; impact of British forest policy, in India, 95–97; impact of German forest policy, in Togo, 56–57, 60–69; impact of Portuguese forest policy, in Brazil, 26; impact on vegetation, in French West Africa, 71–74; impact on world vegetation, 3–6. *See also* Forest legislation: in Tasmania; Forestry agencies
Colonization, state-sponsored: in Brazil, 26; in Kenya, 37–38, 44; in Togo, 61–65. *See also* Environment: impact of settlement
Combretum spp., 272 n.16
Commercial forests: in the United States, 213–14, 218; in the U.S.S.R., 239, 240, 241, 249, 252, 256. *See also* Forest ownership; Timber and wood products trade
Congress of Industrial Organizations (Australia), 178
Coniferous species: in Brazil, 15–16; in India, 280 n.2; in the U.S.S.R., 230, 231, 232, 239, 251, 252, 253, 254
Conservation possibilities: in the Sahel, 77; in the United States, 178; in the U.S.S.R., 233, 257. *See also* Forest management
Conservation Reserve Soil Bank Program (U.S.), 225
Conservation Soil Bank Program (U.S.), 226
Cooperative Forestry Act of 1950 (U.S.), 225. *See also* Forest legislation
Copeland Report (U.S.), 216, 225. *See also* Forest inventories
Cooking practices. *See* Fuel wood
Crops: in Brazil, 19, 26; in Thailand, 113–17; in Togo, 63–67
Crossopteryx africana, 272 n.16
Crown Forests (Kenya), 33–34
Crown Lands Act (Tasmania), 301 n.9. *See also* Forest legislation
Cultivation shifting. *See* Shifting cultivation

Daniella thurifera, 68
Deciduous species: in the U.S.S.R., 230, 231, 252
Deforestation: as a consequence of hardwood exports, 157–60; in Amazonia, 15; in Assam, 91, 105–10; in Brazil, 17–19; in the developing world, 4–5, 147, 157–58; in European colonies, 5; in India, 90, 97; Japanese influence on, 160–62; in Kenya, 39–40, 49; in Niger, 80–81; in the Sahel, 70–72, 79, 82, 83, 274 n.2, 278 n.55; in Senegal, 81–82; in Southeast Asia, 10; in Thailand, 112–113, 117, 123, 126–27, 129–30; in Togo, 57–61; in the United States, 210, 213, 216; in the Upper Volta, 81, 86–88; in the U.S.S.R., 230–61; in the Uttar Pradesh, 91, 97, 102–4; Western European influence on, 161; worldwide trends in, 1–12, 158–63. *See also* Environment; Labor; Overcutting
Deodar cedar (*Cedrus deodara*), 93
Desertification: in Niger, 80–81; in Upper Volta, 86; in West Africa, 70, 79
Detarium senegalense, 272 n.16
Development and Migration Commission (Australia), 199. *See also* Forest legislation: in Tasmania; Forestry agencies
Disease: effect on forests, in the U.S.S.R., 249
Dominion capitalism, 198. *See also* Australia; Colonialism
Douglas fir (*Pseudotsuga menziesii*), 166, 177, 178, 180, 181, 182, 183
Dyewoods, 150

East-African camphor (*Octea usambarensis*), 34

Eastern United States, 223, 229. *See also* United States
Electrolytic Zinc Company (Australia), 199. *See also* Timber and wood products trade
Embu Loca Native Council, 42, 47–49. *See also* Kenya; Kikuyu people
Empire (now Commonwealth) Forestry Association (Australia), 199
English oak, 197
Environment: impact of the metallurgy industry on, in Togo, 58–61; impact of settlement on, in Kenya, 36, 39; impact of timber-cutting on, in Assam, 110; impact of timber-cutting on, in Uttar Pradesh, 103; impact of timber trade on, in Japan, 175, 184–85, 186, 294 n.21; impact of timber trade on, in Tasmania, 208, 209; impact of timber trade on, in the United States, 184–85, 186, 211, 212, 213; impact of timber trade on, in the U.S.S.R., 231–32, 258–59, 260; impact of urbanization on, in Togo, 274 n.2, 278 n.55; impact of urbanization on, in the United States, 222–23; impact of urbanization on, in the U.S.S.R., 232, 259; loss of Sahelian forest, effect on, 71, 73; in Tasmania, 190–93; and traditional Sahel cultivation, 72, 75–76
Estonia, 236, 237, 238, 240, 241
Eucalyptus (*Eucalyptus regnans*), 23, 201, 207
European Economic Community, 206, 208. *See also* World economy
European Russian forests, 230–52 passim
European-Uralian Forests, 231, 250, 252, 253
Extended family: implications for forest resource use, 71, 74–76

Fallowing, 72
Fan palm (*Borassus aethiopum*), 82
Far East (U.S.S.R.), 233, 236, 237, 240, 241, 247, 255, 256
Farmland (U.S.), 223
Finland, 164, 243
Fire: brush fires, in the Sahel, 274 n.2; forest fires: in Brazil, 19–20; in Tasmania, 197; in the United States, 216, 217, 218, 219–22; in the U.S.S.R., 249, 252, 257
Florida, 223
Forest inventories: in the United States, 225; in the U.S.S.R., 248, 252. *See also* Forest management
Forest jobs: changing conditions of, in Japan, 167, 185–86; in the United States, 177, 185–86; in Thailand, 285 n.40
Forest legislation: in Assam, 111; in Australia, 200, 202, 205, 206, 301 n.9; enforcement of, in Thailand, 285 n.40; enforcement of, in the United States, 212, 220, 225, 299 n.38; enforcement of, in the U.S.S.R., 231, 247, 257, 258; in Great Britain, 198; influence of economic legislation on, in Tasmania, 196, 201, 203; in Japan, 169, 174, 175, 176, 178; in Kenya, 40–41, 44, 48; in Tasmania, 197, 200, 202, 206, 208, 301 n.9; in Thailand, 113, 123–29; in Uttar Pradesh, 104. *See also* Forest Police Division (Thailand); Forestry agencies; Land rights
Forest management: in Australia, 191, 200; in Brazil, 28–32; in Europe, 254; in India, 280 n.2; in Japan, 169, 170, 171; in Kenya, 33–34, 38, 43–44, 48–51, 52; in the Sahel, 44–48, 50–53, 72–74, 79; in Tasmania, 197, 200, 201, 209–10; in Togo, 67–68; in the United States, 212, 213, 214, 215–16, 225–27; in the U.S.S.R., 230–60 passim; in Uttar Pradesh, 94–104. *See also* Forestry agencies; Forestry, history of; Overcutting; Sustained yield
Forest ownership: in Canada, 184; in Japan, 169–70, 171, 292 n.11; in Kenya, 44–51; in Togo, 273 n.53;

in the United States, 184. *See also* Land rights; National forests; Private forests; Public forests; State or provincial forests
Forest Police Division (Thailand), 285 n.40. *See also* Forest jobs; Forest legislation; Forestry agencies
Forest products industry. *See* Fuel wood; Paper industry; Pulpwood; Roundwood; Sawmills; Sawnwood; Timber and wood products trade; Woodchip mills
Forest Research Institute (India), 99
Forestry agencies: in Brazil, 27–32; in India, 91, 100, 103; in Japan, 170, 173; in Kenya, 34, 40–44; in Tasmania, 195; in Thailand, 285 n.40; in the United States, 225; in the U.S.S.R., 234, 247, 249, 250, 252, 257, 258, 259. *See also* Forest legislation; Forestry, history of
Forestry, history of: in Australia, 197, 198, 199, 205, 206; in Brazil, 28, 60; in India, 94–95, 97, 99, 101, 103, 104, 107; in Japan, 165; in Tasmania, 197, 198, 205; in the United States, 213, 214, 215, 216; in the U.S.S.R., 230, 233, 249, 258
Foury, P., 81–82
French West Africa: agriculture, 11. *See also* West African Sahel
Fuel wood: impact of consumption on forests, 70, 90; in Japan, 171, 174; in Kenya, 43; in Togo, 272 n.16; in the United States, 227, 228, 242

General Finance Company (Australia), 196
Georgia (U.S.), 224–25. *See also* Carroll County, Georgia (U.S.)
Georgia (U.S.S.R.), 236, 237, 238
Gandhi, M. K., 96, 98
GOSPLAN (U.S.S.R. State Planning Committee), 258. *See also* Timber and wood products trade: in the U.S.S.R.; U.S.S.R.
Great Depression: impact on forestry and timber trade, in India, 100–103; impact on forestry and timber trade, in the United States, 220; impact on the tea industry, 106
Greeley, Thomas, 213, 214, 215, 216, 220, 229. *See also* Deforestation

Hardwoods. *See* Tropical hardwoods
Hinoki (*Chaemaecyparis obtusa*), 166, 180
Honey-hunters: in Kenya, 42–43
Housing industry, impact of timber prices on: in Japan, 180; in the United States, 183
Huon pine, 194
Huon Timber Company (Tasmania), 195, 196, 205. *See also* Commercial forests; Private forests; Timber and wood products trade
Hutchins, David E.: forest report, Kenya, 37–38, 51

Imperial Forestry Institute (Australia), 199
Improvement thinnings: as a forest management technique, in the U.S.S.R., 251, 255, 256. *See also* Forest management
India, 3; agriculture, 92–93; Almora (province), 280 n.2; Assam (province), 5, 6, 91–111; British colonial forestry policy, 94, 95–97; deforestation in, 90, 91, 97, 102–4, 115–20; Kumaon (province), 92–93; Private Forests Act, 111; species composition of forests, 93, 94, 97–98, 280 n.2; Uttar Pradesh (province), 91, 97, 102–4. *See also* Assam; Kumaon; Uttar Pradesh
Indian Forest Act: as model for Kenyan forest regulations, 41. *See also* Forest legislation
Indian Forest Service: description of, 8; establishment of, 91; professionalization of, 100
Indiana, 222
Indonesia, 148, 154–56, 162

Industrialization, 4
Insect infestation: longhorn beetle (*Monochammus alternatus*), in Japan, 171
Interior, Ministry of (Thailand), 285 n.40
Interstate Forestry Congress (Australia), 200
Irkutsk oblast, 260
IUFRO Forest Congress 1981 (Australia), 181
Ivdel-Ob Railroad (U.S.S.R.), 259, 260. *See also* Railroads
Ivory Coast, 79

Japan, 206–8, 231, 291 n.5, 291 n.6; climate, 170; consumption of tropical hardwoods, 160–61; forest stabilization, 2; population changes, 162; role in world deforestation, 3, 6, 174, 184–88, 294 n.21; species composition of forests, 171. *See also* Timber and wood products trade: in Japan; Timber exports; Timber imports
Japanese Ministry for International Trade and Industry, 207
Japanese red pine (*Pinus densiflora*), 171, 293 n.15
Japanese squares, 177, 297 n.29. *See also* Lumber industry; Plywood manufacture
Jhum system. *See* Shifting cultivation
Jujo Paper Company (Japan), 207. *See also* Paper industry

Kail or Blue pine (*Pinus excelsa* or *Wallichiana*), 93
Karamatsu (*Larix leptolepis*), 165–66, 171
Kazakhstan (U.S.S.R.), 238, 239, 240, 241
Kenaf, 117
Kentucky, 227
Kenya: British forest policy, 37–38, 44; Crown Forests, 33–34; deforestation, 36, 39–40, 49; description of, 5–7; European conquest and settlement, 36, 38; forest-related institutions, 40, 44, 45; Kenyan government forest policy, 34; Kikuyu forest policy (Embu Local Native Council), 40–51; Mount Kenya forest, 34; uses of forests by indigenous peoples, 34–35. *See also* Mount Kenya forest
Kenya Land Commission, 36, 44–51
Kenyan Forest Department, 34, 40–44
Kikuyu Central Association, 45–48
Kikuyu people: agriculture in Mount Kenya forest, 39; exclusion from Mount Kenya forest, 40–48; use of Mount Kenya forest, 35–36; Nyagithuci Hill controversy, 48–51
Kirghizia (U.S.S.R.), 238, 239
Knutson-Vandenberg Act of 1920 (U.S.), 225. *See also* Forest legislation
Kumaon (Uttar Pradesh), 91–101; agriculture in, 93; ecological zones, 93; effects of nationalism on, 96–97; forests of, 91, 280 n.2; timber production in, 101; use of the term, 280 n.2. *See also* Assam; Uttar Pradesh

Labor: *begar* labor, in India, 96–97; impact on deforestation, in the West African Sahel, 71, 76; impact of labor organizations on forests, in the United States, 178; migration, to Assam, 105–10. *See also* Forest jobs
Lane-Poole, C. E., 199, 200
Land rights: of Kikuyu, in Kenya, 44–51; in Togo, 63–65; in the Sahel, 72; in Niger, 76–77; in Thailand, 126–29; in the Upper Volta, 77. *See also* Mossi people; Tree property rights
Latvia, 236, 237, 238, 240, 241
Leningrad, 242, 243
Leningrad Oblast (U.S.S.R.), 239
Lithuania, 236, 237, 238, 240, 241, 253
Loblolly pine: in Brazil, 23
Logging industry: in Australia, 205, 206; in Brazil, 20; in Japan, 173, 297 n.29; in Tasmania, 201, 205, 208; in

the United States, 211, 227; in the U.S.S.R., 230, 252, 256, 257, 258, 259, 260. *See also* Timber and wood products trade
Lumber industry: in Japan, 166, 291 n.6; in the United States, 178, 211, 218, 220, 227; in the U.S.S.R., 239, 259. *See also* Forest legislation

Macaranga kilimandscharica, 34
Maize, 117
Malaysia, 154–55, 261. *See also* Tropical hardwoods
Mali, 70, 80, 83; wood supply in, 88–90
Marchal, Jean-Yves (French geographer), 81, 86
Mau Mau rebellion, 33, 36, 51–55; impact on pitsawyers, 53, 54. *See also* Kenya
Mekong, 282 n.10
Metallurgy: impact on environment, in Togo, 58–61; impact on timber trade, 60
Mexico, 148
Middle Atlantic states (U.S.), 229. *See also* United States
Midwest and Plains states (U.S.), 222. *See also* United States
Mining, 130
Mississippi, 227
Mitsubishi Company (Japan), 208. *See also* Private forests
Moldavia, 236, 237, 238, 240, 241
Monocultural planting: as a forest management technique, in Japan, 174, 175. *See also* Forest management; Species composition of forests
Morse amendment (U.S.), 299 n.38; Senator Wayne Morse, 181. *See also* Forest legislation
Moscow, 242
Mossi people (Upper Volta), 73, 75, 86; land tenure, 77; lineage patterns, 74–75. *See also* Extended family
Mountain ash, 201
Mount Kenya forest, 34–55; management and regulation, 39–44, 54–55; species composition, 34. *See also* Kenya
Muthura (*Ocotea renjensis*), 41

National forests: in Australia, 200; in Japan, 170, 172; in the United States, 179, 212, 214, 215, 216; in the U.S.S.R., 232, 234, 249, 256
National Pine Institute (Brazil), 28
Nematode infestation: *Aphelenchoides xylophilus*, in Japan, 171, 174, 293 n.15; *Bursaphelenchus lignicolus*, in Japan, 293 n.15; in the U.S.S.R., 252. *See also* Insect infestation
New England (U.S.), 222, 229. *See also* United States
New Jersey, 222
New South Wales, 191
New York, 222
New Zealand, 198, 205
Niger, 70, 80, 83–85, 275 n.3, 279 n.77; deforestation in, 80–81; forestry and ecological report (1981), 84; history of, 275 n.3; recent cultivation patterns, 84–85. *See also* Club du Sahel
Noncommercial forests: in the United States, 213–14
Northern United States, 233. *See also* United States
North Caucasus (U.S.S.R.), 236, 238, 240, 241
Northern woodchips, 208. *See also* Woodchip mills
North Vietnam, 207
Northwest U.S.S.R., 236, 237. *See also* U.S.S.R.

Ohio, 222
Oklahoma, 227
OPEC, 207
Overcutting: in Japan, 164–65, 175; in Tasmania, 208; in the United States, 211, 216; in the U.S.S.R., 240, 241, 253, 256, 258, 259. *See also* Deforestation; Timber and wood products trade; U.S.S.R.

Overseas Pulp Material Committee (Australia), 207. *See also* Pulpwood

Pacific Rim, 231, 255
Paper industry: in Japan, 208; in Tasmania, 200, 201–6; in the United States, 227, 228; in the U.S.S.R., 239, 242, 253, 258, 259. *See also* Pulpwood
Paraguayan tea (*Ilex Paraguariensis*), 25
Paraná pine (*Araucaria augustifolia*), 5, 15, 148; habitat of, 15, 16; usage of, 17
Paraná state (Brazil): history of forest policy, 28, 29, 32
Pauri Garhwal, 280 n.2. *See also* India
Pennsylvania, 222
Philippines, 150–51, 155
Pinchot, Gifford, 212, 213, 216, 222. *See also* Deforestation
Pine (*Pinus* spp.), 180; as pulpwood, in U.S.S.R., 253. *See also* Chir pine; Huon pine; Japanese red pine; Kail or Blue pine; Paraná pine; Radiata pine
Pitsawing. *See* Sawmills; Sawnwood
Plantation forestry: in Australia, 206; in Japan, 174; in Togo, 65; in the United States, 227. *See also* Forest management
Plywood manufacture: in Japan, 177, 179, 181, 183, 299 n.40; in the United States, 183–84, 227, 228. *See also* Japanese squares
Podo (*Podocarpus gracilor*), 34
Port Orford cedar (*Chamaecyparis lawsoniana*), 177, 178, 181, 182
Private forests: in Japan, 169, 171, 173, 174; in Tasmania, 192, 208; in Togo, 273 n.53; in the United States, 179, 186, 216, 218, 227. *See also* Commercial forests; Forest ownership
Public forests: in Tasmania, 190, 197, 209, 301 n.9. *See also* National forests; Noncommercial forests; State or provincial forests
Pulpwood: in Alaska, 184; in Australia, 206, 207; in Japan, 171, 291 n.5; in Tasmania, 200, 201, 202, 203, 205, 207, 208; in the United States, 185, 227, 228; in the U.S.S.R., 239, 253, 258

Quebracho logs, 150

Radiata pine, 197
Railroads: impact on logging, in Brazil, 20; impact on timber stands, 259; impact on timber trade, in Togo, 273 n.53; in India, 100; in the U.S.S.R., 243, 252, 253, 255
Redwood (*Sequoia sempervirens*), 180
Reforestation, 1; in Brazil, 28–29; in Europe, 254; in Japan, 171; in Kenya, 41–42; in Kumaon, 100; in southern Brazil, 29–32; in Togo, during colonial period, 67–68; in the Upper Volta, 81; in the United States, 2, 9, 184–85, 216, 218, 222–25, 226, 227, 229; in the U.S.S.R., 231, 247. *See also* Afforestation; Forest management
Regeneration: in Australia, 202; in Europe, 254; in the United States, 216–17, 225–27; in the U.S.S.R., 235, 249, 253
Reindeer grazing: in the U.S.S.R., 235. *See also* Sheep herding
Reserved forests: cancellation of, in the Uttar Pradesh, 97; history of, in Assam, 106–10; history of, in India, 97, 103; maintenance of, in Kenya, 8. *See also* Conservation; National forests; Private forests; State or provincial forests
Resin production: in Kumaon, 94–96, 98–99
Roosevelt, Theodore, 212, 216. *See also* Greeley, Thomas; Pinchot, Gifford; Forest legislation
Roundwood, 155, 230, 242, 253
Royal Forest Department (Thailand), 285 n.40. *See also* Forestry agencies
Rubber cultivation, 130
Russia, 3

Russian revolution: impact on forests, 235, 242

Sahel. *See* West African Sahel
Sal (*Shorea robusta*), 93, 100, 103, 108
Salween River, 282 n.10
"Save Our Logs Committee," 299 n.40. *See also* Conservation possibilities; United States
Sawmills: in Japan, 169, 173, 176; in Kenya, 51–54; in Tasmania, 194–209 passim; in the United States, 181, 183, 185
Sawnwood: in Australia, 205, 206; in Brazil, 21, 22, 23, 32; in British Columbia, 179, 184; in developing countries, 148; in Japan, 297 n.29; in Tasmania, 208, 209; in the United States, 179, 181, 184, 213, 214, 227
Selection felling. *See* Improvement thinning
Senegal: deforestation in, 81–82
Sever tribe, 82
S.F.S.R. (U.S.S.R.), 235, 238
Sheep herding: impact on forests, in Tasmania, 191; impact on forests, in the U.S.S.R., 235. *See also* Reindeer grazing
Shifting cultivation: in Assam, 104; in Brazil, 26–27; in Kenya, 41–42; in Togo, 58
Siberia, 230–60 passim
Silver fir (*Abies pindrow*), 94
Silviculture, U.S.S.R., 248
Singapore, 156
Sitka spruce (*Picea sitchensis*), 178
Softwoods. *See* Tropical softwoods
Soil erosion: in Senegal, 82; in Thailand, 128–29; in West Africa, 70–72
South African Cape Forest Ordinance: as model for Kenyan regulations, 41
South America. *See* Argentina; Brazil
South and Southeast Asia. *See* India; Japan; Malaysia; South Korea; Thailand
Southern United States, 222, 224, 227. *See also* Carroll County, Georgia; United States
South Korea, 208
South Seas timber: import trade to Japan, 162
Spanish cedar, 150
Species composition of forests: effect on trade, U.S.S.R., 255. *See also* Monocultural planting
Spruce (*Picea* spp.), 180, 182; as pulpwood, in the U.S.S.R., 233
Starr, Frederick, 212, 216. *See also* Deforestation; Roosevelt, Theodore
State Forestry Committee (U.S.S.R.), 258. *See also* Forestry agencies
State Forests Act (Tasmania), 301 n.9. *See also* Forest legislation
State or provincial forests: in Tasmania, 200, 301 n.9; in Washington state (U.S.), 178; in the U.S.S.R., 234, 236, 237, 238, 239. *See also* National forests
Stumpage prices: in the United States, 179, 181, 183, 184. *See also* Timber and wood products trade: impact of on economic conditions
Sugi (*Cryptomeria japonica*), 166, 180
Sumitomo Company (Japan), 208. *See also* Private forests
Sustained yield: as a management policy, in Tasmania, 200; in the U.S.S.R., 250, 256. *See also* Forest management
Swamp gum, 201
Swidden agriculture. *See* Shifting cultivation

Tadhikistan (U.S.S.R.), 238, 239
Taiga forest region (U.S.S.R.), 249
Tarai belt (Kumaon), 96, 100
Tasmania: British colonial economic policy, 198; climate of, 193; deforestation in, 190–93, 209; environment of, 208, 209; forest legislation, 301 n.9; forest policy, 194–95
Tasmanian Pulp and Forest Holdings, 207, 209

Tasmanian Timber Corporation, 195, 196
Taungya system, 41–42, 122. *See also* Forest management; Reforestation; Shifting cultivation
Tavda-Sotnik Railroad, 260. *See also* Railroads
Tea industry. *See* Assam
Teak, 41, 68, 120, 122–23, 155, 282 n.10
Thailand, 5–6; agricultural expansion in, 112–13, 126–29; agricultural statistics of, 131–43; Chao Phraya River, 282 n.10; climate of, 282 n.4; cropped area, 113–17; deforestation in, 112–13, 117, 123, 126–27, 129–30; exports, 155; forest cover (1913–1980), 118–21; forest policies suggested, 129–30; Forest Protection Acts, 123–25; Ministry of Agriculture, 113–14; population estimates, 286 n.52; reserved forests, 124–27; Royal Forest Department, 122–24, 128; wildlife preservation, 125. *See also* Afforestation; Deforestation; Forest legislation; Forestry agencies
Timber and wood products trade: in Assam, 93, 106–8; in Japan, 164–86; in Kumaon, 93; in Southeast Asia, 150, 164; in Thailand, 117, 120, 122–23, 126–27, 128; in the United States, 178; in the U.S.S.R., 255
—costs of development: in Japan, 170
—government intervention: in Japan, 174, 176; in the U.S.S.R., 231–32, 247, 253, 259
—history of, nineteenth century: in Australia, 301 n.4, 301 n.9; in Japan (Meiji period), 165; in Tasmania, 195; in Thailand, 282 n.10; in the United States, 177, 211, 212, 222
—history of, twentieth century (pre-World War II): in Japan, 165; in South America, 148–50; in Tasmania, 196, 201; in the United States, 177, 183, 212, 213, 217, 220, 222, 225, 229; in the U.S.S.R., 230, 235, 239, 242, 247
—history of, World War II: in Japan, 165
—history of, post-World War II: in Brazil, 20–22, 23, 24, 28; in Indonesia, 154–55; in Japan, 165, 167, 168, 206–8, 299 n.40; in Malaysia, 154–55; in South America, 151; in South and Southeast Asia, 151, 155; in West Africa, 151
—impact of climate on: in the U.S.S.R., 260
—impact of economic conditions on: in Japan, 166–70, 174–75, 178, 185–86; in Tasmania, 207; in the United States, 177, 178, 179, 183, 217, 225, 227, 228, 299 n.38; in the U.S.S.R., 235, 239, 242, 252, 253, 259
—impact of forest accessibility on: in the U.S.S.R., 257
—impact of forest soils on: in Japan, 170
—impact on port cities: in Japan, 176
—internationalization of, 5; in Japan, 186
Timber exports: attempts to limit, in the United States, 299 n.38; from Brazil, 20–23; from British Columbia, 179, 184; destination of tropical hardwoods, 154–55; from developing world, 147–57; from the Far East, 155; illegal exports, through Brazil, 24, 28; from India, to Egypt and Mesopotamia, 95; from Japan, to China, 165; from Japan, to Korea, 165; from Japan, to the United States, 165; from Japan, to the U.S.S.R., 165; postwar expansion of, 151; from South America, 157; from South and Southeast Asia, 156; from Tasmania, 193; from Tasmania, to Japan, 208; from Tasmania, to South Korea, 208; of tropical softwoods, 148; from the United States, to Japan, 176–78, 183, 184; from the U.S.S.R., to China, 231, 255; from the U.S.S.R., to Japan, 231, 255; from the U.S.S.R., to the Pacific Rim, 231; from the U.S.S.R., to Western Europe, 243; from West and West Central Africa, 156–57. *See also* Timber and wood products trade; Timber imports

Timber imports: to Japan, 173, 291 n.5, 297 n.29; of lumber, 291 n.6; to the Pacific Rim, from Siberia, 255; of tropical hardwoods, 160–61, 162–63; from the United States, 165, 179, 207, 291 n.6; from the U.S.S.R., 165, 291 n.6. *See also* Timber and wood products trade; Timber exports
Timber, Pulp, Paper, and Wood-Processing Industry, Ministry of (U.S.S.R.), 257, 259. *See also* Forestry agencies; Timber and wood products trade; in the U.S.S.R.
Timber stands: accessibility of, in the U.S.S.R., 257; age of, in Japan, 174; age of, in Tasmania, 209; growth estimates of, in Brazil, 265 n.20; growth estimates of, in the United States, 215, 216, 217, 218, 222–23, 225, 227; growth estimates of, in the U.S.S.R., 230, 254, 260; juvenile stands, in the U.S.S.R., 253; mature stands, in Europe, 253; mature stands, in the U.S.S.R., 231, 240, 241, 253, 256, 259; middle-aged stands, in the U.S.S.R., 251; mixed-age stands, in the U.S.S.R., 230; overmature stands, in Europe, 253; ponded stands and swamps, in the U.S.S.R., 255, 256
Togo, 3; agricultural expansion in, 58, 60, 61; anti-desertification project, 274 n.61; colonization of, 56–57; German colonial forest policy, 56–57, 60–69; history of deforestation, 57–58; species composition, impact of metallurgy industry on forests, 58–61, 272 n.16; urbanization in, 274 n.2, 278 n.55; vegetation in, 57
Transcaucasia (U.S.S.R.), 236, 237, 240, 241
Trans-Siberian Railway, 247, 253. *See also* Railroads
Tree property rights: control over, in the Sahel, 78–79; description of, 72; division of, 78; of foresters, 78
Trees, types of. *See Acacia albida;* African mahogany (*Khaya senegalensis*); Akamatsu (*Pinus densiflora*); Bamboo (*Arundinaria Alpina*); Birch; Broad-leaved species; Cabinetwoods; Camphor (*Cordia abyssinica*); *Cassipourea malosaria;* Chir pine (*Pinus longifolia*); *Combretum* spp.; Coniferous species; *Crossopteryx africana; Daniella thurifera;* Deciduous species; Deodar cedar (*Cedrus deodara*); *Detarium senegalense;* Douglas fir; Dyewoods; East-African camphor (*Octea usambarensis*); English oak; Eucalyptus (*Eucalyptus regnans*); Fan palm (*Borassus aethiopum*); Hinoki (*Chaemaecyparis obtusa*); Huon pine; Japanese red pine (*Pinus densiflora*); Kail or Blue pine (*Pinus excelsa* or *Wallichiana*); Karamatsu (*Larix leptolepsis*); Loblolly pine; *Macaranga kilimandscharica;* Mountain ash; Paraguayan tea (*Ilex Paraguayiensis*); Paraná pine (*Araucaria augustifolia*); Pine (*Pinus* spp.); Podo (*Podocarpus gracilor*); Port Orford cedar (*Chamaecyparis lawsoniana*); Radiata pine; Redwood (*Sequoia sempervirens*); Roundwood; Sal (*Shorea robusta*); Silver fir (*Abies pindrow*); Sitka spruce (*Picea sitchensis*); Spanish cedar; Spruce (*Picea* spp.); Sugi (*Cryptomeria japonica*); Swamp gum; Teak; Tropical hardwoods; Western hemlock (*Tsugu heterophylla*); Western Himalayan spruce (*Picea smithiana*)
Tropical forests: in Brazil, 15; types of timber yielded, 164
Tropical hardwoods, 183; export from Africa, 156–57; export from Indonesia, 156; export from Latin America, 156–57; export from Malaysia, 155–56; export from the Philippines, 150; export of, 148–49; future demand for, 162–63; importance of, to developing world, 148. *See also* Timber and wood products trade; Timber exports

Tropical softwoods, 22, 148. *See also* Tropical hardwoods
Tulun combine (U.S.S.R.), 260
Turkmenia (U.S.S.R.), 238, 239
Tyumen Oblast, 260

Ukraine, 236, 237, 238, 239, 240, 241, 243, 253. *See also* U.S.S.R.
Upper Volta (Sahel), 70, 80, 83–84, 279 n.77; deforestation in, 81, 86–88; population statistics, 87, 89. *See also* Mossi people; Yatenga region
Uralian forests, 230–60 passim
Urbanization, 4; impact on forests, in the United States, 222; impact on forests, in Uruguay, 198
United States, 207, 211–29; conservation in, 178, 225, 226; consumption of timber products, 227–29, 238, 243; deforestation in, 210, 213, 216; forest acreage/growth estimates, 213, 217, 225–27; forest legislation, 220, 225; forestry institutions, 212; labor institutions, 178; reforestation of, 2, 9; regional forests in the East, 222, 223–29; regional forests in the Midwest and Plains, 227; regional forests in the North, 233; regional forests in the Northwest, 164–68; regional forests in the South, 211–21, 222, 227; regional forests in the West, 229; species composition, 166, 177, 178, 180, 181, 182, 183, 230, 231, 252; timber trade, changes in, 181, 186, 227–29, 300 n.42; timber trade, environmental impact of, 184–85, 186, 211, 212, 213; timber trade, impact on Japan, 186; timber trade, impact on the U.S.S.R., 232, 233, 242–48, 257–61; urbanization, 222–23. *See also* Carroll County, Georgia; Timber and wood products trade
U.S.S.R.: agriculture, 223, 231, 235; Baikal Meridian, 253; Baltics, 236, 237, 240, 241, 248; Belorussia, 236, 237, 238, 240, 241, 243, 253; Black Earth region, 236, 237, 238, 240, 241, 253; Central Asia, 235, 238, 239, 256; Central Industrial Region, 243, 256; Central U.S.S.R., 236, 237, 240, 241; conservation in, 233, 257; deforestation in, 230–60 passim; distribution of forests, 240, 241, 248, 260; economic conditions, 247, 253; economic philosophy, 252, 261; Far East, 233, 236, 237, 240, 241, 247, 255, 256; forest acreage/growth estimates, 230, 231, 232, 234, 235, 248, 249, 250, 252, 253, 255, 256, 260; forest history, 239, 240–48; impact of climate on forest utilization, 255; impact of fires on deforestation, 252, 257; impact of infestation on forest growth, 252; impact of loggers on deforestation, 252, 257; impact of railroads on deforestation, 260; impact of timber trade on forest management, 231–32, 258, 260; Northern Caucasus, 236, 238, 240, 241; Northwest U.S.S.R., 236, 237; Siberia, 230–60 passim; species composition, 230, 231, 232, 233, 251, 253, 254; Taiga forest region, 249; Transcaucasia, 236, 237, 240, 241; Ukraine, 236–41, 243, 253; urbanization, 232, 259; Volga Littoral, 236–37, 240–41, 252; Volga-Vyatka region, 235–37, 240–41, 252, 260. *See also* Timber and wood products trade; Timber exports; Timber imports
Uttar Pradesh (United Provinces): cutting policy in, 103; 1878 forest law, 94; Forest Department, 94, 97, 98–99, 100–101, 103–4; Forest Reserves, 94–95, 97. *See also* Forest legislation; Forest management; Forest Research Institute; Forestry agencies; India
Uzbekistan (U.S.S.R.), 238, 239

Virgin forest: in the United States, 214–15; in the U.S.S.R., 243. *See also* Greeley, Thomas

Virginia, 227
Volga Littoral (U.S.S.R.), 236–37, 240–41, 252
Volgo-Vyatka region (U.S.S.R.), 235, 236, 237, 240, 241, 252, 260

War. *See* Civil War (Russian); Mau Mau rebellion; Russian revolution; World War I; World War II
West African Sahel: agriculture of, 71, 72, 75–76, 83, 274 n.2; conservation in, 77; cooking practices in, 70, 90, 279 n.77; deforestation in, 70–72, 79, 83, 274 n.2, 278 n.55; French colonial forest policy, 71–74; institutions, Club du Sahel, 80, 84, 86, 274 n.1, 279 n.77; species composition, 81
Western hemlock (*Tsuga heterophylla*), 166, 180, 182, 185
Western Himalayan spruce (*Picea smithiana*), 94
Western Himalayas. *See* Kumaon
Western United States, 229. *See also* United States
West Virginia, 222
Weyerhaeuser Timber Company, 179, 186. *See also* Commercial forests; Forest ownership; Timber exports
Wilderness: in the United States, 229
Woodchip mills: in Australia, 207; in Tasmania, 207–8, 209
Woodchip trade: in Australia, 207; international trade in, 207; in Japan, 291 n.5; in Tasmania, 203, 207
Wood stoves: impact on deforestation, in Niger, 279 n.77; impact on deforestation, in the Upper Volta, 279 n.77
World economy, 301 n.1; impact on Australia 190–95, 198, 206; impact on Japan, 206–8; impact on Tasmania, 190–93, 198; impact on the U.S.S.R., 232, 254–55
World War I: impact on forests, in India, 98; impact on forests, in Kenya, 44; impact on forests, in Togo, 273 n.33; impact on forests, in the U.S.S.R., 230; impact on tropical and subtropical hardwoods, 150
World War II: impact on exports of tropical hardwoods, 148–49; impact on forests, in India, 91–92, 95–96; impact on forests, in the U.S.S.R., 230, 243, 247

Yatenga region (Upper Volta), 77, 81, 86; cropping pattern in, 75
Yazoo-Little Tallahatchie Flood Prevention Program (U.S.), 227. *See also* Conservation possibilities
Yurty combine (U.S.S.R.), 260

Zima combine (U.S.S.R.), 260